Topics in
Physical Chemistry

vol 2

Edited by H. Baumgärtel, E. U. Franck, W. Grünbein
On behalf of Deutsche Bunsen-Gesellschaft für Physikalische Chemie

Topics in
Physical Chemistry

E. Illenberger, J. Momigny

Gaseous Molecular Ions

An Introduction to Elementary Processes Induced by Ionization

 Steinkopff Verlag Darmstadt
Springer-Verlag New York

Authors' addresses:
Prof. Dr. Eugen Illenberger
Institut für Physikalische
und Theoretische Chemie
der Freien Universität Berlin
Takustraße 3
D-1000 Berlin 33

Prof. Dr. Jacques Momigny
Institut de Chimie
Département de Chimie
Générale et de Chimie-Physique
Université de Liège
Sart Tilman
B-4000 Liège 1

Edited by:
Deutsche Bunsen-Gesellschaft
für Physikalische Chemie e.V.
General Secretary Dr. Heinz Behret
Carl-Bosch-Haus
Varrentrappstraße 40/42
D-6000 Frankfurt 90

Die Deutsche Bibliothek — CIP-Einheitsaufnahme

Gaseous molecular ions: an introduction to elementary
processes induced by ionization / E. Illenberger; J. Momigny.
— Darmstadt: Steinkopff; New York: Springer, 1992
(Topics in physical chemistry; Vol. 2)
ISBN 3-7985-0870-4 (Steinkopff)
ISBN 0-387-91401-3 (Springer)
NE: Illenberger, Eugen; Momigny, Jacques, GT

ISSN 0941-2646

Chemistry Editor: Dr. Maria Magdalene Nabbe — English Editor: James C. Willis
Production: Heinz J. Schäfer

Printed in Germany

Typesetting: Fotosatz Rosengarten, Kassel — Printing and bookbinding: Druckhaus Darmstadt, Darmstadt

Printed on acid-free paper

Preface

Most of the matter in our solar system, and, probably, within the whole universe, exists in the form of ionized particles. On the other hand, in our natural environment, gaseous matter generally consists of neutral atoms and molecules. Only under certain conditions, such as within the path of lightning or in several technical devices (e. g. gas discharges, rocket engines, etc.) will some of the atoms and molecules be ionized. It is also believed that the chemistry of the earth's troposphere predominantly proceeds via reactions between neutral particles. (The complex system of atmospheric chemistry will be treated in one of the forthcoming volumes to this series.) Why, then, are ions considered so important that hundreds of laboratories all over the world (including some of the most prestigious) are involved in research programs on ions, covering many different facets, from biochemistry to physics? One may obtain as many different answers as there are research groups busy in this field. There is, however, one simple, common feature which makes it attractive to work with ions: since they carry one or more net elementary charges, they can easily be guided, focused or separated by appropriate electric and magnetic fields, and, last but not least, they can easily be detected.

Apart from these advantages, which are welcome and appreciated by the researcher, the study of molecular ions can provide insight into very fundamental aspects of the general behavior of molecules. Moreover, the *ionization process* itself can be used to obtain information on certain properties or quantities of the corresponding *neutral molecule*. Mass spectrometry, for example, is a widely used and established technique to analyze *neutral particles*. This is usually performed by electron-impact ionization of the neutral gas sample, and separation of the generated ions (in space or time) according to their molecular weight.

In photoelectron spectroscopy, on the other hand, the kinetic energy of the ionized electrons is recorded, and the spectrum obtained is an image of the energy levels of the molecular orbitals in the *neutral* molecule. Photoelectron spectroscopy had (and still has) an enormous impact on theory, particularly on quantum chemistry in the development of sophisticated molecular orbital calculations. More refined techniques in photoionization such as coincidence techniques between electrons and ions allow a detailed study of *unimolecular reactions* in that one can follow the evolution of a molecular ion prepared in a definite state (i. e., with a defined amount of internal energy). General concepts to describe unimolecular processes were first developed some decades ago, relating to fragmentation patterns observed in mass spectra. The unimolecular decomposition of energized ions is a key subject of this volume. Today, unimolecular processes in *neutrals* can be studied in great detail by laser techniques (*pump and probe* experiments, i. e., *pumping* a defined amount of energy into a molecule with a first laser, and *probing* the products (identity and state) by means of a second one). Such processes will be treated in detail in one of the forthcoming volumes of this series (Modern Photochemistry). Due to the limited wavelength range of available laser systems, however, these methods are not as generally applicable as photoionization.

Negatively charged ions play a particular role, since they can be formed in the gas phase in large quantities by attachment of very low-energy electrons (sometimes near 0 eV). In contrast, the formation of positive ions requires an energy equal to or greater than the first ionization energy, which is around 10 eV for most organic molecules. Low-energy electrons can be very *reactive*, in that they are effectively captured by many molecules, which then undergo rapid unimolecular decompositions. The cross-section for such processes can be very large, i. e., 4 to 5 orders of magnitude larger than typical photoionization or excitation cross-sections. If a slow electron collides with an excited molecule, dissociative capture processes can have enormous cross-sections, many orders of magnitude larger than the geometrical cross-section of the respective molecule. The role of slow electrons and negative ions has somehow been disregarded, and comparatively few groups are working in that area (at least in Germany). Part II of this volume is particularly dedicated to negative ions and their reactions in low-energy electron collisions.

Removing or adding electrons from or to molecules, along with the reaction of the ionized species, may be regarded as a particular step in relation to a chemical reaction. Of course, a chemical reaction occurring in solution is a much more complex process as it may proceed via energy and charge transfer between *different* molecules (the educts), thereby cleaving bonds and forming new molecules (the products). Approaches toward a *microscopic* study of chemical reactions are currently performed in molecular beam experiments, particularly in supersonic beams containing molecular aggregates (or clusters). Such weakly bound aggregates represent a link between gaseous and condensed matter. Some examples in the case of reactions following electron attachment to van der Waals clusters are discussed in the last sections of this book.

This volume contains three parts and is organized as follows: the first part gives a general overview of the experimental methods used to prepare positive and negative ions, and how their evolution can be studied. In the case of positive ions, it addresses the "classical" techniques of photoionization, i. e., ionization by VUV light sources. Although mentioned occasionally, ionization by multiphoton laser techniques is not treated explicitly. The basic instrumentation (light sources, mass spectrometers, electron energy filters, and detectors) is introduced, with the emphasis placed on the description of the *operation principles* rather than on sophisticated technical details.

Part II (J. Momigny) focuses on general processes which occur when a molecule is subjected to the absorption of energy above its first ionization limit, i. e., *direct* ionization and *auto*ionization, the energy flow via radiative and nonradiative transitions between electronic states, including nonadiabatic interactions, and the decomposition into fragments. Besides photoionization, other techniques for preparing positive ions (charge exchange, Penning ionization, field ionization, etc.) are described. Statistical approaches to calculate the rate constant for a unimolecular reaction and the excess energy distribution among the fragments are introduced. The obtained results are always compared with experimental values in order to test the validity and limits of the approaches.

The last part is devoted to negative ions, their behavior and peculiarities in comparison to positive ions. Electron-capture processes for some prototype systems are presented and discussed. Particular emphasis is placed on the question of how the relevant quantities, e. g., attachment energies, selection rules for populating negative ion states, reaction products and their energy distribution behave on proceeding from an isolated molecule to clusters of increasing size.

Most of the work presented in this last part has generoulsy been supported by the Deutsche Forschungsgemeinschaft and Fonds der Chemischen Industrie which is gratefully acknowledged. Thanks are also due to many colleagues and coworkers for valuable contributions. Their names appear in the original publications cited here.

It is hoped that this text will help to bridge the gap between the body of general knowledge and specific current research, and thus being of benefit in initiating the student and newcomer in the subject of *gaseous molecular ions*.

I would like to thank the publisher Dr. Dietrich Steinkopff Verlag, Darmstadt, particularly the Chemistry Editor Dr. Maria Magdalene Nabbe for her constructive collaboration and the editors for their invitation to contribute to this series.

Special thanks are due to Mrs. L. Brodricks for carefully processing the original manuscript.

Eugen Illenberger   *Berlin, December 1991*

Contents

Part I Preparation and Decomposition of Positive and Negative Ions: Experimental Techniques and Instrumentation

Eugen Illenberger

1 Some General Remarks on Positive and Negative Ions, Photoionization, and Electron Attachment

This volume deals with gaseous molecular ions. It gives an introduction to elementary processes induced in a molecule subjected to the removal or addition of an electron leading to a singly charged cation or anion, respectively. Among the many possibilities to remove or add electrons from or to a neutral molecule, we will focus on reactions induced by photoionization and free electron attachment.

Consider a small organic molecule like halogenated methane $CFCl_3$ which is well known for its chemical inertness in the troposphere. Although it possesses as many as 66 electrons, the stability of the compound may dramatically change on going from the neutral molecule to the ion. The question arises whether the molecular cation or anion formed by photoionization or electron capture exists at all as a stable compound [1,2]. This will generally depend on the amount of energy deposited in the molecule by the respective process. From a more microscopic point of view, it may also depend on the particular influence induced in the molecular system by the ejected or attached electron.

Both the amount of internal energy and the way this energy is initially deposited in the molecule are related to the particular molecular orbital (MO) involved in the ionization or attachment process.

It is the aim of the present volume to introduce basic ideas and concepts used to describe the formation and decomposition of gaseous molecular ions and to present the most important experimental methods and techniques for the study of such reactions.

Photoionization and electron capture may be regarded as isolated and elementary steps in the general field of electron transfer reactions. These processes play a central role in nearly all phenomena of pure and applied science. Formation, dissociation, or recombination of positive and negative ions are basic processes in the earth's upper atmosphere, as well as in the interstellar space, in laboratory plasmas, gaseous dielectrics, or in the various methods and techniques of mass spectrometry [3-6]. A detailed knowledge of these reactions under isolated conditions in the gas phase is important for the understanding of more complex reactions, such as electron transfer in solution (in electrochemical or biological systems [7,8]) or, more generally, for the understanding of elementary steps in chemical reactions [9,10].

Before describing the experimental methods for the study of photoionization and electron attachment and the instrumentation commonly used, we will briefly recall some important facts concerning the formation and decomposition of singly charged positive and negative ions under single collision conditions. Single collision conditions require a mean free path length which is large compared to the dimension of the experimental arrangement. This ensures that secondary collisions of an ion on its way from the source to the detector can be neglected. For nitrogen at $1 \cdot 10^{-5}$ mb the mean free path length is 580 cm.

Consider a molecule M in the gas phase. Ionization can be achieved, for instance, by impact of photons or electrons of sufficient energy

$$h\nu + M \rightarrow M^+ + e^- \qquad\qquad\qquad 1.1$$

$$e^-(1) + M \rightarrow M^+ + e^-(1) + e^-(2). \qquad\qquad\qquad 1.2$$

In the first case, the photon is absorbed by the molecule and the photoelectron carries away the difference between the photon energy and the binding energy of the ejected electron

$$\varepsilon = h\nu - IE(M), \qquad\qquad\qquad 1.3$$

where ε is the kinetic energy of the electron, and $IE(M)$ is the ionization energy of the molecule equal to the binding energy of the ionized electron.

In electron impact ionization the incident electron can lose any amount of its initial energy, so that the excess energy can be shared between the incident ($e^-(1)$) and the ionized electron ($e^-(2)$). This bears important consequences leading to different behavior in the energy dependence of the cross-section between photoionization and electron impact ionization as will be discussed in the second contribution to this volume.

We will restrict our discussion to ionization processes where valence orbitals are involved. These are MOs which are responsible for the formation of chemical bonds. For most organic molecules the energy necessary to remove an electron from the highest occupied molecular orbital (HOMO) (which is equal to the first ionization energy) lies around 10 eV. In that case, the molecular cation in its electronic ground state is created. Accordingly, we speak of the second ionization energy of M when an electron from the next lower lying MO is removed. This creates the cation M^+ in its first electronically excited state.

One should be aware, however, that this electronically excited state does not *necessarily represent the first excited state* of the ion. Ions also possess excited states due to a transition from the HOMO to one of the unoccupied MOs. Such states can be observed in ion absorption spectra [12]. They are normally very weak or absent in photoionization since their population requires a two-electron jump. Although only few ion absorption spectra in the gas phase are known, it appears that the first excited ionic states populated in photoionization (with a single hole in one orbital) generally lie *below* the states due to a HOMO-LUMO transition in the ion.

Photoionization involving valence electrons requires photon energies in the range from about 10 eV to 30-40 eV. This domain of the electromagnetic spectrum is called vacuum ultraviolet (VUV) (or sometimes far-ultraviolet). Spectroscopy in this region necessitates evacuation of the optical path and the design of special optical components, since matter (gaseous or condensed) is opaque for these wavelengths. The photoionization process (Eq.1.1) may be studied either by using a fixed photon wavelength (usually $hv = 21.22$ eV from a He discharge lamp; see Section 3.1.1) and recording the energy distribution of the emitted photoelectrons, or, if a tunable light source is available, by varying the energy of the photons and recording the (parent and fragment) ions formed. The first method is known as photoelectron spectroscopy (PES) [11, 16, 17], the second as photoionization mass spectrometry (PIMS) [13]. In PES the different ionization energies (*IEs*) of a molecule corresponding to different kinetic energies of the electrons are determined (Eq 1.3). Each ionization energy is approximately equal to the negative eigenvalue of the MO from which the electron is ejected. This approximation is known as Koopmans' theorem which is based on the selfconsistent field (SCF) model [14,15]. It is assumed that the interaction of the remaining electrons is the same after one electron has been removed.

Today, preparation of ions is increasingly performed by (multiphoton) *laser techniques* which we will occasionally mention in this context. One of the forthcoming Volumes will particalarly be dedicated to the application of lasers in Physical Chemistry.

From a photoelectron spectrum the particular states of the molecular ion accessible by 21.22 eV photons can be assigned. The method, however, does not provide direct information on the further fate of the ion. If it contains sufficient internal energy, it may undergo different competitive dissociation reactions, such as

$$M^+ \rightarrow R_1 + X^+ \qquad\qquad 1.4$$

$$\rightarrow R_2 + Y^+ \qquad\qquad 1.5$$

$$\rightarrow \ldots\ldots \quad .$$

The obvious question then is to determine the relative decomposition probabilities as a function of the internal energy of the parent ion.

The fate of a molecular ion can be followed in photoionization mass spectrometry (PIMS) by recording the evolution of ions as a function of photon energy.

The energy balance for a dissociative ionization process leading to two fragments

$$
\begin{aligned}
hv + M &\rightarrow M^+ + e^- \\
&\hookrightarrow R + X^+,
\end{aligned}
$$
1.6

requires

$$AE(X^+) = D(R - X) + IE(X) + E^*,$$
1.7

with $AE(X^+)$ the experimentally observed appearance energy of the fragment ion X^+, D the bond dissociation energy in the neutral molecule, and $IE(X)$ the ionization energy of fragment X. E^* is the total excess energy of the process comprised of the kinetic energy of the outgoing electron, and the kinetic and internal energy of the two fragments.

While in PIMS the relative abundance of the different ions is measured vs photon energy, this method does not directly allow information on the internal energy of the parent ions, since the energy of the photoelectron is not known. This can be achieved by measuring (for a given photon energy) the respective ion *in coincidence* with the energy of the ejected photoelectron. This method is called photoelectron-photoion coincidence spectroscopy (PEPICO) [13, 16, 17]. It yields the unimolecular decomposition pathways of M^+ for a defined internal energy and hence a selected state.

The way in which the available excess energy will be distributed among the different degrees of freedom of the fragments represents a rather complex problem and depends on the details of the mechanism of the unimolecular reaction. In the second contribution to this volume some approaches based on a statistical treatment will be given. We will show (in the third contribution to this volume) that measuring the excess kinetic energy distribution of the fragments sometimes allows a description of the decomposition beyond the statistical picture.

We note that the formation of a positive ion M^+ is possible for any photon energy above the ionization threshold. Since the final channel consists of the cation in a discrete state and the electron in the continuum $(M^+ + e^-)$, there is no restriction on the energy of the photoelectron.

The situation becomes different for the formation of negatively charged ions by electron attachment

$$e^- + M \rightarrow M^-.$$
1.8

This process represents a transition from the continuum $(M + e^-)$ to a discrete state of the molecular anion M^-, i.e., a resonant process. Only electrons of appropriate energy can be attached to the neutral molecule. For that reason, the molecular anion is called a "resonance". Synonymously, the term "temporary negative ion" (TNI) is used. Since the additional electron is temporarily trapped in a metastable (or quasi bound) state, M^- is, in principle, unstable with respect to autodetachment[1] which is the reverse of reaction 1.8.

We will not discuss any details of the electronic configuration of resonances at this point. Instead, the interested reader is referred to the third contribution to this volume. To avoid confusion, however, it should be emphasized that under single collision conditions the electronic state of a negative ion which is directly formed by free electron attachment, in every case, lies above the corresponding ground state of the neutral molecule. M^- thus represents a discrete state embedded in the continuum, similar to the autoionizing states of neutral molecules known in photoionization [13].

Within the one electron approximation, electron capture can be pictured as accommodation of an electron into one of the normally empty MOs. It depends on the sign of the electron affinity of the target molecule (positive of negative) whether the lowest unoccupied molecular orbital (LUMO) is involved or one of the higher lying virtual MOs.

The autodetachment lifetime varies over a large scale ranging from 10^{-3} s for larger polyatomic molecules to a few vibrational periods ($\approx 10^{-14}$ s) in smaller systems [3, 18]. If decomposition channels leading to thermodynamically stable fragments are energetically possible at the energy of the TNI, fragmentation according to

$$M^- \rightarrow R_1 + X^-$$
$$\rightarrow R_2 + Y^-$$
$$\rightarrow \ \ldots\ldots \qquad . \qquad\qquad\qquad\qquad\qquad\qquad 1.9$$

strongly competes with autodetachment. The cross-section for electron attachment may exceed the cross-section for photoionization by several orders of magnitude. Since dissociative electron attachment predominantly occurs at low energies (sometimes at zero eV), these processes are of particular interest in phenomena where molecules containing electronegative components and low energy electrons are present.

Electron capture is usually studied by two different techniques, electron transmission spectroscopy (ETS) and electron attachment spectroscopy (EAS). The first method measures the attenuation of an electron beam transmitted through gas as a function of the incident energy [19-21]. Resonances in the transmitted current signalize the formation of TNIs. According to Koopmans' theorem, the attachment energy is then associated with the energy of the involved MO in the anion [22-25]. Whereas PES determines cation energies associated with the removal from filled MOs, ETS yields the energies of anion states formed by the accommodation into normally unfilled MOs.

---

[1] In the case of negative ions, the term detachment is commonly used for the removal of an electron.

In EAS the negative ions (parent and fragment ions) are recorded mass spectrometrically as a function of the incident electron energy. This method additionally yields information on the fate of the parent ion M^-, provided that decomposition channels are available at all.

For a dissociative electron attachment process yielding two fragments

$$e^- + M \rightarrow R + X^-, \hspace{4cm} 1.10$$

energy conservation dictates that

$$AE(X^-) = D(R - X) - EA(X) + E^*, \hspace{3cm} 1.11$$

with AE and D as defined above, and $EA(X)$ the electron affinity of fragment X. Equation 1.11 is analogous to expression 1.7 for dissociative photoionization. In the present case, however, the total excess energy E^* is only shared between translational and internal energy of the two fragments (R and X^-), while in photoionization the energy of the outgoing photoelectron additionally contributes to E^*.

We see that in electron attachment spectroscopy the internal energy of the precursor ion M^- is solely determined by the energy of the incident electron. The method hence represents a convenient way to study energy partitioning in unimolecular reactions. In photoionization, analogous information can only be obtained by the coincidence technique mentioned above.

References

1. Jochims H-W, Baumgärtel H (1976) Photoreactions of Small Organic Molecules. V. Absorption-, Photoion- and Resonancephotoelectron Spectra of CF_3Cl, CF_2Cl_2, $CFCl_3$ in the Energy Range 10-25 eV. Ber Bunsenges Phys Chem 80:130-138
2. Oster T, Kühn A, Illenberger E (1989) Gas Phase Negative Ion Chemistry. Int J Mass Spectrom Ion Proc 89:1-72
3. Christophorou LG (ed) (1984) Electron-Molecule Interactions and Their Applications, vols I and II. Academic Press, Orlando, Florida
4. Scheffler H (1988) Interstellare Materie. Vieweg, Braunschweig
5. Christophorou LG, Pace MO (eds) (1984) Gaseous Dielectrics. Pergamon, New York
6. Budzikiewicz H (1981) Massenspektrometrie negativer Ionen. Angew Chem 93:635-649
7. Marcus RA (1964) Chemical and Electrochemical Electron-Transfer: Theory. Annu Rev Phys Chem 15:155-196
8. Sutin N (1983) Theory of Electron Transfer Reactions: Insights and Hindsights. Prog Inorg Chem 30:441-498
9. Herschbach DR (1987) Molekulare Dynamik chemischer Elementarreaktionen (Nobel-Vortrag). Angew Chem 99:1251-1275
10. Davis HF, Suits AG, Hou H, Lee YT (1990) Reactions of Ba Atoms with NO_2, O_3 and Cl_2: Dynamic Consequences of the Divalent Nature of Barium. Ber Bunsenges Phys Chem 94:1193-1201
11. Turner DW (1970) Molecular Photoelectron Spectroscopy. Wiley, London
12. Maier JP (1991) Approaches to Spectroscopic Characterization of Cations. Int J Mass Spectrom Ion Proc 104:1-22

13. Berkowitz J (1979) Photoabsorption, Photoionization, and Photoelectron Spectroscopy. Academic Press, New York
14. Koopmans T (1934) Über die Zuordnung von Wellenfunktionen und Eigenwerten zu den einzelnen Elektronen eines Atoms. Physica 1:104-113
15. Kutzelnigg W (1978) Einführung in die Theoretische Chemie. Bd 2, Verlag Chemie, Weinheim
16. Rabalais JW (1977) Principles of Ultraviolet Photoelectron Spectroscopy. Wiley, New York
17. Eland JHD (1984) Photoelectron Spectroscopy, II. edn. Butterworths, London
18. Fenzlaff M, Illenberger E (1989) Energy Partitioning in the Unimolecular Decomposition of Cyclic Perfluororadical Anions. Chem Phys 136:443-452
19. Allan M (1989) Study of Triplet States and Short-Lived Negative Ions by Means of Electron Impact Spectroscopy. J Electron Spectrosc Relat Phen 48: 219-351
20. Schulz GJ (1973) Resonances in Electron Impact on Diatomic Molecules. Rev Mod Phys 45: 423-486
21. Hasted JB, Mathur D (1984) Electron-Molecule Resonances. In: Christophorou LG (ed) Electron-Molecule Interactions and Their Applications, vol I. Academic Press, Orlando, Florida
22. Jordan KD, Burrow PD (1978) Studies of the Temporary Anion States of Unsaturated Hydrocarbons by Electron Transmission Spectroscopy. Acc Chem Res 11:341-355
23. Jordan KD, Burrow PD (1987) Temporary Anion States of Polyatomic Hydrocarbons. Chem Rev 87:557-588
24. Heni M, Kwiatkowski G, Illenberger E (1984) Negative Ions in the Gas Phase Observed in Electron Transmission and Electron Attachment Spectroscopy. Ber Bunsenges Phys Chem 88:670-675
25. Simons J, Jordan KD (1987) Ab Initio Electronic Structure of Anions. Chem Rev 87:535-555

2 Experimental Methods

2.1 Photoelectron Spectroscopy (PES)

Figure 2.1 illustrates the schematic experimental arrangement for photoelectron spectroscopy. It consists of a light source, an interaction region where the photon beam collides with the gas under consideration, an electron spectrometer, and an electron detector. In conventional PES, one uses sources emitting light of a fixed wavelength (mostly 584 Å $\cong$ 21.22 eV produced in a He discharge lamp). The target gas is introduced in the reaction volume simply by effusing from a capillary. By scanning the electron energy analyzer, electrons of only one energy at a certain time are transmitted to the detector. By synchronizing the electron energy analyzer with a multichannel analyzer (MCA) a given channel (position on the X axis) is made to correspond to a particular electron energy, yielding a differential photoelectron spectrum, as indicated in Fig. 2.1.

The different components of the experiment are mounted in a high vacuum (or ultra high vacuum) chamber equipped with pumps of sufficient speed. The pressure must be kept low ($\leq 10^{-4}$ mb) in order to ensure the single collision conditions mentioned above. The experimental arrangement may be modified in such a way that

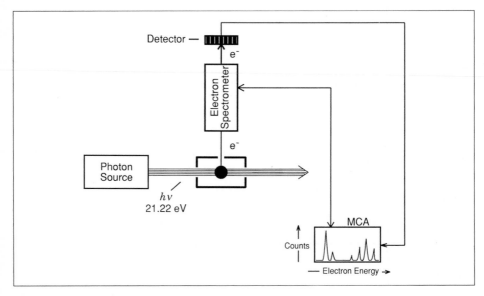

Fig. 2.1. Schematic of the experimental arrangement for photoelectron spectroscopy (PES). MCA: multichannel analyzer.

only electrons emitted at an angle of 54.7° with respect to the photon beam are ana-lyzed. The photoelectrons are not ejected in equal numbers in all directions and their angular distribution varies from one band to the other. If they are collected under the "magic" angle (54.7°) the angular effect vanishes and one obtains a "true" photoelectron energy distribution [1-3].

We will now briefly illustrate PES for the simple case of a diatomic molecule (Fig. 2.2). CO has 10 valence electrons with the configuration $CO \, (\sigma \, 2s)^2 \, (\sigma^* \, 2s)^2 \, (\pi \, 2p)^4 (\sigma \, 2p)^2 \; ^1\Sigma_g^+$. As usual, $2p$, $2s$ denote the atomic orbitals of which the molecular orbitals (π, σ) are composed. Antibonding MOs are assigned by an asterix. 21.22 eV photons are able to ionize electrons from any of the three high-est occupied MOs (Fig. 2.2). With the eigenvalues -14.0, -16.9, and -19.7 eV we expect photoelectrons of 7.2, 4.3, and 1.5 eV kinetic energy, respectively. Removal of an electron from the HOMO $(\sigma 2p)$ generates CO^+ in its electronic ground state, and removal from $\pi 2p$ in its first electronically excited state, etc.

Figure 2.3 shows the experimentally observed photoelectron spectrum and the associated potential energy curves. Since the ionization process is rapid with respect to the time of a molecular vibration, electron ejection can be represented by vertical

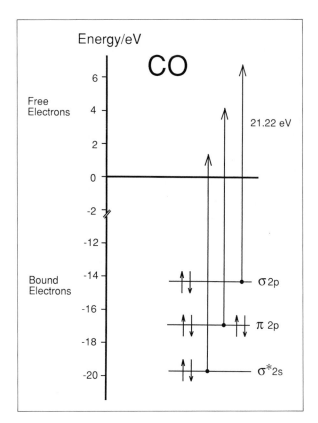

Fig. 2.2. MO diagram for the three highest occupied MOs in CO accessible by HeI radiation.

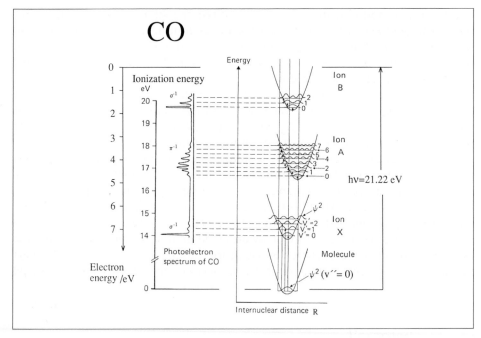

Fig. 2.3. PE spectrum of CO obtained by HeI radiation and potential energy curves for the neutral molecule and the three ionized states (adapted from [48]).

lines in Fig. 2.3[1]). This approximation is known as the *Franck-Condon Principle*. As is obvious from Fig. 2.3, the electronic transition may lead to vibrational excitation in CO^+ in its different electronic states associated with the structures in the photoelectron spectrum. Rotational structures generally cannot be resolved in PES.

The cross section for ionization from the vibrational level v'' of the neutral molecule to a vibrational level v' in the ion is given to a good approximation [3] by the Franck-Condon Factor (FCF) which is equal to the square of the overlap integral between the respective vibrational wave functions

$$\text{FCF} = |\langle \Psi_{v''} | \Psi_{v'} \rangle|^2. \hspace{2cm} 2.1$$

Note that this overlap integral does not vanish by orthogonality because $\Psi_{v''}$ and $\Psi_{v'}$ are vibrational wavefunctions belonging to different electronic states.

Removal of an electron from the strongly bonding $(\pi 2p)$ MO weakens the bond strength in the ion, resulting in a larger nuclear separation and a reduction of the vibrational frequency, as can be seen in the photoelectron spectrum.

---

[1] The time of a vibrational period in CO is $\approx 1.54 \cdot 10^{-14}$ s; during that period the slowest photoelectron escaping from the $\sigma^* 2s$ MO travels over a distance of more than 100 Å. This exceeds the dimension of the molecule by nearly two orders of magnitude.

On the other hand, ejection from the $(\sigma 2p)$ MO leaves the internuclear distance virtually unchanged. According to the Franck-Condon principle, the transition from $v''=0$ to $v'=0$ is then by far the strongest. The energy required to eject an electron from a molecule in its ground vibrational state generating the positive ion in the lowest vibrational level of the respective electronic state is called the *adiabatic ionization energy*. In contrast, the *vertical ionization energy* corresponds to the transition from $v''=0$ to the vibrational level $v'=n$ whose wavefunction gives the largest overlap. For the electronic ground state and the third electronically excited state of CO^+ adiabatic and vertical ionization energies coincide (Fig. 2.3).

In the present case, we have only transitions to bound states of the molecular ion. This is also evident from energetics since the threshold for dissociative photoionization (Eq. 1.7, $E^*=0$) into $C^+ + O$ or $O^+ + C$ is 22.4 eV and 24.7 eV, respectively, [4,5] and thus above the photon energy.

An alternative method to PES is the threshold photoelectron spectroscopy (TPES). This technique uses a tunable light source and only electrons appearing at threshold (near zero eV) are recorded by scanning the wavelength. In TPES additional features due to autoionization resonances appear [6,7]. These are highly excited states of the neutral molecule embedded in the ionization contiuum. TPES has the advantage of high sensitivity. By applying a small electric field across the reaction chamber, nearly 100 % of the threshold electrons are collected. In PES, in contrast, the collection efficiency is controlled by the angular acceptance of the particular spectrometer (see Section 3.2).

2.2 Photoionization Mass Spectrometry (PIMS)

This method requires a tunable light source and a mass filter (Fig. 2.4). Ion yield curves are obtained by measuring the appearance of a specific ion (parent or fragment ion) as a function of the increasing photon energy. The ions created in the interaction volume are focused onto the entrance hole of the mass filter by appropriate electric fields and analyzed according to M/z.

Figure 2.5 illustrates an experimental result for ethylene in the energy range between 10 eV to 15 eV [8]. For this system the energetic thresholds (ΔH_o) for the involved dissociative photoionization processes are well known from different experimental methods [9-12]:

$$hv + C_2H_4 \rightarrow C_2H_3^+ + H \qquad \Delta H_o = 13.22 \text{ eV} \qquad\qquad 2.2$$

$$\rightarrow C_2H_2^+ + H_2 \qquad \Delta H_o = 13.12 \text{ eV}. \qquad\qquad 2.3$$

From Fig. 2.5 it is apparent that the appearance energy of the two fragment ions is very close to the energetic threshold, i.e., the excess energy in reaction 2.2 and 2.3 is close to zero at the experimental appearance energy of $C_2H_3^+$ and $C_2H_2^+$.

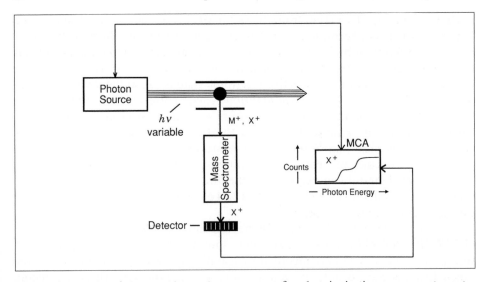

Fig. 2.4. Schematic of the experimental arrangement for photoionization mass spectrometry (PIMS). MCA: multichannel analyzer.

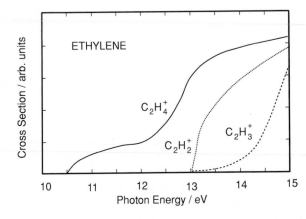

Fig. 2.5. Ion yield curves for C_2H_4 in the energy range 10-15 eV (adapted from [8]).

Determination of appearance energies in photoionization mass spectrometry has frequently been applied to determine unknown thermochemical parameters such as bond dissociation energies or heats of formation of radicals and ions [13-16]. In terms of heats of formation, the energy balance for dissociative photoionization (Eq. 1.7) can be expressed as

$$AE(X^+) = \Delta H_f(R) + \Delta H_f(X^+) - \Delta H_f(M) + E^*, \qquad\qquad 2.4$$

with

$$\Delta H_f(X^+) = \Delta H_f(X) + IE(X).$$

If one of the ΔH_f numbers is unknown, it can be derived from the experimental determination of the respective appearance energy. Such a procedure, however, is always uncertain because the excess energy is not necessarily zero at the experimental threshold as it is in the present example. As mentioned, E^* is composed of the kinetic energy of the photoelectron and the kinetic and internal energies of the fragments.

Figure 2.5 shows that the molecular ion $C_2H_4^+$ is formed with increasing intensity above the ionization threshold. In contrast, various molecules like $CFCl_3$ mentioned in the introduction do not form stable parent ions. It has been shown that only ionic fragments appear in photoionization mass spectrometry of $CFCl_3$ [13].

2.3 Photoelectron-Photoion Coincidence Spectroscopy (PEPICO)

This technique combines PES and PIMS in the way that the dissociation pathways of M^+ for a defined internal energy are determined. Consider photoionization at a fixed photon wavelength

$$hv \ (21.22 \ eV) + M \rightarrow M^+ + e^-$$
$$\longmapsto R_1 + X^+$$
$$\longmapsto R_2 + Y^+$$
$$\longmapsto \ .... \qquad\qquad 2.5$$

If the electron spectrometer is arranged to transmit electrons of a fixed single energy, the corresponding molecular ion is created with a defined amount of internal energy given by

$$E_{int}(M^+) = hv - \varepsilon - IE(M), \qquad\qquad 2.6$$

with $IE(M)$ the first ionization energy. If a fragment ion X^+ is then detected in coincidence with the electron, we know the internal energy of the precursor ion (M^+) from which X^+ is generated. The coincidence technique has to ensure that e^- and X^- arise from the same ionization event. By measuring coincidences at different electron energies, one can measure the appearance of X^+ as a function of the internal energy of M^+.

Figure 2.6 shows a schematic of the experimental arrangment for PEPICO. In contrast to PES a small electric field is applied in the reaction volume in order to draw out electrons and ions in opposite directions. This field will influence the electron energy resolution to some extent since photoelectrons are created across the finite thickness of the photon beam and hence at slightly different potential energies.

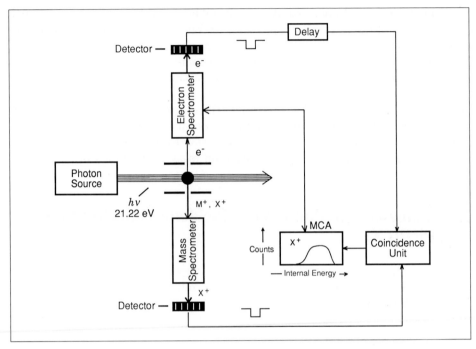

Fig. 2.6. Schematic of the experimental arrangement for photoelectron-photoion coincidence spectroscopy (PEPICO). MCA: multichannel analyzer.

Returning to the ionization process (Eq. 2.5), the electron spectrometer is set to transmit electrons of a single energy, and the mass spectrometer to transmit only fragment ions X^+. Due to its lower mass, the electron will arrive at the detector much faster than the ion X^+ $(t \sim (M)^{1/2})$. The amplified pulse generated by the arriving electron at the detector is delayed to compensate for most of the time difference between e^- and X^+. The coincidence unit is triggered by the delayed electron pulse and is ready for the registration of the ion pulse within a narrow time window. The appropriate setting of time delay and time window ensures that electron and ion indeed arise from the same ionization event. The window must have some finite width since the flight time of the ion to the detector may be influenced by translational excess energy from the unimolecular decomposition. The unit transfers the "coincidence event" to the multichannel analyzer (MCA). Each channel of the MCA (X-axis) corresponds to a certain electron energy. By repeating the procedure at different electron energies, one obtains the formation probability of X^+ as function of the internal energy of M^+. This is called an *ion breakdown curve*. The *breakdown diagram* of a system represents the (normalized) formation probabilities of all ions as function of the internal energy of M^+ (Fig. 2.7).

The basic problem in such an experiment is to minimize "false" or "random" coincidences. These are events which arise when an ion X^+ accidentally appears within the time window, i.e., X^+ originates from a precursor ion M^+ which contains a different amount of internal energy than that selected by the setting of the electron spectrometer.

We will illustrate this problem by an estimation of typical intensities expected in such an experiment.

A reasonable HeI discharge lamp delivers about 10^{11} photons (21.22 eV) per second. Photoionization cross sections of organic molecules are typically of the order of $\sigma = 10^{-16}$ cm^2 ($= 100$ Mbarn). Let the gas density in the reaction volume be $n = 10^{12}$ cm^{-3} and the interaction length between the photon beam and the gas beam be $l = 10^{-1}$ cm. The gas density corresponds to a pressure of some 10^{-5} mb. This guarantees single collision conditions since the pressure decreases rapidly (depending on the pumping speed) outside the interaction zone.

Incident (I_o) and transmitted photon beam (I) are related by the Lambert-Beer Law:

$$I = I_o \exp(-n \cdot \sigma \cdot l). \qquad\qquad 2.7$$

Since the number in the exponent is very small, $n \cdot \sigma \cdot l = 10^{-5}$, we can write

$$I = I_o (1 - n \cdot \sigma \cdot l),$$

and for the number of ionization processes per unit time

$$I_o - I = I_o \cdot n \cdot \sigma \cdot l = 10^6 \text{ s}^{-1}.$$

Note that out of 10^5 initial photons, only one leads to ionization, and

$$I/I_o = 0.999990.$$

The total of 10^6 ionization events per second is generally associated with an electron energy distribution extending over a range of several eV. For a selected single electron energy with the analyzer operating at a resolution of 20 meV and accepting 1% of the total solid angle (see Section 3.2), we expect an electron intensity on the order of 10–100 counts $\cdot$ s^{-1}, depending on the particular Franck-Condon transition selected by the electron energy analyzer. Flight-times of ions are (depending on their mass) on the order of 10 μs, with an accuracy of $\Delta t = 0.5$ μs. In a coincidence experiment, the unit is thus triggered 10-100 times a second by the delayed electron pulse, and is ready for the stopping pulse by fragment ion X^+ expected in the time window $\Delta t = 0.5$ μs. On the other hand, the average production rate of precursor ions M^+ is one per μs. This estimation shows that there is a realistic chance that a fragment X^+ arising from a precursor with internal energy different from that selected by the triggering electron may arrive within the time window of 0.5 μs. These false coincidences comprise a constant average background in the time-of-flight (TOF) win-

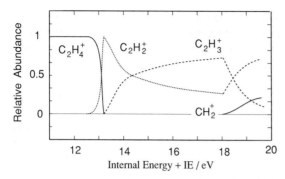

Fig. 2.7. Breakdown diagram for ethylene (adapted from [19]).

dow. The background itself will present no difficulty, as it can be subtracted from the resulting TOF spectrum. Statistical fluctuations, however, will inevitably be present in the background and will result in noise.

The uncertainty in the exact arrival time of fragment ions due to excess translational energy from the unimolecular decomposition is the inherent limitation in distinguishing between true and random coincidences.

The poor data collection rates in coincidence experiments can be improved by the so-called PIPECO method [17, 18], which uses photons of variable energy and detects electrons of zero kinetic energy (threshold photoelectrons) by scanning the wavelength of the light. As mentioned above, detection of zero eV electrons has the advantage of greater sensitivity (nearly 100 % of the electrons are accepted by the spectrometer), but the disadvantage that autoionization resonances may coincide in energy with Franck-Condon allowed transitions to an ionic state.

Figure 2.7 shows the result of a coincidence experiment in the case of ethylene. The breakdown diagram has been obtained in a PIPECO experiment [19]. It has been established that the threshold PES in the region between 11 and 20 eV was indeed almost entirely due to direct ionization (no autoionization).

Figure 2.7 shows that between the first ionization energy (10.5 eV) and 12.5 eV the parent cation is generated exclusively. Increasing the internal energy of $C_2H_4^+$ above 2 eV leads to a decrease of $C_2H_4^+$ in favor of $C_2H_2^+$. The incorporated unimolecular reactions with their thermodynamic limits are as follows

$$h\nu + C_2H_4 \rightarrow C_2H_2^+ + H_2 + e^- \qquad \Delta H_o = 13.12 \text{ eV} \qquad\qquad 2.8$$

$$\rightarrow C_2H_3^+ + H + e^- \qquad \Delta H_o = 13.22 \text{ eV} \qquad\qquad 2.9$$

$$\rightarrow C_2H_2^+ + H + H + e^- \quad \Delta H_o = 17.60 \text{ eV} \qquad\qquad 2.10$$

$$\rightarrow CH_2^+ + CH_2 + e^- \qquad \Delta H_o = 17.90 \text{ eV} \qquad\qquad 2.11$$

The PES spectrum (not presented here) shows different ionic states of $C_2H_4^+$ accessible by Franck-Condon transitions with vertical ionization energies of 10.51, 12.85,

14.66, 15.87, and 19.1 eV [20]. The breakdown diagram is obtained by dividing the number of coincident ions by the number of threshold electrons. It thus represents the relative formation probability of the different product ions as a function of the internal energy of M^+. It should clearly be emphasized that the ionization cross section depends on the Franck-Condon factors as reflected in the PES spectrum. For the present system, C_2H_4, the bands in the photoelectron spectrum associated with the formation of the different electronic states of $C_2H_4^+$ have some overlap so that ionization associated with zero eV electrons indeed occurs in the entire region between threshold and 20 eV, as evident from Fig. 2.7. The breakdown diagram, on the other hand, shows that no abrupt change in fragmentation occurs when the different ionic states are populated. The decomposition of $C_2H_4^+$, in fact, provides a text-book example of the success of the "quasi equilibrium theory" (QET) which will be discussed in more detail in the second Part of this volume. The essential prerequisite of the QET is that energy randomization in the ionized parent molecule occurs rapidly so that the unimolecular fragmentation is solely controlled by the amount of energy and the density of vibrational states, and not by the individual electronic state populated by the Franck-Condon transition. Energy randomization means that the internal energy deposited in the precursor ion (electronic and vibrational) is rapidly converted into vibrational energy of the electronic ground state of the ion.

The experimental breakdown diagram of $C_2H_4^+$ (Fig. 2.7) is, in fact, exactly reproduced by a QET calculation [9,19].

There are, of course, many limitations on the applicability of the statistical picture. In the case when the removal (or addition) of an electron has a specific influence on a localized bond $(R-X)$, the parent ion M^+ may directly dissociate along a repulsive potential energy surface and the statistical description is not applicable. On the other hand, a certain electronically excited state of the ion M^+ may live long enough that radiative transition to the ground state of M^+ (fluorescence) competes with decomposition. Such processes have been studied by photoion/photon coincidence spectroscopy [21,22]. The interested reader is referred to the excellent articles of Baer [23], Bear, Booze and Weitzel [24], and Maier and Thommen [25], who review the dissociation dynamics of energy-selected ions in photoionization.

Although this volume does not deal with laser spectroscopy, we mention here the recent high-resolution studies on state-selected ions excited in resonance-enhanced multiphoton ionization (REMPI) in combination with a particular high-resolution zero kinetic energy photoelectron spectroscopy (ZEKE-PES) [26]. This technique allows the study of unimolecular decay on vibrationally and rotationally resolved ions in molecules accessible in REMPI. These results have been reviewed by Neusser [27].

2.4 Electron Attachment Spectroscopy (EAS)

In electron attachment spectroscopy the appearance of negatively charged ions (parent and fragment ions) as a function of the incident electron energy is recorded

mass spectrometrically (Fig. 2.8). As mentioned above, electron capture is a resonant process which occurs dominantly at low electron energies. For that reason, it is
necessary to use an electron monochromator producing an electron beam of sufficient intensity down to very low electron energies. An ion yield curve is obtained by
synchronizing the MCA with the electron monochromator and scanning the incident electron energy for a fixed mass. The ion yield curve then exhibits pronounced
resonance profiles reflecting the vertical attachment energies in the parent molecule.

Figure 2.9 shows an experimental result in the case of perfluoro-2-butyne (C_4F_6)
[28]. The ion yield curves indicate that C_4F_6 possesses four resonances in the energy
region between 0 and 16 eV, a narrow resonance near 0 eV, and three broader resonances centered around 2, 6, and 11-12 eV. While capture of electrons near zero eV is
exclusively coupled with the formation of a long-lived parent radical anion M^-, the
TNIs formed within the resonances of higher energies immediately decompose into
negatively charged and neutral fragments.

Due to the high electron affinity of the F radical ($EA(F) = 3.399$ eV) [29], F^- formation may represent the most favorable decomposition channel. The energy balance
requires

$$\Delta H_o = D\,(R - F) - EA(F) \qquad\qquad 2.12$$

for the threshold of F^- formation. The C-F bond dissociation energy is not explicitly
known in the present system. It may, however, exceed the electron affinity of F by
1–2 eV [12] so that dissociative electron attachment is only expected for electron
energies above 1 eV. Observation of M^- is then solely controlled by its autodetach-

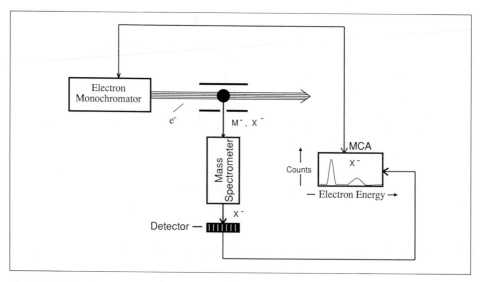

Fig. 2.8. Schematic of the experimental arrangement for electron attachment spectroscopy (EAS).
MCA: multichannel analyzer.

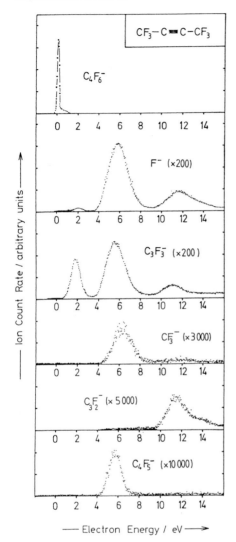

Fig. 2.9. Ion yield curves obtained in electron attachment to hexafluorobutyne (from [28]).

ment lifetime. For the present experiment the flight time of M^- from the source to the detector was 40 μs indicating an autodetachment lifetime at least on the order of 10^{-5} μs.

From energy considerations, it is clear that the 2 eV resonance decomposes according to

$$C_4F_6^- \ (2 \ eV) \rightarrow F^- + C_4F_5$$
$$\rightarrow C_3F_3^- + CF_3 . \qquad\qquad 2.13$$

On the other hand, F^- and $C_3F_3^-$ arising from the resonances of higher energy are likely to be associated with reaction channels consisting of more than one neutral fragment.

Dissociative electron attachment (e.g. F^- formation within the 2 eV resonance in C_4F_6) may be visualized in a simple potential energy diagram (Fig. 2.10). Since the motion of the incoming electron is fast compared to the motion of the atoms in the molecule, M^- is formed by a vertical (Franck-Condon) transition.

In analogy to photoionization, the cross-section for the transition from M to M^- can be approximated by

$$FCF = |\langle \Psi_i | \Psi_f \rangle|^2 . \tag{2.14}$$

In the present case, however, we have a transition from a *bound state* (Ψ_i) to a *continuum state* (Ψ_f), while the Franck-Condon-factor introduced earlier (Eq. 2.1) describes a transition between two bound vibrational states.

For sufficiently large slopes of the ionic potential energy surface $V_f(R)$, Ψ_f can be approximated by a delta function

$$\delta(R - R_T) = \begin{cases} 1, & R = R_T \\ 0, & R \neq R_T \end{cases}, \tag{2.15}$$

with R_T the "classical turning point". Figure. 2.10 shows that R_T varies between R_1 and R_2. Within that approximation, electronic transitions from $V_i(R)$ to $V_f(R)$ are possible for electron energies between ε_1 and ε_2. The approximation says that the

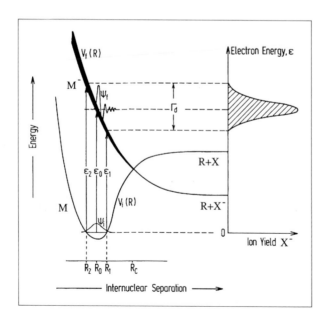

Fig. 2.10. Born-Oppenheimer potential energy diagrams illustrating resonant electron capture.

continuum wave function Ψ_f (related with the positional probability density Ψ_f^2 of the dissociating particles) oscillates rapidly for $R > R_T$, so that only the classical turning points contribute to the overlap integral 2.14.

The parent ion M^- generated by electron capture immediately decomposes into $R + X^-$. The shape of the X^- ion yield curve is thus an *image* of the Gaussian profile of the vibrational wave function Ψ_i^2 *reflected* at the repulsive potential $V_f(R)$. This approximation is known as "reflection principle" [30,31] which holds for transitions between bound and continuuum states.

From Fig. 2.10 it can also be seen that the width of the ion yield curve (the resonance), Γ_d, is directly proportional to the slope of the ionic potential surface

$$\Gamma_d \sim \left| \frac{d}{dR} V_f(R) \right|. \qquad\qquad 2.16$$

Strictly speaking, autodetachment of M^- (which is only possible for $R < R_c$) competes with dissociation. For $R > R_c$, on the other hand, the evolution into the individual particles $R + X^-$ has proceeded to such a stage that autodetachment is no longer possible. Due to the finite lifetime toward autodetachment, M^- possesses some finite width in energy according to Heisenberg's uncertainty principle (see Fig. 2.10).

Figure 2.10 illustrates the simplest case of a *direct electronic dissociation* along a purely repulsive potential energy surface. This picture is generally no longer adequate to describe the decompositions of larger polyatomic molecules. However, regardless of the fragmentation mechanism, the ion yield curve of any product ion is an image of the primary step of electron capture by the parent molecule. For complex fragmentation reactions, in particular, when different competitive fragmentation channels are accessible within the energy of TNI, the line shape of a particular ion yield curve may no longer rigorously be determined by the reflection principle. The profile may be "formed" or "weighted" by the energy-dependence of the individual reaction.

2.5 Electron Transmission Spectroscopy (ETS)

Electron transmission spectroscopy allows access to electron capture processes associated with the formation of short-lived temporary negative ions not observable by mass spectrometric techniques. Figure 2.11 illustrates the schematic experimental arrangement. The electron current transmitted through the gas under consideration is measured as a function of the incident electron energy. Incident (I_0) and transmitted current (I_t) are related by

$$I_t = I_0 \exp(-n l \sigma), \qquad\qquad 2.17$$

with σ now representing the *total* electron scattering cross-section. The formation of temporary negative ions (resonance scattering) is characterized by rapid changes of

the cross-section with electron energy. By that rapid variation, negative ion formation can be identified on the slowly varying background due to non resonant (direct)
scattering, as indicated in Fig. 2.11.

Any scattering event will cause the electron to deviate from its initial axial direction and will, hence, not contribute to the transmitted current measured at the electron collector.

Figure 2.12 gives an experimental example in the case of the nitrogen molecule
[32]. The top curve represents the incident electron current (no target gas in the
absorption cell), below that is the transmitted current through the chamber filled
with 10^{-2} mb N_2, and the bottom curve is a measure of the total electron scattering
cross-section:

$$\ln \frac{I_o}{I_t} = n \sigma l. \tag{2.18}$$

Figure 2.13 presents the transmission spectrum in its common form, namely, as
the *derivative* of the transmitted current $dI_t/d\varepsilon$. The derivative representation
enhances the sharp variations due to resonant scattering and suppresses the nearly
constant background of direct scattering.

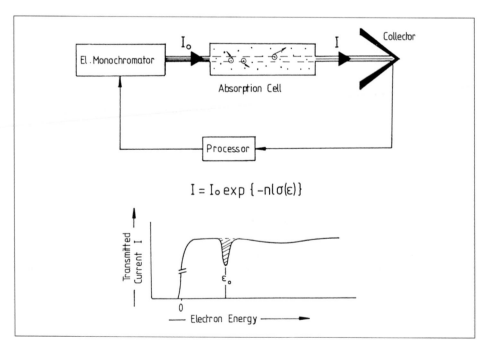

Fig. 2.11. Schematic of the experimental arrangement for electron transmission spectroscopy
(ETS).

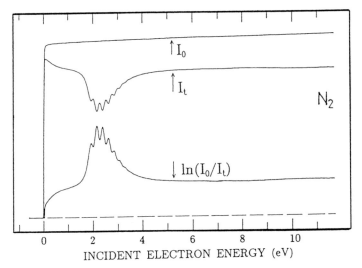

Fig. 2.12. Electron transmission in N_2. I_o, I_t: incident and transmitted electron current (see text, adapted from [32]).

The pronounced structures between 2–3 eV are due to resonant electron capture into the lowest unoccupied MO in N_2 (π^*) generating N_2^- in its different vibrational states as indicated by the potential energy diagram in Fig. 2.13. The present system is an instructive example that illustrates the complementary of PES and ETS with respect to electron removal from occupied or electron attachment to unoccupied MOs, respectively.

The short lifetime of N_2^- with respect to autodetachment is related with a potential energy curve having a finite width due to Heisenberg's uncertainty principle

$$\Gamma \approx \frac{h}{\tau},$$

2.19

with τ the autodetachment lifetime and h the Heisenberg constant ($h = 4.14 \cdot 10^{-15}$ eV s).

A detailed analysis of electron scattering on N_2 suggests that the autodetachment lifetime of N_2^- is on the order of a few vibrational periods, i.e., on the order of some 10^{-14} s [33,34].

N_2 represents a large class of molecules where the respective anion in its electronic ground state is unstable with respect to autodetachment. By definition, the electron affinity of such molecules is considered *negative*. For a great many molecules, including such important prototypes as ethylene, acetylene, or benzene, the ground states of the anions are known to be unstable in the gas phase. These temporary anions have comprehensively been studied in electron transmission spectroscopy as reviewed in the article of Jordan and Burrow [35].

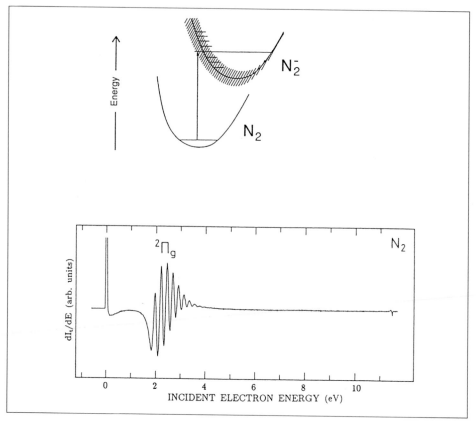

Fig. 2.13. Electron transmission in N_2. Derivative of the transmitted electron current (adapted from [32]) and potential energy diagrams of N_2 and N_2^-. The derivative representation enhances the structures due to negative ion formation.

On the other hand, atoms or molecules which can bind an excess electron in a thermodynamically stable state are considered to possess a *positive* electron affinity. The electron affinity is the basic quantity in the relation between a neutral molecule and its anion, as it is the ionization energy in the case of the cation.

From the energy balance (Eq. 1.11) electron affinities of fragments formed in dissociative electron attachment can be derived in the limit of the unknown excess energy. Exact determination of electron affinities is performed in photodetachment spectroscopy of negative ions. This method is the analogue to photoionization with one more electron in the system.

2.6 Photodetachment Spectroscopy

The photodetachment process

$$h\nu + M^- \rightarrow M + e^- \qquad\qquad 2.20$$

can either be studied with a fixed photon energy and by recording the electron energy distribution or by measuring the *decrease* of the ion intensity as a function of the photon energy. Since the binding energy of the extra electron is typically below 3–4 eV, lasers are frequently used. Laser photodetachment is the most accurate technique to determine electron affinities, or more precisely, vertical detachment energies.

For experiments using a light source of fixed wavelength the principal arrangement corresponds to that of PES (Fig. 2.1, p. 8), the target gas now consisting of a beam of negative ions [29,36].

An often applied technique to monitor the decrease of M^- as a function of the wavelength of the incident light uses the ion cyclotron resonance (ICR) spectrometry [37]. An ICR spectrometer allows to generate gas phase ions under controlled conditions and trap them for periods up to several seconds. When the instrument is operated in a steady-state mode (continuous ion formation), the photodetachment cross-section is determined by the decrease of the ion signal upon irradiation.

The cross-section behavior of electron impact ionization and photoionization will be discussed in the second Contribution to this volume. Here, we will briefly consider how the photodetachment cross-section behaves near threshold.

In a very general treatment, Wigner showed in a landmark paper [38] that the collision cross-section for a process with two final particles is only controlled by their long-range interaction. In photoionization of neutrals, the long-range force is always dominated by the Coulomb interaction between the photoelectron and the positive ion. This results in a photoionization cross-section which is independent of energy near threshold, yielding a step function. In polyatomic molecules the step function is often smeared out due to the population of different successive and overlapping vibrational levels, resulting in a more gradual increase of the cross section than in atoms, as can be seen in Fig. 2.5.

In photodetachment from anions, however, the departing electron interacts with the remaining neutral with a potential falling off much faster than r^{-1}. The long-range interaction will then be controlled by the centrifugal term in the effective potential

$$V(r) = \frac{\hbar l\,(l + 1)}{2m\,r^2}, \qquad\qquad 2.21$$

with $\hbar l$ the angular momentum of the detached electron, $\hbar = h/2\pi$.

For this interaction the Wigner formalism predicts the form of the cross-section as

$$\sigma \sim \Delta E^{l + 1/2}, \qquad\qquad 2.22$$

with ΔE the energy above threshold. Equation. 2.22 is known as "Wigner Law".

In the case of atoms, the usual dipole selection rule $\Delta l = \pm 1$ connects the angular momentum of the photoelectron with the atomic orbital (AO) from which it is ejected.

Photodetachment from an s orbital (e.g., in H^-) leads to $l=1$ allowed for the photoelectron and $\sigma \sim \Delta E^{3/2}$. Ejection from an atomic p orbital (e.g., in O^-) results in $l=0$ and $l=2$, and the cross-section near threshold will be dominated by $l=0$ and $\sigma \sim \Delta E^{1/2}$ (Fig. 2.14).

The Wigner Law can be extended to polyatomic systems by applying a simple algorithm based on group theory to find the "effective angular momentum" quantum number l^* of the MO from which the photoelectron is detached [39]. This algorithm has successfully been applied to describe photodetachment in many polyatomic molecules [37,39-43], including systems of very high angular momentum like the perinaphtenyl anion. The HOMO in the perinaphtenyl anion (D_{3h}) transforms like an atomic g orbital, corresponding to $l^* = 4$. This results in an extremely shallow increase of the cross-section near threshold, namely $\sigma \sim \Delta E^{7/2}$ [43].

Photodetachment spectroscopy has very recently gained particular interest as a *transition state spectroscopy* to probe the transition state in hydrogen transfer reactions of the form [44-46]

$$X + HY \rightarrow XHY \rightarrow XH + Y \qquad\qquad 2.23$$

by performing photoelectron spectroscopy on the anion XHY^-.

The transition state refers to the region of the *reaction coordinate* near the top of the barrier which separates reagents and products (Fig. 2.15). The aim of state-to-state chemistry over the last 15 years has been to gain information on *asymptotic properties*, and to conclude from that on the region of the potential energy surface in the vicinity of the transition state [47]. The most interesting challenge in the field of reaction dynamics is to *directly* measure the nature (energy, structure, lifetime) of a transition state in a chemical reaction. A transition state does not have stationary energy levels, but probably has metastable states with lifetimes that enable access by spectroscopic techniques. In the case of the hydrogen transfer reaction 2.23 one can use

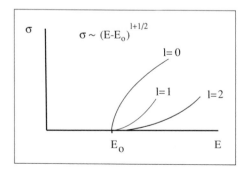

Fig. 2.14. Behavior of photodetachment cross section near threshold (E_o) according to Wigner's law, l denotes the angular momentum quantum number of the detached electron.

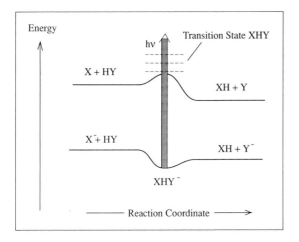

Fig. 2.15. Potential energy diagrams illustrating photodetachment from molecules of the type XHY⁻.

the anion (XHY)⁻ which exists as a thermodynamically stable molecule when X and Y are halogen atoms. Upon photodetachment the neutral system is generated in a geometry close to the transition state (Fig. 2.15) of reaction 2.23. If the photon energy is fixed, the signature of the photoelectron spectrum contains information on the transition state. This has been demonstrated in photodetachment studies of a series of XHY⁻ compounds [46].

References

1. Turner DW (1970) Molecular Photoelectron Spectroscopy. Wiley, London
2. Eland JHD (1984) Photoelectron Spectroscopy, II. edn. Butterworths, London
3. Rabalais JW (1977) Principles of Ultraviolet Photoelectron Spectroscopy. Wiley, New York
4. Radzig AA, Smirnov MB (1975) Reference Data of Atoms, Molecules and Ions, Springer Series in Chemical Physics, vol 31. Springer, Berlin
5. Herzberg G (1950) Spectra of Diatomic Molecules. Van Nostrand, New York
6. Eland JHD (1980) Branching in Electronic Autoionization of CS_2 and N_2O. Mol Phys 40:917-931
7. Guyon PM, Baer T, Ferreira LFA, Nenner I, Tabche-Fouhaille A, Botter R, Govers T (1979) Non-Franck-Condon Transitions in Resonant Autoionization of N_2O. J Chem Phys 70:1585-1592
8. Chupka WA, Berkowitz J, Refaey KMA (1969) Photoionization of Ethylene with Mass Analysis. J Chem Phys 50:1938-1941.
9. Berkowitz J (1979) Photoabsorption, Photoionization and Photoelectron Spectroscopy. Academic, New York
10. Franklin JL, Dillard JG, Rosenstock HM, Herron JT, Draxl K, Field FH (1969) Ionization Potentials, Appearance Potentials and Heats of Formation of Positive Ions. National Bureau of Standards, NSRDS-NBS 26, Washington
11. Lide DR, Jr (ed) (1985) JANAF Thermochemical Tables, 3rd edn. American Chemical Society, New York

12. CRC Handbook of Chemistry and Physics, 66th edn. CRC Press, Boca Raton, Florida 1985
13. Jochims H-W, Baumgärtel H (1976) Photoreactions of Small Organic Molecules. V. Absorption-, Photoion- and Resonancephotoelectron Spectra of CF_3Cl, CF_2Cl_2, $CFCl_3$ in the Energy Range 10-25 eV. Ber Bunsenges Phys Chem 80:130-138
14. Jochims H-W, Lohr W, Baumgärtel H (1979) Photoreactions of Small Organic Molecules. VI. Photoionization Processes of Difluoroethylenes. Nouveau Journal de Chimie 3:109-113
15. Lohr W, Jochims H-W, Baumgärtel H (1975) Photoreaktionen kleiner organischer Moleküle. IX. Absorptionsspektren, Photoionen- und Resonanzphotoelektronenspektren von Vinylbromid. Ber Bunsenges Phys Chem 79:901-906
16. Brutschy B, Bisling P, Rühl E, Baumgärtel H (1987) Photoionization Mass Spectrometry of Molecular Clusters Using Synchrotron Radiation. Z Phys D5:217-231
17. Stockbauer R (1973) Threshold Electron-Photoion Coincidence Mass Spectrometric Study of CH_4, CD_4, C_2H_6 and C_2D_6. J Chem Phys 58:3800-3815
18. Werner AS, Tsai BP, Baer T (1974) Photoionization Study of the Ionization Potentials and Fragmentation Paths of the Chlorinated Methanes and Carbon Tetrabromide. J Chem Phys 60:3650-3657
19. Stockbauer R, Inghram MG (1974) Threshold Photoelectron-Photoion Coincidence Mass Spectrometric Study of Ethylene and Ethylene-d_4. J Chem Phys 62:4862-4870
20. Kimura K, Katsumata S, Achiba Y, Yamazaki T, Iwata S (1981) Handbook of HeI Photoelectron Spectra of Fundamental Organic Molecules. Japan Scientific Society Press, Tokyo.
21. Maier JP (1980) Spectroscopic Studies of Open-Shell Organic Cations in the Gas Phase. Chimia 34:219-231
22. Maier JP (1982) Open-Shell Organic Cations: Spectroscopic Studies by Means of Their Radiative Decay in the Gas Phase. Acc Chem Res 15:18-23
23. Baer T (1979) State Selection by Photoion-Photoelectron Coincidence. In: Bowers MT (ed) Gas Phase Ion Chemistry, vol 1. Academic Press, New York
24. Baer T, Booze J, Weitzel K-M (1991) Photoelectron Photoion Coincidence Studies of Ion Dissociation Dynamics. In: Ng CJ (ed) Vacuum Ultraviolet Photoionization and Photodissociation of Molecules and Clusters. World Scientific, London
25. Maier JP, Thommen F (1984) Relaxation Dynamics of Open-Shell Cations Studied by Photoelectron Photon Coincidence Spectroscopy. In: Bowers MT (ed) Gas Phase Ion Chemistry, vol 3. Academic Press, New York
26. Chewter LA, Sander M, Müller-Dethlefs K, Schlag EW (1987) High Resolution Zero Kinetic Energy Photoelectron Spectroscopy of Benzene and Determination of the Ionization Potential. J Chem Phys 86: 4737-4744
27. Neusser HJ (1989) Lifetimes of Energy and Angular Momentum Selected Ions. J Phys Chem 93:3897-3907
28. Süzer S, Illenberger E, Baumgärtel H (1984) Negative Ion Mass Spectra of Hexafluoro-1,3-butadiene, Hexafluoro-2-butyne and Hexafluoro-cyclobutene. Identification of the Structural Isomers. Org Mass Spectrom 19:292-293
29. Mead RD, Stevens AE, Lineberger WC (1984) Photodetachment in Negative Ion Beams. In: Bowers MT (ed) Gas Phase Ion Chemistry, vol 3. Academic Press, New York
30. Taylor HS (1970) Models, Interpretations, and Calculations Concerning Resonant Electron Scattering Processes in Atoms and Molecules. In: Prigogine I, Rice SA (eds) Advances in Chemical Physics, vol XVIII. Interscience Publishers, New York
31. O'Malley TF (1966) Theory of Dissociative Attachment. Phys Rev 150:14-29
32 Allan M (1989) Study of Triplet States and Short-Lived Negative Ions by Means of Electron Impact Spectroscopy. J Electron Spectrosc Relat Phen 48:219-351
33. Birtwistle DT, Herzenberg A (1971) Vibrational Excitation of N_2 by Resonance Scattering of Electrons. J Phys B4: 53-70
34. Bernan M, Estrada H, Cederbaum LS, Domcke W (1983) Nuclear Dynamics in Resonant Electron-Molecule Scattering Beyond the Local Approximation: The 2.3 eV Shape Resonance in N_2. Phys Rev A28:1363-1381

35. Jordan KD, Burrow PD (1987) Temporary Anion States of Polyatomic Hydrocarbons. Chem Rev 87:557-588
36. Coe JV, Snodgrass JT, Freidhoff CB, McHugh KM, Bowen KH (1987) Photoelectron Spectroscopy of the Negative Cluster Ions $NO^-(N_2O)_{n=1,2}$. J Chem Phys 87: 4302-4309
37. Drzaic PS, Marks J, Brauman JI (1984) Electron Photodetachment from Gas Phase Molecular Ions. In: Bowers MT (ed) Gas Phase Ion Chemistry, vol 3. Academic Press, New York
38. Wigner EP (1948) On the Behavior of Cross Sections Near Thresholds. Phys Rev 73:1002-1009
39. Reed KJ, Zimmermann AH, Andersen HC, Brauman JI (1976) Cross Sections for Photodetachment of Electrons from Negative Ions Near Threshold. J Chem Phys 64:1368-1375
40. Janousek BK, Brauman JI (1979) Electron Affinities. In: Bowers MT (ed) Gas Phase Ion Chemistry, vol 1. Academic Press, New York
41. Wetzel DM, Brauman JI (1987) Electron Photodetachment Spectroscopy of Trapped Negative Ions. Chem Rev 87:607-622
42. Illenberger E, Comita PB, Brauman JI, Fenzlaff H-P, Heni M, Heinrich N, Koch W, Frenking G (1985) Experimental and Theoretical Investigation of the Azide Anion (N_3^-) in the Gas Phase. Ber Bunsenges Phys Chem 89:1026-1031
43. Gygax R, McPeters HL, Brauman JI (1979) Photodetachment of Electrons from Anions of High Symmetry. Electron Photodetachment Spectra of the Cyclooctatetraenyl and Perinaphthenyl Anions. J Am Chem Soc 101:2567-2570
44. Schatz GC (1990) Quantum Theory of Photodetachment Spectra of Transition States. J Phys Chem 94:6157-6164
45. Metz RB, Weaver A, Bradforth SE, Kitsopoulos TN, Neumark DM (1990) Probing the Transition State with Negative Ion Photodetachment: The Cl+HCl and Br+HBr Reactions. J Phys Chem 94:1377-1388
46. Bradforth SE, Weaver A, Arnold DW, Metz RB, Neumark DM (1990) Examination of the Br+HI, Cl+HI, and F+HI Hydrogen Abstraction Reactions by Photoelectron Spectroscopy of BrHI$^-$, ClHI$^-$, and FHI$^-$. J Chem Phys 92:7205-7222
47. Levine RD, Bernstein RB (1987) Molecular Reaction Dynamics and Chemical Reactivity. Oxford University Press, New York
48. Baker AD, Brundle CR (1977) An Introduction to Electron Spectroscopy. In: Brundle CR, Baker AD (eds) Electron Spectroscopy. Theory, Techniques and Applications, vol 1. Academic Press, London

3 Instrumentation

The experimental methods described in Chapter 2 require light sources in the VUV region, electron monochromators and analyzers, mass spectrometers and finally detectors for the registration of the electrons and ions involved in the corresponding processes.

This chapter is not intended to give a comprehensive overview on the many different instruments and experimental configurations described in the literature. Instead, we will focus on the basic operation principles and illustrate them in the case of some prototype devices.

3.1 Light Sources and Monochromators

Today lasers have supplanted traditional light sources in many branches of spectroscopy [1-3], and they may eventually also replace the sources used in VUV spectroscopy. The short wavelength limit of standard commercial laser systems has already been pushed into the UV region by excimer lasers. The available energy, however, is still below the ionization energy of most molecules [4, 5, 7]. A frequently applied approach is to use (tunable) lasers producing visible light (2-3 eV) and generate ions by multiphoton ionization (MPI). This can be achieved either by absorption of several photons via virtual states or through a real excited state of the molecule [6,8,9]. The latter technique is the resonant enhanced multiphoton ionization (REMPI) which requires frequency doubling of the laser light in order to have access to the first electronically excited state of the neutral molecule. Although laser technology is a rapidly developing field, the traditional VUV light sources are still important components for the study of photoionization processes [10].

3.1.1 Line Sources

The most commonly used photon source in conventional PES is a discharge in pure He which dominantly emits the HeIα resonance line at 584 Å $\cong$ 21.22 eV. It corresponds to the transition between the first excited state, He* $(1s2p, \, ^1P)$ and the ground state, He $(1s^2, \, ^1S)$.

Figure 3.1 shows a simple performance of such a resonance lamp [11]. The whole device is cylindrically symmetric, the aluminum cylinder acting as anode and the small tantalum tube as hollow cathode. The quartz capillary provides that the pressure inside the lamp (≈ 1 mb) is efficiently reduced to $\approx 10^{-6}$ mb in the main chamber so that differential pumping is not necessary. The capillary additionally acts as a

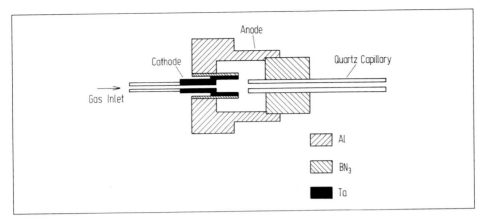

Fig. 3.1. He discharge lamp for the production of HeI radiation (adapted from [11]).

light tube and ensures that metastable (He* $(2^1S, 2^3S)$) and other highly excited He atoms are effectively quenched by collisions inside the capillary. Typical operation voltage is 500 V between cathode and anode with an output of 10^{10}–11^{11} photons·s^{-1}.

Higher members of the series He* $(1snp)$ are also present in the output, however, with an intensity normally less than 1 %. Positive ions which are generated in the discharge are rejected by electrodes or grids of appropriate potentials between the source and the ionization zone.

By reversing the polarity and/or changing the pressure in the discharge it is often possible to generate the HeIIα line (303 Å $\cong$ 40.81 eV) which corresponds to the transition between He$^+$ $(2p)$ and He$^+$ $(1s)$.

Of course, such a discharge lamp can be run by gases other than helium producing lines of lower energy. NeIα, for example, gives two lines at 16.67 and 16.84 eV (due to spin orbit splitting in the excited state), which complicates the interpretation of the PES spectra.

Many different designs of such line sources have been described in the literature including hot filament discharges [12,13], microwave discharges [14,15] or particular devices for the HeII production [16,17]. The interested reader is referred to the monograph of Samson [18].

3.1.2 Continuum Sources

Photoionization mass spectrometry requires photon sources of variable energy in the range above 8 eV. In this domain only a few continuum sources are available. The emitted radiation must then be monochromatized by appropriate dispersive elements. The most useful continuum source is synchrotron radiation originating from electrons or positrons accelerated in storage rings.

Laboratory VUV sources have also been developed [18,19]. One type uses the H_2 many line spectrum between 900 Å (13.8 eV) and 1650 Å (8.75 eV) and another uses He and Ar discharges. These discharges emit continua in the range 600 Å–1000 Å and 1000 Å–1400 Å, respectively. The hydrogen lamp operates at a pressure near 1 mb. The emitted light is due to transitions from excited H_2^* to the repulsive potential energy curve $^3\Sigma_u^+$ correlated with the two hydrogen atoms in their ground states (the antibonding combination of $H + H$). The noble gas continuum lamp operates at substantially higher pressures (100 mb) which requires extensive differential pumping to reduce the pressure in the monochromator and the main chamber. While the H_2 lamp works continuously, the noble gas lamp needs repetitively high power pulses in order to maintain the discharge. It is believed that transitions from weakly bound excited noble gas molecules to the repulsive potential energy curve of the ground state are responsible for these continua [18].

While the laboratory sources emit light in a comparatively narrow region, synchrotron radiation covers a much wider spectral range, as can be seen from Fig. 3.2.

In a storage, ring electrons of 100 MeV to 10 GeV are forced into circular orbits by strong magnetic fields. This motion is associated with the emission of light. The

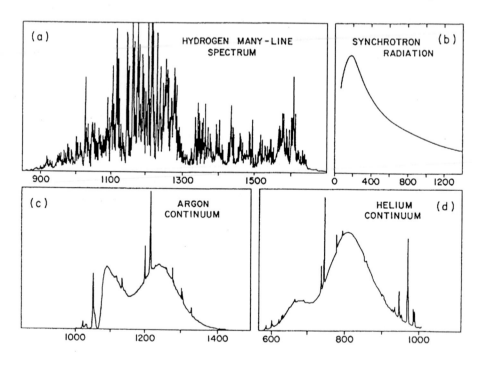

Wavelength / Å

Fig. 3.2. Comparison between laboratory continuum sources and synchrotron radiation (adapted from [19]).

energy loss is equalized by a corresponding acceleration in an RF cavity on every circuit. Synchrotron radiation is characterized by some unique features:
a) an absolute smooth continuum,
b) 100% polarization of the light in the plane of the ring, and
c) its time structure, i.e., light coming in short pulses (≈ 1 ns), separated by some 10 or 100 nanoseconds apart, according to the size of the ring and the particular operation mode.

In the meantime several storage rings as dedicated sources for synchrotron radiation have been built all over the world.

3.1.3 Dispersive Elements / Monochromators

Light from continuum sources has to be monochromatized in order to study photoionization processes at defined photon energies. The heart of a monochromator is its dispersive element which separates light in space according to its wavelength. As mentioned, in VUV spectroscopy the optical path has to be evacuated in order to avoid absorption by air which occurs for wavelengths below 2000 Å. Dispersive elements used in traditional spectroscopy like interferometers or prisms can no longer be used since even particular material like LiF is opaque for wavelengths below 1040 Å ($\hat{=}$ 11.9 eV).

Monochromators for the VUV domain operate exclusively with gratings as dispersive elements. A diffraction grating consists of a large number of parallel and equidistant grooves ruled on a hard material, the ruled surface being coated with a reflecting film. The fundamental theory for light dispersion on a reflecting grating is the same as for light passing through a series of narrow and parallel slits and may be found in elementary textbooks of optics or spectroscopy [23-26].

The condition for two rays reflected from two adjacent grooves to reinforce is that their path difference must equal an integer number of wavelength. If light strikes the surface under the angle α (angle of incidence, Fig. 3.3), this condition is fulfilled for

$$m \cdot \lambda = a \left(\sin \alpha - \sin \beta \right). \qquad\qquad 3.1$$

β is the angle of emergence, a the distance between two grooves, and m the order of the spectrum, m $= 0,1,2...$. Both angles are measured with respect to the normal of the macroscopic surface and are defined positive.

For a fixed order and constant angle of incidence, images of different wavelengths will be formed at different angles of emergence. However, for a constant value of β, Eq. 3.1 requires m $\cdot \lambda = $ const, which may be achieved for the combinations λ, $2\lambda/2$, $3\lambda/3$,.... Thus, the spectrum of 900 Å in first order will be observed under the same emergence angle as 450 Å in second order, 300 Å in third order, etc.. Separation of overlapping orders is indeed a serious problem when using light sources of wide spectral range like synchrotron radiation. It may be overcome by the use of an LiF

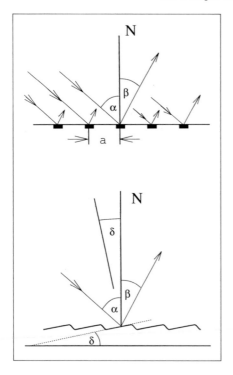

Fig. 3.3. a) Reflecting diffraction grating, α: angle of incidence, β: angle of emergence, a: distance between two grooves. b) Blazed grating (see text).

cutoff filter when operating in the range below 11.9 eV (LiF cutoff) or by the use of materials whose reflectance effectively drops down at shorter wavelengths.

We will now briefly consider a few important properties of diffraction gratings. The *dispersive power* is defined as $|\Delta\beta/\Delta\lambda|$. If the angle of incidence in arranged constant, Eq. 3.1 yields

$$\left|\frac{\Delta\beta}{\Delta\lambda}\right| = m\;\frac{1}{a\cos\beta}\,.$$
3.2

The dispersive power thus increases linearly with the number of lines per unit width and the order of the spectrum.

The *resolving* power $(\lambda/\Delta\lambda)$ can be expressed [18] as

$$\frac{\lambda}{\Delta\lambda} = m\cdot N\,,$$
3.3

with N the total number of ruled lines exposed to the incident radiation.

To obtain a resolution of $\Delta\lambda = 0.1$ Å at a wavelength of $\lambda = 500$ Å, one needs a grating of 5000 ruled lines. In terms of energies $(E(\mathrm{eV}) = 12.4\cdot10^{3}/\lambda(\text{Å}))$ this corresponds to $\Delta E = 4.96$ meV at a photon energy of $E = 24.8$ eV. This number should be kept in mind when the resolving power of electron monochromators and analyzers is discussed. At present, gratings with more than 10 000 lines per cm are routinely available.

The main disadvantage of a diffraction-grating is that it disperses the incident energy over a large number of orders. Modern gratings are often machined in the way that the grooves have a definite shape in order to concentrate the diffracted light into a definite direction. Such a *blazed grating* is illustrated in Fig. 3.3. The principle of concentrating energy into a given wavelength is that this wavelength must be diffracted into a direction corresponding to the speculary reflected beam with respect to the surface of the facet, as illustrated in Fig. 3.3

This condition requires

$$\alpha - \delta = \delta + \beta. \qquad\qquad 3.4$$

With Eq.3.1 at normal incidence ($\alpha = 0$) an elementary calculation leads to

$$m \cdot \lambda_{bl} = a \sin 2\delta \qquad\qquad 3.5$$

for the wavelength λ_{bl} for which the grating is blazed.

The *reflectance* of a grating depends on the material used and the method of producing the reflecting surface.

The reflectance of a metal surface generally decreases with energy in the VUV region. A gold surface, for example, possesses 12 % reflectance at 1000 Å, decreasing to 3 % at 400 Å in normal incidence [18].

However, from the theory of metals it follows that there is a critical angle Θ_e and a minimum wavelength λ_{min} related by

$$\sin \Theta_e = \Gamma \cdot \lambda_{min}, \qquad\qquad 3.6$$

for which total reflection occurs. Θ_e is the "crazing angle" (measured with respect to the surface of the grating) and Γ a constant which essentially depends on the electron density of the material.

As an example, an Al surface reflects a few percent of the incident light at $\lambda = 8$ Å and $\Theta = 10°$. This is increased to nearly 100 % for $\Theta \leq 5°$! Thus, by operating in crazing incidence, it is possible to use reflecting gratings deep in the X-ray region.

In practice, nearly all VUV monochromators are realized by combining the principle of diffraction with the focusing properties of a concave mirror. In 1882, Rowland discovered that if a concave grating placed tangentially to a circle of diameter D, which equals the radius R of the curvature of the grating, the spectrum of an illuminated point lying on this "Rowland circle" will be focused on this circle (Fig. 3.4).

For monochromators in use with synchrotron radiation it is essential that the entrance and exit slit remains fixed when the wavelength is scanned. Otherwise, the whole equipment has to be moved, maintaining vacuum conditions.

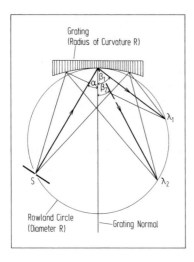

Fig. 3.4. Concave grating and Rowland circle: the spectrum of an illuminated point (S) lying on the Rowland circle will be focused on the circle.

The simplest scanning mechanism of a monochromator is the rotation of the grating over a vertical axis through the center of the grating which is known as Seya-Namioka mounting. When the angular separation between entrance and exit beam is 70° a simple rotation of the concave grating varies the wavelength and maintains the image in approximate focus [27].

More complicated motions involving rotation as well as translation of the grating on scanning the wavelength are used in high resolution monochromators. Resolutions better than 0.1 Å in the VUV are availabe by standard commercial instruments. (For more details the interested reader is referred to the relevant monographs [18-20,22].)

3.2 Electron Energy Analyzers and Monochromators

An electron energy analyzer is used when electrons arising from a process like photoionization are analyzed according to their energies. An electron monochromator is applied to generate a beam of (nearly) monoenergetic electrons of variable energy. Monochromatized electron beams are used in electron-scattering experiments like electron transmission and electron attachment spectroscopy.

Free electrons are usually produced by emission from an electrically heated metal filament. Due to the high temperature and the electric potential drop across the filament, such electrons possess a broad energy distribution, usually between 0.5 and 1 eV. From this distribution the electron monochromator cuts out a small width, typically in the range 20–50 meV. These monochromatized electrons are then accelerated or decelerated to the desired energy by appropriate electrostatic fields and focused into the reaction volume.

Many devices can be operated as analyzers as well as monochromators. In such cases, we may speak of an electron energy filter. The most interesting parameter of an electron energy filter is its energy resolution $\Delta\varepsilon$. It is usually described in terms of the full width at half maximum (FWHM) of the energy pass band. A further property of interest is the angular acceptance when used as analyzer and the available electron current when used as monochromator.

Most filters are of the *dispersive* type, applying static electric or magnetic fields either separately or in combination.

It should be noted that a pure electrostatic filter can also be used to analyze the *kinetic energies of ions*, since the trajectory of a charged particle depends solely on its kinetic energy. (In the case of positive ions the direction of the electrostatic field has to be reversed.) In PES, electron energy analysis can also be achieved through time-of-flight (TOF) techniques, provided that a pulsed light source is available [28,29] (see Section 3.2.9).

To obtain high energy resolution, electrostatic filters require Mu metal shielding to block the earth's magnetic field.

3.2.1 Retarding Field Analyzer

The principle of this method is to apply a retarding potential between two or more grids, permitting only those electrons with energies higher than the retarding potential to reach the detector (Fig. 3.5). Spectra are produced by recording the electron current at the detector as a function of the stopping potential. The spectrum obtained is called an *integral spectrum* since any step is superimposed of the current of all electrons of higher energy. Fig. 3.5a illustrates a configuration with spherical grids and 3.5b illustrates one with plane retarding grids [30]; the latter one is rotationally symmetric. While the spherical configuration analyzes electrons ejected into the whole solid angle [31,32], the angular acceptance of the plane configuration is determined by the aperture of the exit hole in the ionization chamber. The diverging electrons leaving the ionization chamber are made parallel by an einzellens before entering the plane retarding grids. The electrons are then conveniently detected by an electron multiplier.

For an exit hole of 2 mm diameter at a distance of 1 cm from the reaction volume, we have an angular acceptance of $d\Omega/\Omega = 2.5 \cdot 10^{-3}$, i.e., 0.25% of the total solid angle. Figure 3.6a shows schematically the integral spectrum one expects in photoionization of Ar at a fixed wavelength, yielding the ion in its two different states: $Ar^+ (^2P_{3/2})$ and $Ar^+ (^2P_{1/2})$; they are separated in energy by 0.18 eV, thus creating two steps when scanning the stopping potential.

A differential spectrum (Fig. 3.6b) can be obtained, either numerically by differentiating the integral spectrum, or experimentally by sweeping the retarding voltage between ε and $\varepsilon + d\varepsilon$ and taking the difference of the current registered between ε and $\varepsilon + d\varepsilon$. The disadvantage of this experimental technique is the comparatively high statistical noise which depends on the *total* electron current (on the lefthand

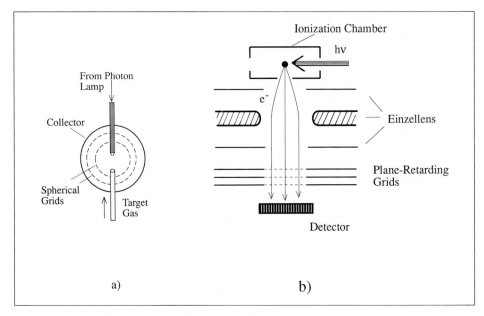

Fig. 3.5. Retarding field analyzers: a) spherical grids, b) plane retarding grids.

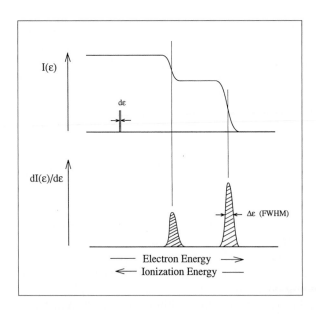

Fig. 3.6. Integral and differential photoelectron spectrum obtained in photoionization of Ar by HeI radiation. $\Delta\varepsilon$ is the width (first width at half maximum, FWHM) of the photoelectron peak which represents the energy resolution of the spectrometer, $d\varepsilon$ is the sweeping amplitude for the differentiation procedure (see text).

side of the spectrum one has to take the difference of two large numbers, giving a number near zero). Energy resolutions on the order of 20–50 meV can routinely be obtained in retarding field analyzers.

The principle of differentiating retarding potentials has also been applied to generate nearly monoenergetic electron beams. This technique is known as retarding potential difference (RPD) method [33,34].

3.2.2 Parallel Plate Analyzer

The parallel plate analyzer is the simplest design in the group of *deflection analyzers*. These are instruments which use electric and/or magnetic fields to make electrons of different kinetic energy follow different paths.

In the parallel plate analyzer (Fig. 3.7) a homogeneous electric field E is produced by a potential difference ΔV between the two plates, separated by a distance d.

If electrons of velocity v_o enter the analyzer at an angle Θ, the electric field causes an acceleration in the $(-y)$ direction while it has no effect on their motion in x direction. The differential equations for motion are

$$\frac{d^2y}{dt^2} = -a \qquad\qquad 3.7$$

$$\frac{dx}{dt} = v_{ox}, \qquad\qquad 3.8$$

with $a = eE/m$. The problem is analogous to a ballistic trajectory in the gravitational field. Because of its simplicity, we present the explicit solution for this case. Note that for a ballistic trajectory the acceleration is independent of mass $(a = g)$. The components of the initial velocity vector (v_o) are

$$v_{oy} = v_o \sin \Theta$$

$$v_{ox} = v_o \cos \Theta \qquad\qquad 3.9$$

Integration of 3.7 and 3.8 yields

$$y = v_{oy} t - \frac{1}{2} at^2 \qquad\qquad 3.10$$

$$x = v_{ox} t . \qquad\qquad 3.11$$

This is the parametrized representation of the trajectory. Elimination of t yields

$$y = x \frac{v_{oy}}{v_{ox}} - \frac{ax^2}{2 v_{ox}^2}, \qquad\qquad 3.12$$

which is the expression for a parabolic curve.

From $dy/dt = 0$, we obtain the time an electron needs to reach the maximum,

$$t_m = \frac{v_{oy}}{a}.$$

     3.13

The second half of the trajectory is the mirror image of the first. So the time until an electron returns to the lower plate is $2\, v_{oy}/a$. The distance they have travelled in the x direction is then

$$R = \frac{2\, v_{ox}\, v_{oy}}{a}$$

or, according to 3.9

$$R = \frac{2\, v_o^2 \sin \Theta \cos \Theta}{a} = \frac{v_o^2 \sin 2\Theta}{a}.$$

     3.15

Replacing v_o^2 by $2eV_o/m$ (eV_o is the initial energy of an electron in electronvolts) and the acceleration a by eE/m, we obtain

$$R = \frac{2eV_o}{eE} \sin 2\Theta.$$

     3.16

Equations 3.15 and 3.16 show that the trajectory is independent of the mass of the charged particle so that the analyzer can also be used for kinetic energy analysis of

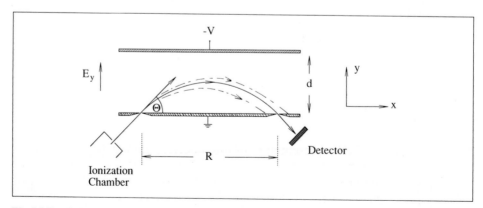

Fig. 3.7. Parallel plate analyzer. The three electron trajectories correspond to electrons entering the analyzer under the same angle ($45°$), but with different kinetic energies.

ions. For a ballistic trajectory ($a = g = 9.81 \text{ m s}^{-2}$) the range is solely determined by the initial velocity, leading to

$$R = \frac{2T}{mg} \sin 2\Theta,$$ 3.16a

with T the initial kinetic energy of mass m moving in the gravitational field.
    Expanding Eq. 3.16 in a Taylor series as a function of $\Delta\Theta$

$$R = R_o + \frac{\partial R}{\partial \Theta} \Delta\Theta + \frac{1}{2!} \frac{\partial^2 R}{\partial \Theta^2} (\Delta\Theta)^2 + \dots,$$ 3.17

we see that the linear term

$$\frac{\partial R}{\partial \Theta} = \frac{4 V_o}{E} \cos 2\Theta$$ 3.18

vanishes for $\Theta = 45°$, while the quadratic term does not. We say the $45°$ parallel plate analyzer has *first order focusing*. In the case when the quadratic term also vanishes, we have second order focusing. Equation 3.17 (and related equations expanding the trajectory in a Taylor series as a function of various perturbations) applies to all deflection analyzers. Note that R has a maximum for $\Theta = 45°$.
    The parallel plate analyzer focuses electrons in *one plane*, which is therefore called single focusing. Other instruments like the hemispherical filter (see Section 3.2.4) focus in *two planes* and are said to have double-focusing properties.
    In practice, the distance R is fixed by the position of the entrance and exit slits and electrons of different energies are brought into focus (on the exit slit) by varying the electric field. If the lower plate is grounded and V is the potential of the upper plate, then $E = V/d$ and Eq. 3.16 becomes

$$\frac{2 V_o}{V} = \frac{R}{d} \quad \text{or,}$$

$$V_o = V \cdot \frac{1}{2} \frac{R}{d}.$$ 3.19

We have thus a linear dependence between the electron energy $(e V_o)$ and the potential applied to the upper plate. If the configuration is arranged such that the distance R is twice the distance between the plates, the potential applied to focus electrons (in volts) is numerically equal to the energy of these electrons (in electronvolts). By scanning the potential of the upper plate, we directly obtain a differential spectrum.

In a real instrument, the entrance and exit slits will have finite widths, ΔS_1 and ΔS_2. The energy resolution is then given by an equation of the form [35]

$$\frac{\Delta \varepsilon}{\varepsilon} = \frac{S_1 + S_2}{R} + \text{higher order terms}.\qquad\qquad 3.20$$

The last term includes the deviation of the electron entry angle from 45°. An equation of the form 3.20 holds for any electrostatic energy analyzer.

3.2.3 127° Cylinder Filter

This instrument is often used as electron energy analyzer and as electron monochromator. The electric field E between the plane concentric analyzer segments (Fig. 3.8) is given by

$$E(R) = \frac{\Delta V}{R \ln \dfrac{R_2}{R_1}},\qquad\qquad 3.21$$

with R_1 and R_2 the radii of curvature of the inner and outer cylinder, respectively, and ΔV the potential difference between them.

For electrons travelling along a circular orbit, $R = R_o$, centrifugal and radial forces must be equal:

$$\frac{mv_o^2}{R_o} = eE(R_o).\qquad\qquad 3.22$$

With $mv_o^2 = 2\,eV_o$ (as above), we obtain

$$\frac{2\,V_o}{R_o} = \frac{\Delta V}{R_o \ln \dfrac{R_2}{R_1}},$$

which gives

$$\Delta V = 2\,V_o \ln \frac{R_2}{R_1}.\qquad\qquad 3.23$$

The potential difference required for a circular motion depends linearly on the energy of the electrons and is independent of R_o.

The differential equations of motion and their solution are not as simple as in the case of the parallel plate analyzer, and we will present only the results.

It can be shown that for $\alpha = \pi/\sqrt{2} = 127°\,17'$ (Fig. 3.8) the instrument has strong first order focusing properties in one plane [36,37]. This means that a diverging beam entering the analyzer through the entrance slit is focused first order onto the exit slit.

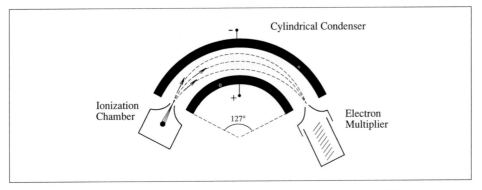

Fig. 3.8. 127° cylinder analyzer. Electrons of the same energy injected under different angles are focused (first order) onto the exit slit (see text).

One problem is that the slits are not in a plane of constant potential and may thus disturb the electric field between the plates. This effect can be minimized by fringing field corrector elements or specially shaped slits as indicated in Fig. 3.8.

3.2.4 Hemispherical Analyzer

In this configuration electrons travel in the field between the sections of two concentric spheres. The electric field between the spheres is

$$E(R) = \frac{\Delta V\, R_1 R_2}{(R_2 - R_1)} \cdot \frac{1}{R^2}, \qquad\qquad 3.24$$

with R_1 the radius of the inner sphere, R_2 that of the outer sphere, and ΔV the electric potential difference.

In analogy to the cylindric analyzer, the condition for circular motion of electrons with energy eV_o along a plane of constant potential, $R = R_o$, leads to

$$\Delta V = \frac{2 V_o (R_2 - R_1)}{R_1 R_2} \cdot R_o. \qquad\qquad 3.25$$

Note that in this case ΔV depends on R_o. For the median ray, $R_o = \dfrac{R_1 + R_2}{2}$, Eq. 3.25 becomes

$$V_o = \frac{\Delta V}{\left(\dfrac{R_2}{R_1} - \dfrac{R_1}{R_2} \right)}. \qquad\qquad 3.36$$

It has been shown [38] that for 180° (two half hemispheres) the configuration has *first order double focusing* properties. This means that conical diverging electrons entering the entrance aperture are focused (first order) onto the exit aperture.

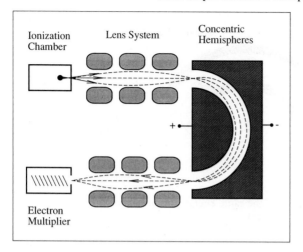

Fig. 3.9. 180° hemispherical analyzer with an electrostatic lens system (virtual entrance and exit slits).

Kuyatt and Simpson [39] developed a most elegant device by replacing real entrance and exit slits with "virtual slits" represented by an electrostatic lens system which provides images of real physical apertures (Fig. 3.9). Apart from photoelectron spectrometry, the Kuyatt-Simpson design has often been adopted as combined monochromator-analyzer system in electron scattering experiments such as electron energy loss spectroscopy (EELS) [40-42]. Electron energy resolutions below 20 meV and electron currents in the range of 10^{-9} - 10^{-8} A have been reported.

3.2.5 Cylindric Mirror Analyzer

In the 127° analyzer the electrons travel in a radial plane around the cylinder axis, while the trajectories in the cylindrical mirror analyzer occur in a plane containing

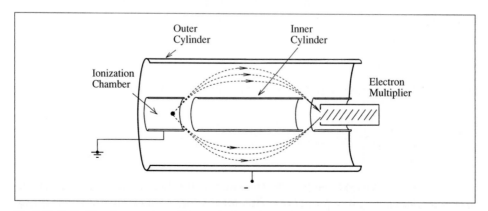

Fig. 3.10. Cylindric mirror analyzer. Electrons ejected over the whole radial angle (2π) are accepted.

the cylinder axis (Fig. 3.10). This instrument has strong second order focusing properties for angular deviations in the planes of the trajectories and, due to its geometry, perfect focusing for variations of the radial angle [43,44]. Since focusing is good over a wide range of angles [45,46], the magic angle (54°44′) introduced in Section 2.1 can conveniently be used. Because of its particular geometry the CMA has exclusively been applied as analyzer.

3.2.6 Trochoidal Electron Filter

The trochoidal electron monochromator employs a combination of an electric and a magnetic field (Fig. 3.11). The use of an axial magnetic and a perpendicular electrostatic field for energy dispersion was first realized by Stamatovic and Schulz [47,48]. The instrument has subsequently often been used to generate monochromatized electron beams in electron transmission [49-52] and electron attachment [53-55] studies.

In the trochoidal monochromator the electrons emitted by a heated filament are aligned by the axial magnetic field and enter the deflection region defined by the crossing of the magnetic field $\vec{B}$ with the electric field $\vec{E}$. The latter is generated by a potential difference between the two parallel plates C_1 and C_2 (Fig. 3.11). Under the influence of the combined electric and magnetic field the electrons describe a trochoidal or cycloidal motion, depending on whether or not they possess velocity components perpendicular to the z axis when entering the deflection region. The trochoidal or cycloidal path implies motion of its guiding center with the constant velocity

$$\vec{v}_x = \frac{[\vec{E} \times \vec{B}]}{|\vec{B}|^2} \qquad\qquad 3.37$$

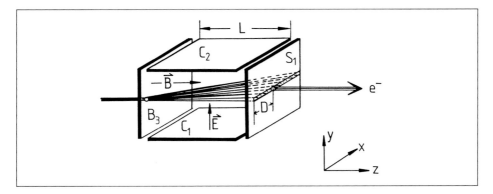

Fig. 3.11. Trochoidal electron monochromator. The electrons are dispersed according to their z-velocity along the x axis.

along the x direction, i.e., along a plane of constant electrostatic potential. This results in a dispersion of the electrons according to their z velocity, and only those electrons which reach the exit aperture (displaced by some distance D with respect to the entrance aperture) are transmitted. These energy selected electrons are then accelerated or decelerated by subsequent electrodes before entering the reaction volume. A calculation of the electron trajectories and the electron energy distribution has been carried out by several authors [56-58].

Attempts have been made to use the trochoidal electron filter also as an analyzer in electron energy loss spectrocopy [59]. Several background problems, however, prevented a broader use. These shortcomings have recently been overcome with an instrument designed by Allan et al. [49,60,61]. They obtained excellent results using two analyzers in series. The trochoidal electron monochromator has shown to be particularly suited to the study of (dissociative) electron attachment processes in the very low energy region [53]. The axial magnetic field prevents spreading of the electron beam so that reasonable intensities can be achieved even near zero eV electron energy. Beam spreading at low energies is a major problem when electrostatic devices are used. In addition, the alignment of the electron beam by the magnetic field allows a separation of electrons from negative ions thus preventing the electrons from reaching the ion detector.

3.2.7 Wien Filter

The Wien filter also uses a combination of a static electric and magnetic field. In that case, electric field, magnetic field, and electron momentum are mutually perpendicular (Fig. 3.12). The instrument is operated in such a way that the electrostatic force (eE) is balanced by the velocity dependent magnetic force $(e[\vec{v} \times \vec{B}])$. The condition for electrons travelling on a linear trajectory is then

$$v = \frac{E}{B}.$$ 3.38

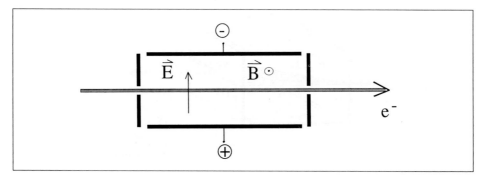

Fig. 3.12. Schematic of the arrangement for the Wien filter. The configuration can be used as velocity selector for electrons and as mass analyzer for ions (see section 3.3.6).

Some decades ago the Wien filter was used as electron monochromator and electron analyzer in high energy scattering experiments [62,63]; energy resolutions below 0.01 eV have been reported.

The instrument can also be used as a mass filter [64,65] provided the ions enter the analyzer with the same constant energy eV_o. The condition for a linear trajectory of ions is

$$\frac{E}{B} = \left(\frac{2\,e\,V_o}{M}\right)^{1/2}$$ 3.39

3.2.8 Threshold Photoelectron Analyzer

The threshold analyzer detects just those photoelectrons ejected within a few millielectronvolts of zero. As mentioned in Section 2.1 threshold photoelectron spectroscopy (TPES) is used in combination with a continuum light source. A small electric field is applied to accelerate the photoelectrons towards an electrode with parallel tubular apertures (Fig. 3.13). Then, only electrons having velocity vectors along the axis of the apertures, i.e., electrons initially created near zero eV, can penetrate them to reach the detector.

Of course, a small amount of energetic electrons having initial velocity vectors along the axis will also be transmitted. This depends a) on the aspect ratio ($d/2l$, Fig. 3.13) of the cylindrical channels, and b) on the ratio of initial kinetic energy of the photoelectrons with respect to the energy gain in the extraction field, to what degree the background of energetic electrons will contribute to a threshold photoelectron spectrum. A comprehensive geometrical analysis of this problem is given by Spohr et al. [66].

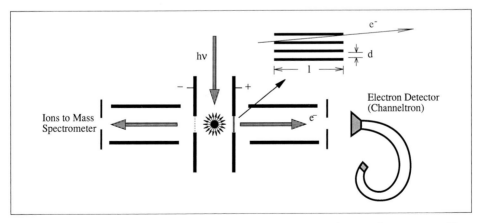

Fig. 3.13. Threshold photoelectron analyzer. Only electrons with initial energies near zero eV are transmitted through the array of tubular apertures.

Several attempts have been made to suppress these "hot" electrons. Peatman et al. [67] use a geometry with the reaction volume having no direct "line of sight" access to the apertures. A particular choice of potentials in the ionization chamber provides that only zero eV electrons have velocity vectors normal to the aperture while *all* electrons having initial kinetic energy are not transmitted.

Baumgärtel et al. [68] applied a 90° electrostatic deflection "postmonochromator" for a crude separation of zero eV electrons from hot electrons. Electron energy resolutions below 10 meV have been obtained in TPES.

Since the static electric field extracts photoelectrons and photoions in opposite directions, coincidences between electrons and ions can conveniently be measured [69-71]. As mentioned, the zero eV analyzer detects nearly 100% of the threshold electrons formed in the reaction volume.

3.2.9 Further Spectrometers and Limiting Effects on Energy Resolution

Many more electron spectrometers than those introduced above have been described in the literature.

We here mention the Bessel box (also known as the pill box) which consists of a cylindrical cavity with grounded end plates containing circular slits. Interestingly, its design was optimized by numerical calculation of the electric field and the electron trajectories, instead of an analytical solution of the equations of motion [72,73].

Considerable enhancement of the collection efficiency can be achieved in configurations where retarding fields and deflection of the electrons are combined (hybrid analyzers) [74,75]. Inhomogeneous fields to parallelize the electrons emerging from a source will also improve the collection efficiency [76,77].

When a TOF analysis is used, flight time t and electron energy are connected by

$$ t = s \left(\frac{m}{2\varepsilon} \right)^{1/2} , \qquad\qquad 3.40 $$

where s is the travelling distance. For a typical distance, say $s = 10$ cm, a 10 eV electron travels $t = 54$ ns.

To apply the TOF principle for photoelectron energy analysis, one needs a light source that delivers sufficiently narrow pulses, and a detector system with an equivalent time resolution.

Today, pulsed laser systems or synchrotron radiation can provide light pulses well below 1 ns. Standard electron multipliers and channeltrons have time resolutions on the order of 10 ns, while the resolution of channel plates is generally well below 1 ns (see Section 3.4.2).

For an estimated overall time resolution of $\Delta t = 1$ ns and a nominal electron energy of 10 eV, the TOF analysis (Eq. 3.40) is associated with an energy resolution of $\Delta\varepsilon = 370$ meV. This is poor compared to any electrostatic instrument. From Eq. 3.40, however, it follows immediately that

$$\Delta\varepsilon \sim \varepsilon^{3/2} \Delta t . \qquad\qquad 3.41$$

The energy resolution is thus considerably improved at low energies. For the same given time resolution and a nominal energy of 1 eV, we have $\Delta\varepsilon = 12$ meV!

At this point we again mention the "zero kinetic energy photoelectron spectroscopy" (ZEKE-PES) which uses a TOF analysis of threshold electrons formed in laser multiphoton ionization [78-80]. This technique can provide electron energy resolutions below 0.1 meV, allowing ionizing transitions with full rotational state selection at threshold [81].

We will finally discuss some effects which limit the energy resolution of an energy analyzer. The theoretical resolution can be made arbitrarily high by narrowing the entrance and exit slits and keeping the divergence of the entering electrons low (Eq. 3.20). However, due to local variations of the surface potential of the spectrometer and the aperture electrodes and due to unwanted electric and magnetic fields, the experimental resolution is often worse than predicted by theory. Low energy electrons are particularly sensitive to these perturbations so that $\Delta\varepsilon/\varepsilon$ is generally not a constant below 5 eV as predicted by Eq. 3.20.

In addition to these experimental shortcomings there are three physical effects which principally limit the electron energy resolution in photoelectron spectroscopy. Although in practice only one effect contributes, we will briefly discuss all three of them.

a) Recoil Energy Between Electron and Ion

The fraction of energy carried off by the ion is equal to the mass ratio between electron and ion. In the most unfavorable case, ionization of atomic hydrogen by HeI radiation, the H^+ ion carries away 4 meV of the 7.6 eV electron energy. For molecules the recoil energy is thus completely negligible.

b) Line Width of the Light Source

The natural width of a spectral line (ΔE) is related to the lifetime of the associated excited state (τ) by Heisenberg's uncertainty principle

$$\Delta E \approx \frac{h}{\tau} \qquad\qquad 3.42$$

($h = 4.14 \cdot 10^{-15}$ eV s).

Dipole allowed electronic states (like He* ($1s2p^1P$) leading to HeI radiation) have lifetimes on the order of 10^{-8} s resulting in a natural line width of $\Delta E \approx 10^{-6}$ eV. The real linewidth of a He resonance lamp is additionally affected by i) collisions between the atoms in the discharge lowering the average lifetime of an excited atom (pressure broadening). For a HeI lamp operating near 1 mb, pressure broadening is less than 10^{-7} eV, and ii) the thermal velocity of the atoms in the discharge causes a Doppler shift in the emitted radiation, its contribution is $\Delta E \approx 10^{-4}$ eV. Further effects which may influence the linewidth (Stark effect, self absorption, etc.) are discussed by Samson [82]. In any case, the linewidth of a resonance lamp is generally much narrower than the resolution of an electron energy analyzer.

c) Motion of the Target Gas (Doppler Effect)

The electron energies analyzed by the spectrometer are modified by the motion of the target molecules. The degree to which this motion influences the energy of a photoelectron can easily be quantified: Consider a photoelectron ejected with velocity v_e from a target molecule moving with velocity $+ v_t$ towards the spectrometer. The kinetic energy of the photoelectron in the laboratory frame is then $\varepsilon^+ = m \left(v_e + v_t\right)^2/2$. Accordingly, we have $\varepsilon^- = m \left(v_e - v_t\right)^2/2$ when the target moves away from the detector. The resulting energy difference $\Delta\varepsilon = \varepsilon^+ - \varepsilon^-$ becomes

$$\Delta\varepsilon = 2mv_ev_t . \tag{3.43}$$

Assuming a Maxwell-Boltzmann distribution for the target molecules, its projection on the relevant axis (connecting the ionization origin with the spectrometer entrance slit) is

$$f(v_t) \sim \exp\left(\frac{-Mv_t^2}{2kT}\right), \tag{3.44}$$

with M the mass of the target molecule. The width (FWHM) of the Gaussian distribution is

$$v_t = \pm \left(\frac{2kT \ln 2}{M}\right)^{1/2} . \tag{3.45}$$

Replacing v_e by $(2\varepsilon/m)^{1/2}$, Eq. 3.43, results in

$$\Delta\varepsilon = 4((m/M) \, \varepsilon \, kT \, \ln 2)^{1/2} . \tag{3.46}$$

If M is expressed in atomic units and ε in electronvolts, we can write

$$\Delta\varepsilon \, (\text{meV}) = 0.72 \, (\varepsilon T/M)^{1/2} . \tag{3.47}$$

Equation 3.47 holds when the analyzer accepts only electrons emitted along the axis. The numerical factor changes slightly when the analyzer accepts electrons over a wider angular range.

In the unfavorable case of H_2 ionization by HeI radiation at 300 K, the electron energy at threshold ($\varepsilon = 5.8$ eV) is Doppler broadened by 22 meV. This is more than the rotational constant of H_2 (7.5 meV) or H_2^+ (4 meV). This illustrates that resolving of rotational states in PES of H_2 is more limited because of Doppler broadening than by the resolving power of an analyzer. The Ar^+ doublet which is often used as a standard test of resolution and energy calibration has a Doppler width of 5 meV.

An attractive solution to reduce the thermal motion is to provide the target gas in the form of a supersonic jet generated by adiabatic expansion of the gas. By this technique, the relevant translational temperature can easily by reduced to a few Kelvin [83-85]. See also Section 4.1 in Part III of this volume.

3.3 Mass Spectrometers

Parent ions and fragment ions formed in photoionization and electron capture are analyzed according to their M/z value by mass spectrometers. In addition to identifying the different ionic species, one may also wish to record their relative abundances as a function of the energy of the ionizing beam. Although this volume deals with singly charged ions $(z = 1)$, we will always explicitly use the number of charges in the equations of motion in order to demonstrate that common mass spectrometers indeed analyze according to M/z.

Of course, sufficiently high photon energies can generate multiply charged cations due to electron correlation in the target molecule. In the case of electron attachment, single collision conditions necessarily bring about only singly charged temporary anions (resonances). From the thermodynamic point of view, a neutral molecule can generally bind *one* extra electron, provided the molecule possesses a positive electron affinity. Only molecules of considerable size may be able to bind two excess charges in a thermodynamically stable state [86].

3.3.1 Principles of Mass Analysis

Mass analyzers operate by the behavior of the ion in electric and magnetic fields, either separately or in combination. In static fields, ions of different M/z values can be distinguished in space through their different trajectories (static mass analyzers), or in time through their different flight times (dynamic mass analyzers). A further class of dynamic instruments uses the different periodic motions in oscillating fields.

Before considering practical analyzers, we will briefly recall the motion of charged particles in homogeneous electric and magnetic fields.

Consider a uniform electric field which is realized between two parallel metallic plates separated by a distance d. The strength of the field is $E = \Delta V / d$ when ΔV is

the potential difference between the two plates. An ion formed at rest will be accelerated along the electric field E according to

$$a = z \, e \, E/M = \text{const} \qquad\qquad 3.48$$

$(e = 1.6 \cdot 10^{-19} \text{ C})$.

After travelling a certain distance s (e.g., by passing the aperture of one of the plates), the ion has gained the velocity

$$v = (2 \, a \, s)^{1/2}, \qquad\qquad 3.49$$

and with $V = E s$, we can write

$$v = \left(2 \, V \, e \, \frac{z}{M}\right)^{1/2}. \qquad\qquad 3.50$$

If M is expressed in atomic units and V in volts, we have

$$v\,(\text{m/s}) = 1.39 \cdot 10^{4} \left(\frac{V}{M}\right)^{1/2}. \qquad\qquad 3.51$$

For example, an N_2^+ ion accelerated through a potential difference of 100 V will
gain the velocity $v = 2.63 \cdot 10^{4}$ m/s.

If an ion of velocity v is injected into a uniform magnetic field $\vec{B}$, it experiences
the Lorentz force

$$\vec{F} = z e \, [\vec{v} \times \vec{B}]. \qquad\qquad 3.52$$

For velocity vectors parallel to the magnetic field, there will be no force acting on the
ion. On the other hand, if $\vec{v}$ is perpendicular to $\vec{B}$, the Lorentz force is perpendicular
to both $\vec{v}$ and $\vec{B}$, with the magnitude

$$F = z e v B .$$

This results in a circular motion of the ion. The radius R for the motion follows from
the condition that the Lorentz force Eq. 3.52 must equal the centrifugal force
$(M v^2/R)$ leading to

$$R = \frac{M \, v}{z \, e \, B}. \qquad\qquad 3.53$$

Replacing v by $(2 z e V/M)^{1/2}$, the radius becomes

$$R = \frac{1}{B}\left(\frac{2 \, M \, V}{z e}\right)^{1/2}. \qquad\qquad 3.53\,a$$

For constant V and B the radius is proportional to $(M/z)^{1/2}$. Expressing B in Teslas (1 Tesla $= 1$ Vs/m^2) and M and V as above, we have

$$R(\text{m}) = 1.44 \cdot 10^{-4} \frac{1}{B} \left(\frac{MV}{z} \right)^{1/2}.$$

The time for traveling one circular orbit $(2\pi R)$ is given by

$$T = \frac{s}{v} = 2\pi \frac{M}{zeB}.$$

The reciprocal of this time, $f = 1/T$ is called the Larmor frequency

$$f = \frac{zeB}{2\pi M} \qquad\qquad 3.54$$

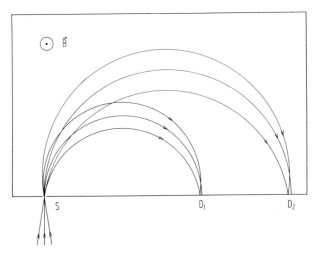

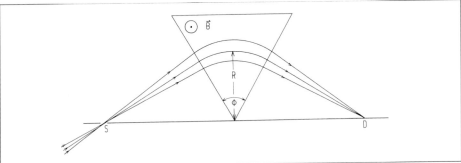

Fig. 3.14. a) Focusing of ions in a magnetic field after travelling $180\,°$. For an angular deviation of α, the width of the focused image is $R \cdot \alpha^2/4$. b) Focusing of ions in a $60\,°$ magnetic sector.

Again, if B is expressed in Teslas and M in atomic units,

$$f(\text{MHz}) = 15.25 \ B \ \frac{z}{M}.$$

The radius of an N_2^+ ion injected perpendicular into a magnetic field of 1 Tesla at an energy of 1000 eV will experience a circular motion with $R = 2.41$ cm, its Larmor frequency is 0.545 MHz, independent of the kinetic energy of the ion.

From Eq. 3.53a it is clear that the radius is equally sensitive on mass *and* energy. For ions injected in a magnetic field with the same kinetic energy, first order focusing is obtained after travelling 180°, as illustrated in Fig. 3.14a. Ions of different M/z ratios are separated according to their different radii. For an angular deviation of α of the injected ions, the width of the focused image at 180° is $R \cdot \alpha^2/4$ [87,88]. For $R = 20$ cm and $\alpha = 5°$, for example, the image is broadened by 1.5 mm.

In fact, 180° focusing for total immersion was used in Dempster's apparatus [89], one of the early mass spectrometers. 180° focusing is a special case for first order magnetic focusing [87]. Since the 180° design requires immersion of both the ion source and the detector, it is generally more convenient to use magnetic sector analyzers with smaller sector angles such as 90° or 60° (Fig. 3.14b). The smaller the sector angle, the greater is the distance of the source and image point from the boundary of the magnetic field.

It has been shown by Kerwin [87] that *perfect focusing* is obtained for ions travelling with radius R in a uniform magnetic field whose field boundary is defined by

$$y = \frac{x(A - x)}{(R^2 - x^2)^{1/2}},\tag{3.55}$$

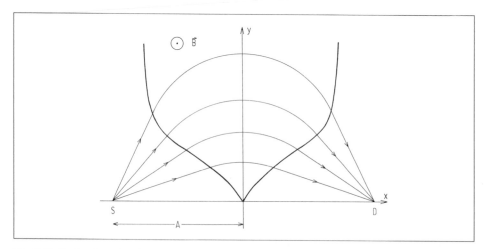

Fig. 3.15. Perfect focusing in a magnetic field with boundary defined by Eq. 3.55 with $R = 0.75 \ A$.

with A the distance of the source and image point from the symmetry plane of the magnetic field (Fig. 3.15). The magnetic sectors introduced above are straight line approximations with *first order* focusing properties [87].

3.3.2 Double Focusing Magnetic Sector Mass Analyzers

The resolving power $(M/\Delta M)$ of a magnetic sector mass analyzer is limited by the initial spread of ion translational energies (Eq. 3.53). This spread is caused by
a) the Maxwell-Boltzmann distribution of the target molecules in the ionization source,
b) fragment ions may be formed with some additional kinetic energy from the respective unimolecular decomposition, and
c) by field inhomogeneities in the acceleration of the ions.

Kinetic energy spread of ions can be overcome using electrostatic energy analyzers prior to mass analysis. In Section 3.2.3, it has been shown that a cylindrical analyzer focuses (first order) for 127°. Focusing can also be obtained for smaller deflection angles, the conjugate foci then no longer lie at the field boundary (as in the case of 127°), but lie outside the field.

Instruments with an electrostatic energy analyzer are designated as *double focusing mass spectrometers*, since the ions are focused in *energy* and *direction*.

Figure 3.16a shows a geometry with 90° electric and magnetic deflection. It can be shown [90, 91] that a radial electrostatic analyzer can always be combined with a magnetic analyzer in such a way that the *velocity dispersion* is *opposite* and approximately *equal* in magnitude. The slit between the two sectors can be widened and the velocity aberration of the ion image is substantially eliminated.

Double focusing instruments are commercially available and many different machines have been built, from miniature devices used in space research, to large machines with a magnetic sector radius of over 2 m and capable of mass resolutions $(M/\Delta M)$ on the order of 10^6.

Double focusing mass spectrometers are often used in the way that ions are transmitted by the magnetic sector *prior to entering* the electric sector. Such instruments are said to have *reversed geometry* (Fig. 3.16b); they can be used to elucidate the decomposition channels of a given ion [92].

Consider a particular ion transmitted through the magnetic sector. In the absence of decomposition, the ion enters the electric analyzer with kinetic energy zeV and is transmitted to the detector since the potential energy difference between the cylinders is adjusted to that energy. If, however, the ion (M^+) decomposes in the field free region between the mass filter and the kinetic energy analyzer, the daughter ion (m^+) enters the electric sector with kinetic energy $zeV\,(m/M)$. Thus, by scanning ΔV downwards from ΔV_o, all daughter ions of a given parent ion can be detected. This procedure is called *mass analyzed ion kinetic energy spectroscopy* (MIKES), or direct analysis of daughter ions (DADI).

Ions which decompose between the two sectors of a mass analyzer are called *metastable* ions and the associated process a *metastable decomposition*. The ions are usually accelerated by a few keV before entering the electric or magnetic sector having velocities on the order of $\approx 10^5$ m s^{-1} (Eq. 3.51). Thus, an unstable ion surviving the flight time from the source to the region between the two sectors must have a

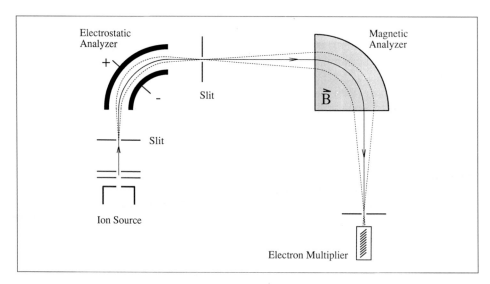

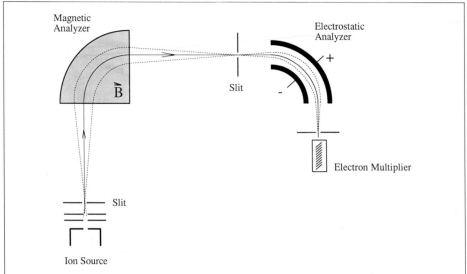

Fig. 3.16. a) Double focusing magnetic sector mass analyzer in conventional geometry with 90° electric and magnetic deflection. b) Double focusing mass analyzer in reversed geometry.

lifetime on the the μs scale. This is long, compared to a direct electronic dissociation along a repulsive potential energy surface which occurs on the time scale of vibrational periods (10^{-13}–10^{-14} s).

We mention that analogue information on metastable decompositions can be obtained using machines of conventional geometry (Fig. 3.16a). In that case, the data are acquired in the opposite sense: The magnetic sector is now tuned to transmit a given daughter ion and all metastable dissociations occurring in the field free region prior to the electric sector resulting in this daughter ion can be found by scanning the electric sector voltage [92]. All parent ions which dissociate to yield the given daughter ion are thus documented. Note that by repeating these scans for the different daughter ions, one obtains the same information as from the MIKE spectra.

The conventional or reversed geometry can also be used to study the *kinetic energy release* of daughter ions arising from metastable decompositions occurring in the region between the two sectors [92-96]. According to Eq. 3.53a, the radius of an ion in the magnetic sector depends on both its M/z ratio and its kinetic energy. Thus, if in conventional geometry a metastable decomposition is associated with some release of kinetic energy, the mass peak of the daughter ion will be broadened giving "flat topped" or "dish topped" metastable peaks from which the kinetic energy release can be calculated [92,95,96]. This technique, sometimes assigned by the acronym KERD (from kinetic energy release distribution), allows determination of mean kinetic energies as low as in the 10–100 meV region. It has recently been applied to study metastable decompositions of cluster ions [97,98].

3.3.3 Time-of-Flight Mass Spectrometers and Reflectrons

Time-of-flight (TOF) instruments belong to the group of dynamic mass analyzers. There are different options to define the start (time zero) for the TOF measurement: One can use
a) a pulsed primary beam (which is obvious for pulsed lasers and easily realized for electrons),
b) the signal of a photoelectron (in the case of photoionization), or
c) simply pulse the ion draw out field.

The bunch of ions formed at time zero is then further accelerated by appropriate fields and projected through a field free space toward a detector (Fig. 3.17). The flight time of an ion is proportional to $(M/z)^{1/2}$ (Eq. 3.50), thus, if the signal at the detector is recorded vs time one observes a series of peaks, each representing ions of a particular M/z value.

For acceleration energies on the order of keV the flight times of ions fall in the μs range (see below), so the repetition cycle can be operated in the 10–100 kHz region. Of course, the width of the ionizing pulse or ion draw out pulse must be sufficiently short (i.e. on the ns scale).

Although the ionizing period in a TOF spectrometer covers only a small fraction of the cycle, virtually all ions generated in the source are transmitted to the detector.

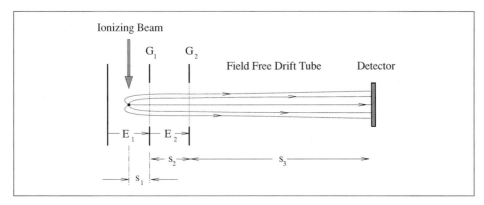

Fig. 3.17. Two -field time-of-flight (TOF) mass spectrometer. E_1: draw-out field, E_2: acceleration field.

By contrast, in other instruments like the magnetic sector analyzer or the quadrupole mass filter ions of only one mass (M/z) are detected at a given moment.

Mass spectrometers working on the TOF principle have been reported for more than 40 years [99,100,103]. The mass resolution achieved depends on the spread of the flight times for ions of the same M/z value. Spreading of flight times is principally caused by two effects:

a) The ionizing beam possesses a finite thickness associated with ion formation at different potential energies across the beam, which generally results in different flight times, and

b) the ions are not created at rest, but have a Maxwell-Boltzmann distribution.

In addition, fragment ions may possess some distribution of kinetic energy from the unimolecular decomposition of the parent ion.

The first limiting effect is referred to as *space resolution* and the second as *energy resolution*. Space focusing can easily be obtained by adjusting the electric field(s) in such a way that an ion created closer to the detector (and hence at lower potential energy) will be overtaken by another ion created at higher potential energy in the drift space just at the detector.

In a "one-field TOF spectrometer" (one acceleration region followed by a drift space) the requirement for space focusing is that the drift space must be twice the acceleration space (see below). The ions then spend 50% of their total flight time in the acceleration region. The instrument can then be operated such that relatively large variations in acceleration energy (some 10^{-2}) result in only small variations in flight time (less than 10^{-4}). Ions which exactly match the space focus condition have minimum flight times.

While space focus is easily obtained, the second effect inherently limits the mass resolution of TOF spectrometers. If two ions start with axial velocities of equal magnitude but opposite direction, the antiparallel ejected ion will be decelerated, reversed, and then accelerated toward the detector. This *turn around* ion will pass the plane of origin with the same velocity as the *direct ion*, but with some time delay which will persist all the way to the detector. This time delay must be smaller than the time difference between two adjacent masses.

The time difference between adjacent masses can always be enlarged by using longer drift spaces, which in turn violates the condition for space focusing. A detailed analysis shows that for ions having a Maxwell-Boltzmann velocity distribution (300 K) the simple one-field TOF arrangement has a resolving power on the order of $M/\Delta M \approx 100$.

This defect was partly remedied by the introduction of the "two field" TOF spectrometer [101] which allows more flexibility in adjusting the experimental parameters.

Ions formed in the interaction volume are drawn out by the first electric field E_1, further accelerated by the second filed E_2 before they enter the field free drift tube and arrive at the detector. With this configuration, space focusing can be obtained for conditions where the flight time in the acceleration regions is considerably less than 50 % of the total flight time. This helps to increase the mass resolution with respect to initial translational energies of the ions.

With the symbols from Fig. 3.17 the flight times t_{s_1}, t_{s_2}, and t_{s_3} for the three segments s_1, s_2, and s_3 are

$$t_{s_1} = \frac{(2M)^{1/2}}{zeE_1} \left\{ (W_o + zeE_1s_1)^{1/2} \pm W_o^{1/2}) \right\} \tag{3.56}$$

$$t_{s_2} = \frac{(2M)^{1/2}}{zeE_2} \left\{ W^{1/2} - (W_o + zeE_1s_1)^{1/2} \right\} \tag{3.57}$$

$$t_{s_3} = \frac{(2M)^{1/2}}{W^{1/2}} s_3, \tag{3.58}$$

where W_o is the initial kinetic energy of a (fragment) ion and W the kinetic energy after acceleration,

$$W = ze(E_1s_1 + E_2s_2) + W_o. \tag{3.59}$$

The + and − signs in Eq. 3.56 refer to initial velocity components directed, respectively, away and toward the detector.

Note that W is independent of the initial orientation of the velocity vector since turn-around and direct ions leave the acceleration region with the same kinetic energy.

In the ideal case where all ions begin their flight in the plane located at a distance s_1 from grid G_1 with zero kinetic energy, $W_o = 0$, the total flight time $t = t_{s_1} + t_{s_2}, + t_{s_3}$ can be written in the form

$$t = \left(\frac{M}{ze\,(E_1s_1 + E_2s_2)} \right)^{1/2} \left(2k^{1/2}s_1 + \frac{2k^{1/2}s_2}{k^{1/2} + 1} + s_3 \right), \tag{3.60}$$

with constant k defined as

$$k = \frac{E_1 s_1 + E_2 s_2}{E_1 s_1}.$$

3.61

To find the condition at which ions formed across the finite thickness of the ionizing beam are focused (in time) at the detector, one has to set

$$\frac{\partial t}{\partial s_1} = 0,$$

which leads to

$$s_3 = 2k^{3/2} s_1 \left(1 - \frac{1}{k + k^{1/2}} \frac{s_2}{s_1}\right).$$

3.62

According to the instrumental parameters, Eq. 3.62 specifies a maximum, a minimum, or an inflection point for the flight time of ions vs s_1. Once the dimensions of s_1, s_2 and s_3 are fixed, k (and hence the ratio E_1/E_2) is uniquely determined by Eq. 3.62.

Note that for a one-field configuration ($k = 1$, $E_2 = s_2 = 0$), Eq. 3.62 yields $s_3 = 2s_1$ and Eq. 3.60 then specifies a minimum.

As mentioned, kinetic energies of the ions will cause some spread in their arrival time. On the other hand, peak broadening in TOF spectra can in turn be used to obtain information on translational excess energies in unimolecular decomposition [102]. This question is considered in more detail in Part III of this volume.

Space focusing and energy focusing place opposite requirements on several system parameters. An exact mathematical treatment of this problem becomes prohibitively difficult and we will instead illustrate this in an explicit computational example.

For the arbitrary dimensions $s_1 = 1$ cm, $s_2 = 3$ cm, and $s_3 = 45$ cm space focusing, Eq. 3.62, is obtained for $k = 9.55$ and hence $E_2/E_1 = 2.85$. If we use $E_1 = 200$ V/cm and $E_2 = 570$ V/cm, ions created exactly in the plane at $s_1 = 1.0$ cm will be accelerated to 1910 eV. The flight time of an ion with $M/z = 100$ is $t = 9.195$ µs. The time difference of adjacent atomic masses M and $M + 1$ can be obtained from Eq. 3.60, yielding

$$t_{M+1} - t_M = \left\{\left(1 + \frac{1}{M}\right)^{1/2} - 1\right\} t_M;$$

3.63

for $M \gg 1$ we can write

$$t_{M+1} - t_M \approx \frac{t_M}{2M}.$$

At $M = 100$ the time between two adjacent masses is thus approximately 50 ns. The explicitly calculated times are 9.148 μs ($M = 99$) and 9.240 μs ($M = 101$).

Let the thickness of the ionizing beam be $\Delta s_1 = 6$ mm in diameter. We then have ionization across the beam between $s_1 - 0.3 = 0.7$ cm and $s_1 + 0.3 = 1.3$ cm. An ion ($M = 100$) formed closer to the detector (0.7 cm) will be accelerated to $W = 1850$ eV and will strike the detector after 9.188 μs, while an ion formed on the far end of the beam diameter (1.3 cm) will gain 1970 eV energy and reach the detector after 9.191 μs. Both ions have, by a few nanoseconds, shorter flight times than that created in the plane at $s_1 = 1.0$ cm. Equation 3.62 hence specifies a maximum. This maximum is flat, so flight time differences across the ionizing beam are small (4 and 7 ns). Compared to the time difference between two adjacent masses (≈ 50 ns). Note that the variation in energy of about 6 % corresponds to a variation in time of $6 \cdot 10^{-4}$. It can be shown [101] that the maximum mass, M_{max}, for which adjacent mass resolution can be obtained is given by

$$M_{max} \approx 16 \ k \left(\frac{s_1}{\Delta s_1} \right)^2 ,$$

which leads to $M_{max} \approx 420$ amu for the present example. Target molecules having a Maxwell-Boltzmann velocity distribution yield Gaussian peak shapes in the TOF spectrum (see Section 2.3.1 in Part III), with a first width at half maximum given by

$$t_{1/2} = \frac{(8kTM \ln 2)^{1/2}}{zeE_1}. \qquad\qquad 3.64$$

For the present example, one calculates $t_{1/2} = 19$ ns at $T = 300$ K. This is considerably more than spreading of flight times due the finite thickness of the ionizing beam.

Figure 3.18 illustrates that ions ($M = 100$) created with (discrete) excess translational energy of $W_o = 0.3$ eV and isotropically distributed velocity vectors will generate a rectangular TOF peak having a width of 79 ns. Thus, direct and turn-around ions with 0.3 eV initial energy have flight times (9.155 μs and 9.234 μs, respectively) close to those of adjacent masses created at rest. This arbitrary example demonstrates that the mass resolution in a two field TOF spectrometer adjusted to space focusing is also inherently limited by initial kinetic energies of the ions. The limiting effect, however, is considerably smaller than in a one field TOF spectrometer. In the present example the ions spend approximately 25 % of their total flight time in the two acceleration regions. Since the time spread due to initial kinetic energies solely depends on the magnitude of the first draw out field (E_1) one can always minimize this effect by using a longer drift space thus increasing the time between two adjacent masses which in turn violates the condition for space focusing. In practice, mass resolutions up to 300 are feasible.

In such a TOF configuration metastable decompositions can only be seen if they occur in the acceleration region. If a decomposition occurs in the drift space, parent (precursor) and daughter ions have the same velocity, so precursor and daughter

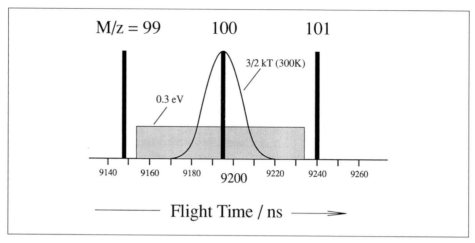

Fig. 3.18. Calculated TOF spectra illustrating the effect of initial kinetic energies on the peak shape. The shaded area is for a discrete initial energy of 0.3 eV.

ions cannot be distinguished. On the other hand, if the decay occurs in the acceleration region, the daughter ion will gain more velocity and can eventually be distinguished from the parent.

It has been shown by Neusser, Schlag and coworkers [104] that metastable decays occurring in the drift region can clearly be identified in a reflectron time-of-flight (RETOF) mass spectrometer. The reflectron was invented in the early 1970s by Mamyrin et al. as a high resolution TOF spectrometer [105]. Figure 3.19 shows the schematic design. After acceleration to a certain energy, W, the ions enter a drift region (2), are then decelerated, reflected and accelerated again ((3) and (4)) until they travel through a second drift region (5) and arrive at the detector.

The idea is that variations in energy W (due to ion formation at different potential energies across the ionizing beam and due to initial translational energies) are compensated by corresponding variations in the transit time through the reflector.

Mamyrin showed that such a system provides *second order time focusing* for ions with respect to *energy and angle of injection.*

It should be grasped that second order energy focusing is obtained for variation of ion energies entering the first drift space. The derivation, however, does not account for the time spread of the ions when leaving the acceleration region. Such a time spread is caused by ions created with translational excess energy, as well as for ions created across the finite thickness of the ionizing beam. As for the Wiley-McLaren design (two field TOF spectrometer) discussed above, a turn-around ion will leave the draw-out field with the same energy as the direct ion, but with some time delay which will, of course, persist in the reflectron all the way to the detector. Accordingly, ions created at different positions across the ionizing beam will enter the first drift space at different times (and with different energies).

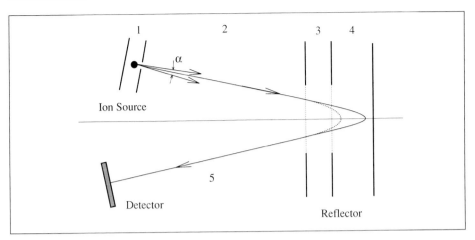

Fig. 3.19. Schematic of the arrangement for a reflectron time-of-flight (RETOF) mass spectrometer.

The point of the reflectron is that it focuses on variations of ion energies and that the time the ions spend in the acceleration region is small compared to the overall flight time. It thus provides the condition that the time difference between adjacent masses is large compared with the time spread of the ions when leaving the acceleration field. Note that in the conventional TOF arrangement, the length of the drift space with respect to the acceleration region was fixed by the space focus conditions.

Apart from the possibility of observing metastable decompositions, the power of the reflectron as a high resolution TOF mass spectrometer has also been shown by Neusser and Schlag [104]: at an acceleration voltage of 540 V and a draw out field of 500 V/cm, the time difference between two adjacent masses at $M = 78$ is 400 ns at a total flight time of 60 µs. With a peak width of 8 ns the mass resolution becomes $M/\Delta M = 3900$. The time the ion is subject to acceleration (Eq. 3.56) is on the order of 0.56 µs, i.e., less than 1% of the total flight time. Ionization across the laser focus (100 µm) leads to an energy spread of 5 eV and a time spread of only 3 ns. The energy spread is completely compensated by the reflectron. The calculated width for ions having a Maxwell-Boltzmann energy distribution (Eq. 3.64) becomes $t_{1/2} = 6.9$ ns at 300 K, and even at 0.3 eV the time difference between a direct and turn around ion is only 27 ns. This clearly demonstrates that the resolving power of a RETOF mass spectrometer is much less limited by translational excess energies than in the conventional TOF design.

Returning to the problem of metastable decompositions, it is clear that a daughter ion (m^+) formed in the first drift space will stay a shorter time in the reflecting field in comparison to the parent ion (M^+). By that, they can be distinguished in the mass spectrum. If the shift in kinetic energy between parent and daughter ion:

$$\Delta E = z e V_o \left(1 - \frac{m}{M}\right)$$

falls in the range which is well corrected by the reflectron, the daughter ion appears at the correct mass.

Reflectrons and modifications of it have proved to be versatile instruments for the study of metastable decays, particularly in connection with pulsed lasers as ionizing sources [104, 106, 107]. Mass resolutions of 35 000 have been reported [108].

3.3.4 Quadrupole Mass Spectrometers

The quadrupole mass spectrometer was invented in the 1950s by Paul and coworkers [109–111] and has since proved to be the most successful competitor to the traditional magnetic sector instruments. Quadrupole mass filters evolved from techniques used in high-energy physics where proton beams were focused by alternating quadrupole fields. It was found that focusing could be made mass selective, which led directly to the development of the quadrupole mass filter.

Figure 3.20 depicts the arrangement of electrodes and an end-on view of the analyzer. Ions from the ionization source are accelerated only modestly by 10-20 eV prior to injection to the quadrupole array where they are separated according to M/z.

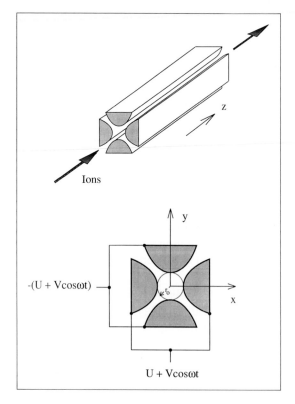

Fig. 3.20. Electrode arrangement of a quadrupole mass spectrometer.

A two-dimensional quadrupole field is established between the four rods of hyperbolic cross-section, and opposite rods are electrically connected pairwise.

The voltage applied to the array consists of a constant component (U) and a radio-frequency component $(V \cos \omega t)$. In cartesian coordinates (Fig. 3.20) the potential at any point between the rods is

$$\Phi(x,y) = \frac{(U + V \cos \omega t)(x^2 - y^2)}{r_o^2}, \qquad\qquad 3.64$$

where $2r_o$ is the spacing between the electrodes. The equations of motion for an ion are then

$$\frac{M \, d^2 x}{dt^2} + \frac{2 \, ze \, (U + V \cos \omega t) \, x}{r_o^2} = 0 \qquad\qquad 3.65\,a$$

$$\frac{M \, d^2 y}{dt^2} - \frac{2 \, ze \, (U + V \cos \omega t) \, y}{r_o^2} = 0 \qquad\qquad 3.65b$$

$$\frac{M \, d^2 z}{dt^2} = 0 \qquad\qquad 3.65\,c$$

The only nontrivial solution for Eq. 3.65c is $v_z = $ const. By introducing the dimensionless parameters

$$A = \frac{8 \, z \, e \, U}{M \, r_o^2 \, \omega^2}$$

$$Q = \frac{4 \, z \, e \, V}{M \, r_o^2 \, \omega^2}$$

$$\xi = \frac{\omega t}{2},$$

the equations of motion for the x and y axes become

$$\frac{d^2 x}{d\xi^2} + (A + 2 \, Q \cos 2 \, \xi) \, x = 0 \qquad\qquad 3.66\,a$$

$$\frac{d^2 y}{d\xi^2} - (A + 2 \, Q \cos 2 \, \xi) \, y = 0, \qquad\qquad 3.66\,b$$

which are known as Mathieu equations. The properties of the Mathieu equations are well known [112] and the theory of the quadrupole mass filter based on these equations has been treated in detail [113-115].

The solutions designate oscillations by an ion in x and y directions. For certain values of A and Q these oscillations are stable (finite amplitude at any time) while for other values A,Q the oscillations are unstable, i.e., the amplitude increases exponentially.

For an ion oscillating stable in both x and y directions, the solutions of the two Mathieu equations must simultaneously be stable.

Figure 3.21 indicates the range of the values A and Q for which this is the case. If the ratio $A/Q=2\,U/V$ is kept constant, the points for different M/z values fall on a straight line pointing through the origin of the (A,Q) space. For a mass such as M_1 ($Q=0.64$ / $A=0.16$) the ion will describe a stable oscillatory path, while for a mass such as M_2 ($Q=0.4$ / $A=0.1$) the ion is eventually lost due to the exponentially increasing amplitude of its oscillating trajectory. Note that $M_2 > M_1$ and that a mass spectrum can be obtained by varying U and V for a constant U/V ratio. U and V are then directly proportional to M/z (for constant ω, $M/z \sim V/Q = 2U/A$). In other words, by performing a mass scan, ions of increasing M/z will successively pass the stability region along the straight line of constant A/Q. The mass resolution increases with the ratio $A/Q=2\,U/V$ until it becomes theoretically infinite at the apex of the stability region ($A=0.237$ / $Q=0.706$).

For a real quadrupole, mass separation does not rigorously depend on whether an ion is subject to a stable or unstable oscillation, but whether it is able to travel through the quadrupole of finite length without striking a rod.

In practice, the ideal quadrupole field is often approximated by cyclindrical electrodes. Care must then be taken of the proper choice of the spacing between the electrodes in order to minimize the high-order terms in the power series expansion of the non-hyperbolic field [115,116]. This is achieved for $r_o = r/1.1486$ where r is the radius of the cylindrical rod.

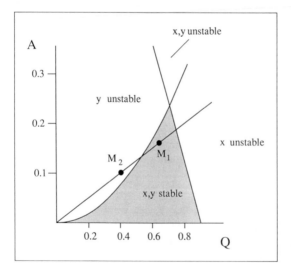

Fig. 3.21. Stability diagram for the quadrupole mass spectrometer.

Nowadays, hyperbolic electrodes can be manufactured by computer- controlled machining, and quadrupole mass filters utilizing such electrodes are commercially available. One main difficulty in the construction of a high-resolution quadrupole remains: the electrodes must be adjusted parallel with a tolerance of mm over a distance of 0.2–0.5 m.

Commercial instruments operate in the MHz frequency region and at transmission energies of 10–20 eV. Resolving powers in the 10^3 domain are possible with a maximum M/z value of some 10^3. An ion ($M/z=100$, 10 eV) will spend 45.5 μs in a quadrupole of 20 cm length; it will oscillate 91 times when the instrument operates at 2 MHz.

3.3.5 Ion Cyclotron Resonance (ICR) Mass Spectrometers

Figure 3.22 shows an overview of the ICR spectrometer. The ions of interest are generated in a vacuum can, which is placed between two poles of an electromagnet. According to Eq. 3.53 and Eq. 3.54, the ions describe a circular motion with radius

$$R = \frac{Mv}{zeB},$$

and frequency

$$\omega = 2\pi f = \frac{zeB}{M}.$$

Note that R depends on the velocity of the ion, but not on the Larmor (or cyclotron) frequency. At thermal energy the radii are quite small. An N_2^+ ion at 1 Tesla describes a radius of 0.15 mm, having a cyclotron frequency of 545 kHz.

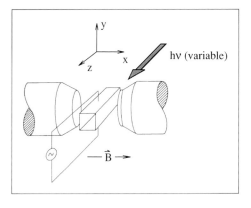

Fig. 3.22. Ion cyclotron resonance (ICR) mass spectrometer.

If now a radio frequency is applied in the x-y plane (Fig. 3.22) which oscillates just at the cyclotron frequency the ion will be accelerated and hence absorb energy from the rf field. In the absence of collisions, it will spiral until it strikes one of the plates. If the irradiating oscillator is sufficiently sensitive to energy absorption, the presence of ions can readily be detected. By scanning either the magnetic field or the frequency, a mass spectrum of the ions present in the cell can be obtained. In an ICR the ions are observed without physically collecting and detecting them.

In many ICR spectrometers ionization region and resonance region are spatially separated (drift cell ICR spectrometers) [117,118]. This is performed by applying a constant electric field across the ion source. The combined interaction of the electric and magnetic field causes trochoidal trajectories for the ions, the guiding center moving with the constant velocity

$$v_x = E/B$$

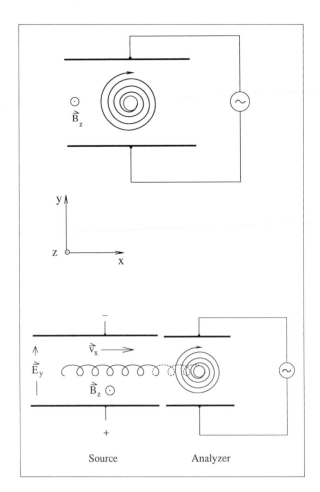

Source   Analyzer

Fig. 3.23. Spiral trajectory of an ion excited by radiofrequency (top), and trochoidal trajectory in the drift cell ICR spectrometer (bottom, see text).

in the x direction (Fig. 3.23). In the analyzer region the upper and lower plates of the cell are placed into the capacitance bridge circuit of an oscillator, bringing the ions into resonance and causing an imbalance in the bridge. A mass spectrum is obtained by sweeping the magnetic field. The drift cell ICR spectrometers represent extremely versatile instruments which have been used in many fields of gas phase ion chemistry [119].

Presently, ICR spectrometers are mostly used with the *Fourier transform* technique (FT-ICR) [120-122] in combination with super conducting magnets allowing magnetic field strengths between 5 and 10 Teslas [123, 124]. In FT-ICR all ions in the cell are coherently excited by a short RF pulse (typically ms), the frequency of which is linearly swept over the range of interest. The irradiated ions induce an image current at the cell plates containing the cyclotron frequencies of the ions. This image current is amplified, digitized, and stored in a computer. The transient signal is then Fourier transformed to give the frequency spectrum which corresponds to the mass spectrum of the ions present in the cell.

Mass resolution in FT-ICR is linearly related to the duration of signal acquisition, i.e., how long the transient signal can be observed. High resolution thus requires low pressure in the cell ($< 10^{-9}$ mb), otherwise deactivation of the excited ions by collisions restricts the time for signal acquisition. Allemann, Kellerhals, and Wanczek [125] report an ultra-high resolution spectrum of H_2O^+ with $M/\Delta M > 10^8$! This was achieved at $8 \cdot 10^{-11}$ mbar and an acquisition time of 51 s.

The reader should realize that this ultra-high mass resolution is close to separate electronically excited species from their ground states by Einstein's law $\Delta E = \Delta M\, c^2$! The mass of the long-lived excited helium atom He* $(2^3 S)$ (excitation energy 19.8 eV), for example, is by $\Delta M = 2 \cdot 10^{-8}$ amu higher than that of the neutral atom, corresponding to $M/\Delta M = 2 \cdot 10^8$.

3.3.6 Other Mass Analyzers

Apart from the four classes of commonly used mass spectrometers (magnetic, TOF, ICR, quadrupole), many other designs of dynamic and static mass analyzers have been reported in the literature over the last 60 to 70 years.

We mention here the *Wien filter* which can be used as energy analyzer for electrons and as mass analyzer for ions. This instrument was briefly described in Section 3.2.7. In the Wien filter the electric field, the magnetic field, and the direction of ion beam are mutually perpendicular. The electric field is adjusted such that electric and magnetic deflection just balance for a given M/z value. The ion beam then passes the analyzer undeflected. The Wien filter has excellent transmission, but comparatively low mass resolution. It is often used for the generation of mass selected ion beams, e.g., in photodetachment experiments. [126,127].

A perpendicular electric/magnetic field configuration is also used in the *cycloidal focusing mass analyzer* [128]. Here, the ions enter the analyzer parallel to the electric field and the cycloidal trajectories focus on some image which is a replica of the entrance slit.

Prior to the development of the quadrupole mass spectrometer, various radiofrequency (rf) techniques were applied to mass spectrometry. This was done for linear or circular motion of the ions. In the first case ions travelling down a tube can be selected by subjecting them to a series of electrodes which are periodically switched from (−) to (+) polarity (traffic light mass analyzers) [129]. In the *omegatron* [130,131] the spiral trajectory of an ion in a crossed rf and magnetic field is used to analyze ions (similar to the ICR spectrometer).

3.4 Detectors

The experimental methods described in Chapter 2 require detectors for the registration of ions and photoelectrons arising from photoionization or electron attachment. In addition, for normalization procedures or determination of reaction cross sections it is necessary to measure (or monitor) the intensity of the primary (photon or electron) beam.

A flux of charged particles (positive/negative ions, electrons) represents an electric current. This current can, in principle, be measured by an ampèremeter connected to a metallic plate at which the charged particle beam is neutralized. Photons, on the other hand, can be detected by their various interaction with matter, e.g., by the photoelectric effect.

3.4.1 Faraday Cup Collectors

In the Faraday cup the electric current carried by a charged particle beam is directly measured. It is the simplest and cheapest way to detect ion or electron beams. Faraday detectors have been used for nearly a century [132,133]. Figure 3.24 shows a typical design. An inner cylinder receives the beam and the resulting current is directly measured with an electrometer.

However, the charged particles may not only be neutralized directly at the cylinder. Rather, their interaction with the surface is a complex scattering event which

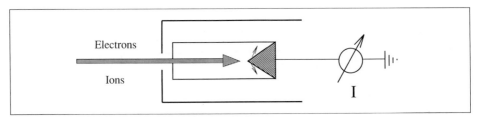

Fig. 3.24. Faraday cup collector for the detection of charged particle beams.

may lead to backscattered primary particles and/ or secondary electrons from the surface.

Of course, these reactions depend on the nature and energy of the primary beam, the surface material and, last but not least, on the surface contamination.

In spite of the microscopically complex situation, the Faraday cup has only to ensure that once the beam enters the cup no significant fraction of primary or secondary charged particles can escape from it. This is achieved by making the cup deep relative to the size of the entrance hole and machining the back surface at a cute angle so that no direct backscattering can occur (Fig. 3.24). Keeping the outer cylinder at an appropriate electric potential may also help to prevent charged particles from escaping the cup. Under optimum conditions (minimizing leakage currents and using an electrometer with a minimum of noise level) current as low as 10^{-15} A can be detected; this corresponds to a particle flux of about 10^4 s^{-1}. Thus, Faraday cups can easily be used to detect *primary* electron or ion beams. The registration of ions or electrons resulting from photoionization, however, requires detectors of considerably higher sensitivity.

3.4.2 Electron Multipliers, Channeltrons, Channelplates

An enhancement of sensitivity by several orders of magnitude can be obtained by electron multipliers. Such a multiplier converts the primary particle current into an electron current which is further amplified by means of the secondary emission effect. *A discrete electron mulitplier* consists of a series of dynodes with secondary emission coefficients greater than unity. The particle beam is directed to the first (conversion) dynode, the resultant secondary electrons are accelerated to the next dynode until the electron pulse arrives at the final dynode (anode). The overall voltage between first and final dynode is usually in the range of 2–4 kV. For a 17- stage multiplier and an emission coefficient of 3, for example, the current gain is $\approx 10^8$. Common dynode materials are Be-Cu or Ag-Mg alloys.

An electron multiplier can be operated in the analogue mode as a dc current amplifier or in the digital mode as a single event counter.

If it is used to detect positive ions (Fig. 3.25), the first dynode is at negative high voltage and the electron current at the last (collector) dynode can conveniently be measured at ground potential.

For the detection of negative ions or electrons, however, the last dynode is far off ground and a dc operation with an electrometer at that level is inconvenient. In that case one can use a blocking capacitor which brings the electron pulse at the collector to ground potential (Fig. 3.25). The pulses can then be amplified and stored by standard pulse counting electronics.

Of course, positive ions can also be detected in the single pulse counting mode by keeping the first dynode at some negative voltage.

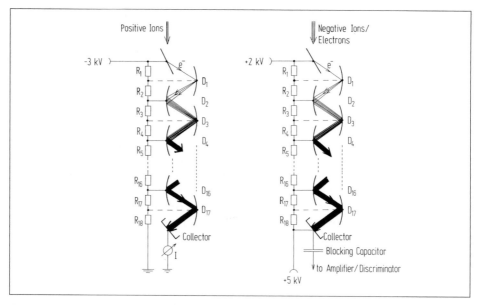

Fig. 3.25. Discrete electron multiplier operating in the analogue mode (lefthand side), and in the pulse counting mode (righthand side).

The noise level of a well designed electron multiplier can be considerably less than one event per minute, corresponding to a particle flux of less than 10^{-20} A.

A continuous dynode multiplier (channeltron) consists of a hollow glass tube which is covered inside with a semiconducting layer (Fig. 3.26). The layer acts as material for secondary electron production and as resistor to establish the necessary voltage along the multiplier tube. Channeltrons need less space than do discrete multipliers. They are, however, less resistive to larger current bursts and contamination with reactive molecules. The total resistance is on the order of 10^9 Ω. Channeltrons are commercially available in curved and linear forms (Fig. 3.26).

When the beam of particles is not well collimated and extends over a larger area as it is the case in TOF or RETOF mass spectrometers, *channelplate detectors* can be used. A micro channel plate (MCP) is an array of many microscopic channels (10-100 μm diameter) etched in a lead glass plate by modern chemical etching techniques [134]. The length to diameter ratio of a channel is between 50 and 100. The channels are covered with a semiconducting material, each channel thus acting as a micro continuous dynode electron multiplier. Parallel electric contact to each channel is provided by a metallic coating of the input and output surface of the plate. A typical MCP with an area of 10 cm^2 and a 20 μm spacing between the channels has a total number of channels on the order of 10^6. Two or more MCPs can be used in series in order to increase the current gain (Fig. 3.26).

When the collector plate consists of a resitive sheet, the micro channel plate can be operated as a *position sensitive detector* [135,136]. Such detectors are used to

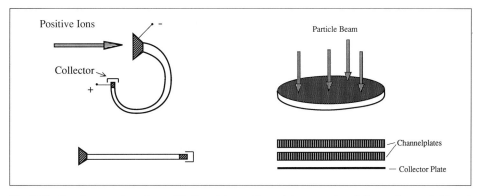

Fig. 3.26. Continuous dynode multipliers (channeltrons) in curved and linear forms (lefthand side), and microchannel plate detectors (righthand side).

obtain information on the spatial distribution of products [137,138]. The spatial resolution is only limited by the channel dimensions and their spacing.

In contrast to the discrete dynode multiplier and the conventional channeltron, the MCP possesses an ultrahigh time resolution (< 100 ps) which makes it an excellent detector for TOF experiments.

3.4.3 VUV Light Detectors

The electron multipliers described above can detect any particle provided its energy is considerably higher than the work function of the dynode material (the work function of a solid is the threshold energy for the photoelectric effect). Thus, charged particles as well as electronically excited neutral (metastable) particles and, of course, VUV photons can be detected by electron multipliers. However, since the conversion factor depends on the contamination of the dynode material and since the material itself may suffer from this contamination, it is inconvenient to use *open* electron multipliers as described above (particle multipliers) for photon detection.

There are many classical ways to detect photons in the visible to VUV region which are all based on their various modes of interaction with matter. The interested reader can find a comprehensive description of photographic plates, fluorescent materials and detectors based on photoelectric emission in the monograph of Samson [18].

VUV photons can most conveniently be detected by a (closed) photomultiplier tube operating in the visible range and a sodium salycilate coating. Sodium salycilate possesses an approximately constant fluorescence efficiency in the VUV with an emission spectrum in the visible range. The VUV radiation (which cannot enter the closed multiplier tube) is converted into visible light, which is then detected by the photomultiplier (by electron emission and multiplication, as in the particle electron multiplier described above).

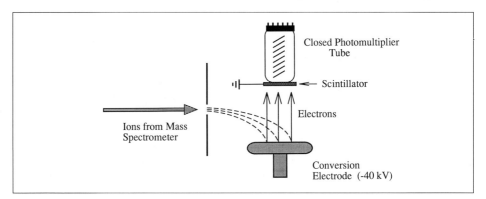

Fig. 3.27. Schematic of the Daly detector. Ions strike the high voltage cathode. The emitted secondary electrons strike the scintillator surface and emit light pulses which are amplified by the closed photomultiplier.

Closed photomultiplier tubes are sometimes also used in mass spectrometry to avoid exposure of the electron multiplier to corrosive gases. In that case ions from the mass spectrometer are accelerated to a conversion dynode. The ejected secondary electrons are then accelerated to the scintillator and the emitted photons are detected by the closed photomultiplier. (Fig. 3.27). This arrangement is known as Daly detector [139].

Today, the familiar photomulitplier tube is challenged by new devices fabricated from silicon and using modern semiconductor technology, mostly as array detectors, and with the option of position sensitive detection [140].

References

1. Bisner J, Schnieder J, Ahlers G, Xie X, Welge KH, Ashford MNR, Dixon RN (1989) State Selective Photodissociation Dynamics of Ã State Ammonia, II. J Chem Phys 91:2901-2911
2. Welge KH, Schmiedl R (1981) Doppler Spectroscopy of Photofragments. In: Jortner J, Levin RD, Rice SA (eds) Advances in Chemical Physics, vol XLVII. Wiley, New York
3. Demtröder W (1991) Laser Spektroskopie, Grundlagen und Technik. Springer, Berlin
4. Hutchinson MHR (1987) Excimer Lasers. In: Mollenauer LF, White JC (eds) Topics of Applied Physics, vol 59. Springer, Berlin
5. Kneubühl FK, Sigrist MW (1989) Laser. Teubner Studienbücher Physik, Teubner, Stuttgart
6. Lin SH, Fujimura Y, Neusser H, Schlag EW (1984) Multiphoton Spectroscopy of Molecules. Academic Press, Orlando
7. Parker DH, Berg JO, El-Sayed MA (1978) Multiphoton Ionization Spectroscopy of Polyatomic Molecules. In: Zewail AH (ed) Advances in Laser Chemistry, Springer Series in Chemical Physics, vol 3. Springer, Berlin
8. Letokhov VS (1987) Laser Photoionization Spectroscopy. Academic Press, New York
9. Brutschy B (1990) Reactions in Molecular Clusters Following Photoionization. J Phys Chem 94:8637-8647

10. Baer T (1989) Vacuum UV Photophysics and Photoionization Spectroscopy. Annu Rev Phys Chem 40:637-669

11. Morgner H (1976) Ionisation von atomarem Sauerstoff durch metastabiles Helium. Dissertation, Fakultät für Physik, Universität Freiburg

12. Hartman PL (1961) Improvements in a Source for Use in the Vacuum Ultraviolet. J Opt Soc Am 51:113-114

13. Brehm B, Siegert H (1965) Eine intensive H_2-Entladungslampe für das ferne Vakuum Ultraviolett. Z Angew Phys 19:244-246

14. Schlag EW, Comes FJ (1960) Intense Light Sources for the Vacuum Ultraviolet. J Opt Soc Am 50:866-867

15. Fehsenfeld FC, Evenson KM, Broida HP (1965) Microwave Discharge Cavities Operating at 2450 MHz. Rev Sci Instrum 36:294-298

16. Burger F, Maier JP (1979) "Charged Particle Oscillator" He(II) Photon Source. J Electron Spectrosc Rel Phen 16:471-474

17. Potts AW, Williams TA, Price WC (1973) Ultraviolet Photoelectron Data on the Complete Valence Shells of Molecules Recorded Using Filtered 30.4 nm Radiation. Discuss Faraday Soc 54:104-115

18. Samson JAR (1967) Techniques of Ultraviolet Spectroscopy. Wiley, New York

19. Berkowitz J, (1979) Photoabsorption, Photoionization and Photoelectron Spectroscopy. Academic Press, New York

20. Winick H, Doniach S (eds) (1980) Synchrotron Radiation. Plenum Press, New York

21. Marr GV (ed) (1987) Handbook of Synchrotron Radiation, vol 2. North Holland, Amsterdam

22. Kunz C (ed) (1979) Synchrotron Radiation, Topics in Current Physics, vol 10. Springer, Berlin

23. Walker S, Straw H (1961) Spectroscopy, vols I and II. Chapman and Hall, London

24. Bauman RP (1962) Absorption Spectroscopy. Wiley, New York

25. Born M (1972) Optik, 3. Auflage. Springer, Berlin

26. Klein MV (1970) Optics. Wiley, New York

27. Namioka T (1959) Theory of the Concave Grating. III. Seya Namioka Monochromator. J Opt Soc Am 49:951-961

28. Ganteför G, Gansa M, Meiwes-Broer KH, Lutz HO (1988) Photoelectron Spectroscopy of Jet-Cooled Aluminum Cluster Anions. Z Phys D 9:253-261

29. Rademann K (1989) Photoionization Mass Spectrometry and Valence Photoelectron-Photoion Coincidence Spectroscopy of Isolated Clusters in a Molecular Beam. Ber Bunsenges Phys Chem 93:653-670

30. Spohr R, von Puttkamer E (1967) Energiemessung von Photoelektronen und Franck-Condon-Faktoren der Schwingungsübergänge einiger Moleküle. Z Naturforsch 22a:705-710

31. Frost DC, McDowell CA, Vroom DA (1967) Photoelectron Kinetic Energy Analysis in Gases by Means of a Spherical Analyser. Proc R Soc, Lond A296:566-579

32. Huchital DA, Rigden JD (1972) Resolution and Sensitivity of the Spherical-Grid Retarding-Potential Analyzer: Application to Auger Electron Spectroscopy and x-Ray Photoelectron Spectroscopy. In: Shirley DA (ed) Electron Spectroscopy. North Holland, Amsterdam

33. Fox RE, Hickam WM, Groove JD, Kjeldaas Jr T (1955) Ionization in a Mass Spectrometer by Monoenergetic Electrons. Rev Sci Instr 26: 1101-1107

34. Stockdale JAD, Compton RN, Reinhardt PW (1969) Studies of Negative-Ion-Molecule Reactions in the Energy Region from 0 to 3 eV. Phys Rev 184:81-93

35. Harrower GH (1955) Measurement of Electron Energies by Deflection in a Uniform Electric Field. Rev Sci Instrum.26:850-854

36. Hughes AL, Rojansky V (1929) On the Analysis of Electronic Velocities by Electrostatic Means. Phys Rev 34:284-290

37. Arnow M, Jones DR (1972) Reanalysis of the Trajectories of Electrons in 127° Cylinder Spectrometers. Rev Sci Instrum 43:72-75

38. Purcell EM (1938) The Focusing of Charged Particles by a Spherical Condenser. Phys Rev 54:818-826

39. Kuyatt CE, Simpson JA (1967) Electron Monochromator Design. Rev Sci Instrum 38:103-111
40. Boness MJW, Schulz GJ (1974) Vibrational Excitation in CO_2 via the 3.8 eV Resonance. Phys Rev A9:1969-1979
41. Daviel S, Wallbank B, Comer J, Hicks PJ (1982) Electron Energy Loss Spectroscopy of Carbon Monoxide Using a New Position-Sensitive Multidetector Spectrometer. J Phys B15:1929-1937
42. Sohn W, Jung K, Erhardt H (1983) Threshold Structures in the Cross Sections of Low-Energy Electron Scattering of Methane. J Phys B16:891-901
43. Hefner H, Simpson JA, Kuyatt CE (1968) Comparison of the Spherical Deflector and the Cylindrical Mirror Analyzers. Rev Sci Instrum 39:33-35
44. Aksela S (1971) Analysis of the Energy Distribution in Cylindrical Electron Spectrometers. Rev Sci Instrum 42:810-812
45. Sar-el HZ (1967) Cylindrical Capacitor as an Analyzer. Rev Sci Instrum 38:1210-1216
46. Sar-el HZ (1971) Cylindrical Mirror Analyzer with Surface Entrance and Exit Slits. Rev Sci Instrum 42:1601-1606
47. Stamatovic A, Schulz GJ (1968) Trochoidal Electron Monochromator. Rev Sci Instrum 39:1752-1753
48. Stamatovic A, Schulz GJ (1970) Characteristics of the Trochoidal Electron Monochromator. Rev Sci Instrum 41:423-427
49. Allan M (1989) Study of Triplet States and Short-Lived Negative Ions by Means of Electron Impact. J Electron Spectrosc Relat Phen 48:219-351
50. Jordan KD Burrow PB (1987) Temporary Anion States of Polyatomic Hydrocarbons. Chem Rev 87:557-588
51. Giordan JC, Moore JH, Tossell JA, Weber J (1983) Negative Ion States of 3d Metallocenes. J Am Chem Soc 105:3431-3433
52. Modelli A, Scagnolari F, Distefano G, Guerre M, Jones D (1990) Electronic Structure of Tert-Butyl Halides and Group 14 Derivatives: Electron Affinities and Dissociative Electron Attachment. Chem Phys 145:89-99
53. Oster T, Kühn A, Illenberger E (1989) Gas Phase Negative Ion Chemistry. Int J Mass Spectrom Ion Proc 89:1-72
54. Tronc M, Azria R, Ben Arfa M (1988) Differential Cross Section for H^- and NH_2^- Ions in NH_3. J Phys B: At Mol Phys 21:2497-2506
55. Olthoff JK, Moore JH, Tossell JA (1986) Electron Attachment by Chloro- and Bromomethanes. J Chem Phys 85:249-254
56. Roy D (1972) Characteristics of the Trochoidal Monochromator by Calculation of Electron Energy Distribution. Rev Sci Instrum 43:535-541
57. Verhaart GJ, Brongersma HH (1980) Electronic Excitation (5-9 eV) in Ethylene and Some Haloethylenes by Threshold Electron-Impact Spectroscopy with an Improved Energy Resolution. Chem Phys 52:431-442
58. McMillan MR, Moore JH (1980) Optimization of the Trochoidal Electron Monochromator. Rev Sci Instrum 51:944-950
59. Tam WC, Wong SF (1979) Magnetically Collimated Electron Impact Spectrometer. Rev Sci Instrum 50:302-307
60. Allan M (1982) Forward Electron Scattering in Benzene: Forbidden Transitions and Excitation Functions. Helv Chim Acta 65:2008-2023
61. Dressler R, Allan M (1987) A Dissociative Electron Attachment, Electron Transmission, and Electron Energy Loss Study of the Temporary Negative Ion of Acetylene. J Chem Phys 87:4510-4518
62. Boersch H, Geiger J, Topchowsky M (1965) Rotational Structure in the Energy Loss Spectrum of H_2. Phys Lett 17:266-269
63. Boersch H, Geiger J, Stickel W (1964) Das Auflösungsvermögen des elektrostatisch-magnetischen Energieanalysators für schnelle Elektronen. Z Physik 180:415-424
64. Herzog R (1934) Ionen- und elektronenoptische Zylinderlinsen und Prismen. Z Physik 89:447-473

65. von Ardenne M (1975) Tabellen zur Angewandten Physik, Band I. VEB Verlag, Berlin

66. Spohr R, Guyon PM, Chupka WA, Berkowitz J (1971) Threshold Photoelectron Detector for Use in the Vacuum Ultraviolet. Rev Sci Instrum 42:1872-1879

67. Peatman WB, Kasting GB, Wilson DJ(1975) The Origin and Elimination of Spurious Peaks in Threshold Electron Photoionization Spectra. J Electron Spectrosc Relat Phen 7:233-246

68. Lohr W, Jochims HW, Baumgärtel H (1975) Photoreaktionen kleiner, organischer Moleküle IV. Absorptionsspektren, Photoionen- und Resonanzphotoelektronenspektren von Vinylbromid. Ber Bunsenges Phys Chem 79:901-906

69. Stockbauer R, Inghram MG (1975) Threshold Photoelectron-Photoion Coincidence Mass Spectrophotometric Study of Ethylene and Ethylene-d_4. J Chem Phys 62:4862-4870

70. Baer T (1979) State Selection by Photoion-Photoelectron Coincidence. In: Bowers MT (ed) (1979) Gas Phase Ion Chemistry, vol 2. Academic Press, New York

71. Baer T (1986) The Dissociation Dynamics of Energy Selected Ions. In: Prigogine I, Rice SA (eds) Adv Chem Phys, vol LXIV. Wiley, New York

72. Allen JD, Durham JD, Schweitzer GK, Deeds WE (1976) A New Electron Spectrometer Design. J Electron Spectrosc Relat Phen 8:395-410

73. Allen JD, Grimm FA (1979) High-Resolution Photoelectron Spectroscopy through Deconvolution. Chem Phys Lett 66:72-78

74. Lindau I, Helmer JC, Uebbing J (1973) A New Retarding Field Electron Spectrometer with Differential Output. Rev Sci Instrum 44:265-274

75. Hotop H, Hübler G (1977) Photoelectron and Penning Ionization Electron Spectrometry with a Differential Retarding Field Analyzer. J Electron Spectrosc Relat Phen 11:101-121

76. Beamson G, Porter HQ, Turner DW (1981) Photoelectron Spectromicroscopy. Nature 290:556-561

77. Kruit P, Reed FH (1983) Magnetic Field Paralleliser for 2 π Electronspectrometer and Electron-Image Magnifier. J Phys E: Sci Instrum 16:313-324

78. Chewter LA, Sander M, Müller-Dethlefs K, Schlag EW (1987) High Resolution Zero Kinetic Energy Photoelectron Spectroscopy of Benzene and Determination of the Ionization Potential. J Chem Phys 86:4737-4744

79. Müller-Dethlefs K, Sander M, Schlag EW (1984) A Novel Method Capable of Resolving Rotational Ionic States by the Detection of Threshold Photoelectrons with a Resolution of 1.2 cm^{-1}. Z Naturforsch A39:1089-1091

80. Müller-Dethlefs K, Sander M, Schlag EW (1984) Two-Colour Photoionization Resonance Spectroscopy of NO: Complete Separation of Rotational Levels of NO$^+$ at the Ionization Threshold. Chem Phys Lett 112:291-294

81. Neusser HJ (1989) Lifetimes of Energy and Angular Momentum Selected Ions. J Phys Chem 93:3897-3907

82. Samson JAR (1969) Line Broadening in Photoelectron Spectroscopy. Rev Sci Instrum 40:1174-1177

83. Fricke J (1973) Kondensation in Düsenstrahlen. Physik in unserer Zeit 4:21-27

84. Anderson JB (1974) Molecular Beams for Nozzle Sources. In: Wegener PP (ed) Molecular Beams and Low Density Gas Dynamics. Dekker, New York

85. Miller DR (1988) Free Jet Sources. In: Scoles G (ed) Atomic and Molecular Beam Methods, vol I. Oxford University Press, New York

86. Maas WPM, Nibbering NMM (1989) Formation of Doubly Charged Negative Ions in the Gas Phase by Collisionally Induced "Ion Pair" Formation from Singly Charged Negative Ions. Int J Mass Spectrom Ion Proc 88:257-266

87. Kerwin L (1963) Ion Optics. In: McDowell CA (ed) Mass Spectrometry. McGraw-Hill, New York

88. Morrison JD (1986) Ion Focusing, Mass Analysis and Detection. In: Futrell JH (ed) Gaseous Ion Chemistry and Mass Spectrometry. Wiley, New York

89. Dempster AJ (1918) A New Method of Positive Ray Analysis. Phys Rev 11:316-325

90. Johnson EG, Nier AO (1953) Angular Aberrations in Sector Shaped Electromagnetic Lenses for Focusing Beams of Charged Particles. Phys Rev 91:10-17
91. Mattauch J, Herzog RFK (1934) Über einen neuen Massenspektrographen. Z Physik 89:786-795
92. Howe I, Williams DR, Bowen RD (1981) Mass Spectrometry, Principles and Applications, 2rd ed. McGraw-Hill, New York.
93. Beynon JH, Saunders RA, Williams AE (1965) Dissociation of Meta-Stable Ions in Mass Spectrometers with Release of Internal Energy. Z Naturforsch 20a:180-183
94. Beynon JH, Cooks RG, Amy JW, Baitinger WE, Ridley TE (1973) Design and Performance of a Mass-Analyzed Ion Kinetic Energy (MIKE) Spectrometer. Anal Chem 45:A1023-A1027
95. Beynon JH, Fontaine AE (1967) Mass Spectrometry: The Shapes of "Meta-Stable Peaks", Z Naturforsch 20a:334-346
96. Cooks RG, Beynon JH, Caprioli RM, Lester GR (1973) Metastable Ions. Elsevier, Amsterdam
97. Iraqi M, Lifshitz C (1989) Studies of Ion Clusters of Atmospheric Importance by Tandem Mass Spectrometry. Neat and Mixed Clusters Involving Methanol and Water. Int J Mass Spectrom Ion Proc 88:45-47
98. Stace AJ, Shukla AK (1982) A Measurement of the Average Kinetic Energy Release During the Unimolecular Decomposition of CO_2 Ion Clusters. Chem Phys Lett 85:157-160
99. Cameron AE, Eggers DF (1948) An Ion "Velocitron". Rev Sci Instrum 19:605-607
100. Wolf HM, Stephens WE (1953) A Pulsed Mass Spectrometer with Time Dispersion. Rev Sci Instrum 24:616-617
101. Wiley WC, McLaren IH (1955) Time-of-Flight Mass Spectrometer with Improved Resolution. Rev Sci Instrum 26:1150-1157
102. Franklin JL (1979) Energy Distribution in the Decomposition of Ions. In: Bowers MT (ed) (1979) Gas Phase Ion Chemistry, vol 1. Academic Press, New York
103 Farmer JB (1963) Types of Mass Spectrometers. In: McDowell CA (ed) Mass Spectrometry. McGraw-Hill, New York
104. Boesl U, Neusser HJ, Weinkauf R, Schlag EW (1982) Muliphoton Mass Spectrometry of Metastables: Direct Observation of Decay in a High-Resolution Time-of-Flight Instrument. J Phys Chem 86:4857-4863
105. Karataev VI, Mamyrin BA, Shmikk DV (1972) New Method for Focusing Ion Bunches in Time-of-Flight Mass Spectrometers. Soviet Physics-Technical Physics 16:1177-1179
106. Kühlewind H, Kiermeier A, Neusser HJ, Schlag EW (1987) Laser Measurements of the Unimolecular Kinetics of Energy Selected Molecular Ions: Isotope Effects in Benzene. J Chem Phys 87:6488-6498
107. Kühlewind H, Neusser HJ, Schlag EW (1983) Metastable Fragment Ions in Multiphoton Time-of-Flight Mass Spectrometry: Decay Channels of the Benzene Cation. Int J Mass Spectrom Ion Proc 51:255-265
108. Bergmann T, Martin TP, Schaber H (1989) High Resolution Time-of-Flight Mass Spectrometer. Rev Sci Instr 60:792-793
109. Paul W, Steinwedel H (1953) Ein neues Massenspektrometer ohne Magnetfeld. Z Naturforsch 8a:448-450
110. Paul W, Raesler M (1955) Das elektrische Massenfilter. Z Physik 140:262-273
111. Paul W, Reinhard HP, von Zahn U (1958) Das elektrische Massenfilter als Massenspektrometer. Z Physik 152:143-182
112. McLachlan NW (1951) Theory and Application of Mathieu Functions. Oxford University Press, New York
113. Dawson PH (ed) (1976) Quadrupole Mass Spectrometry and its Application. Elsevier, Amsterdam
114. Campana JE (1980) Elementary Theory of the Quadrupole Mass Filter. Int J Mass Spectrom Ion Phys 33:101-117

115. Dayton IE, Shoemaker FC, Mozley RF (1954) The Measurement of Two-Dimensional Fields. Part II: Study of a Quadrupole Magnet. Rev Sci Instrum 25:485-489

116. Denison DR (1971) Operating Parameters of a Quadrupole in a Grounded Cylindrical Housing. J Vac Sci Technol 8:266-269

117. Baldeschwieler JD (1968) Ion Cyclotron Resonance Spectroscopy. Science 159:263-273

118. Beauchamp JL, Anders LR, Baldeschwieler JD (1967) The Study of Ion-Molecule Reactions in Chloroethylene by Ion Cyclotron Resonance Spectroscopy. J Am Chem Soc 89:4569-4577

119. Farrar JM, Saunders WH (eds) (1988) Techniques for the Study of Ion-Molecule Reactions. Wiley, New York

120. Comisarow MB, Marshall AG (1974) Fourier Transform Ion Cyclotron Resonance Spectroscopy. Chem Phys Lett 25:282-283

121. Hartmann H, Wanczek KP (eds) (1982) Ion Cyclotron Resonance Spectrometry II, Lecture Note in Chemistry Series, vol 31. Springer, Berlin

122. Wanczek KP (1984) Ion Cyclotron Resonance Spectrometry - A Review. Int J Mass Spectrom Ion Proc 60:11-60

123. Allemann M, Kellerhals HP, Wanczek KP (1980) A New Fourier Transform Mass Spectrometer with a Superconducting Magnet. Chem Phys Lett 75:328-331

124 Irion MP, Selinger A, Wendel R (1990) Secondary Ion Fourier Transform Mass Spectrometry: A New Approach Towards the Study of Ion Metal Clusters Chemistry. Int J Mass Spectrom Ion Proc 96:27-47

125. Allemann M, Kellerhals HP, Wanczek KP (1983) High Magnetic Field Fourier Transform Ion Cyclotron Resonance Spectroscopy. Int J Mass Spectrom Ion Phys 46:139-142

126. Ervin KM, Hoe J, Lineberger WC (1989) A Study of the Singlet and Triplet States of Vinylidene by Photoelectron Spectroscopy of $H_2C=C^-$, $D_2C=C^-$ and $HDC=C^-$. Vinylidene-Acetylene Isomerization. J Chem Phys 91:5974-5992

127. Leopold DG, Murray KK, Stevens Miller AA, Lineberger WC (1985) Methylene: A Study of the $\tilde{X}^3B_1$ and $\tilde{a}^1A_1$ States by Photoelectron Spectroscopy of CH_2^- and CD_2^-. J Chem Phys 83:4849-4865

128. Bleakney W, Hipple JA (1938) A New Mass Spectrometer with Improved Focusing Properties. Phys Rev 53:521-533

129. Bennett WH (1950) Radiofrequency Mass Spectrometer. J Appl Phys 21:143-149

130. Berry CE (1954) Ion Trajectories in the Omegatron. J Appl Phys 25:28-31

131. Sommer H, Thomas HA, Hipple JA (1951) Measurement of e/M by Cyclotron Resonance. Phys Rev 82:697-702

132. Kuyatt CE (1968) Measurement of Electron Scattering from a Static Gas Target. In: Bederson B, Fite WL (eds) Atomic and Electron Physics - Atomic Interactions. Academic Press, New York

133. Hasted JB (1964) Physics of Atomic Collisions. Butterworths, London

134. Wiza JL (1979) Microchannel Plate Detectors. Nucl Instr Methods 162:587-601

135. Gao RS, Robert WE, Smith GJ, Stebbings RF (1988) High-Resolution Position-Sensitive Detector. Rev Sci Instr 59:1954-1956

136. Gao RS, Gibner PS, Newman JH, Smith KA, Stebbings RF (1984) Absolute and Angular Efficiencies of a Microchannel-Plate Position-Sensitive Detector. Rev Sci Instr 55:1756-1759

137. Kalamarides A, Walter CW, Lindsay BG, Smith KA, Dunning FB (1989) Post-Attachment Interactions in K(nd)-CF$_3$I Collisions at Intermediate n. J Chem Phys 91:4411-4413

138. Kalamarides A, Marawar RW, Ling X, Walter CW, Lindsay BG, Smith KA, Dunning FB (1990) Negative Ion Production in Collisions Between K(nd) Rydberg Atoms and CF$_3$Br and CF$_2$Br$_2$. J Chem Phys 92:1672-1676

139. Daly RN (1960) Scintillation Type Mass Spectrometer Ion Detector. Rev Sci Instrum 31:264-267

140. Grossmann WEL (1989) A Comparison of Optical Detectors for the Visible and Ultraviolet. J Chem Ed 66:697-700

Part II The Monomolecular Decay of Electronically Excited Molecular Ions

Jacques A. G. Momigny

Introduction

This contribution to "Topics in Physical Chemistry" will introduce the reader to the present experimental and theoretical study of the processes involved when electronically excited molecular ions lose their energy by decaying into more stable states. As the lowest ionization energy of the molecules has been chosen here as the origin of the energy scale, we will mainly deal with positive ionization processes without neglecting the possible occurrence of excited states of the molecules lying above the energy scale origin.

In the first Chapter of this work we will describe how ions, either molecular or fragmentary, are produced in gases in such a pressure range that their decay behaves unimolecularly. Such a pressure range is realized when the mean free path of the molecules is larger than the dimension of the chamber where the ionization takes place, or under molecular beam conditions.

In Chapter 2, the different ways used by electronically excited molecular ions to lose their energy either by radiative, but more frequently, by nonradiative processes leading to dissociation processes, will be examined. This will lead to a classification of the dissociative decay of molecular ions, the dissociation phenomena being produced from successively isolated electronic states, or through radiationless transitions that completely transform the electronic energy in internal energy of the ground state molecular ion which dissociates.

In Chapter 3, the energy balance characterizing the dissociation process of molecular ions will be considered. As such thermochemical considerations are possible only for completely defined states of the ions, the problem of the structural identity of the appearing ions will be treated. Theoretical and experimental methods giving the best possible solution to this problem will be described.

In Chapter 4, the additional hypothesis of the complete randomization of internal energy in the ground state molecular ions examines how statistical theories of unimolecular decay of internally excited molecules can be applied for a priori calculations of the rate constant, characterizing the dissociation processes of the molecular ions. These calculations will lead, through the use of an appropriate kinetic scheme of concurrent and consecutive unimolecular dissociation reactions, to the determination of an a priori calculated breakdown diagram, to be compared with the experimental one.

In Chapter 5, miscellaneous useful topics in molecular dissociation phenomena will be discussed in order to check, in some definite cases, the application limits of the pure statistical theories.

This contribution ends with two appendices that illustrate the realization of a limited direct count of quantum states in a rather simple molecule (CO_2), and the calculation of rate constants as function of energy for two dissociation channels of $C_2H_3Cl^+$.

1 Ionization Processes in Gaseous Phase

The enrichment in energy of gaseous molecules from their first ionization energy will be considered as resulting from the following interactions:
— Resonant photon absorption and photoionization
— Resonant electron impact and electroionization
— Charge exchange
— Field ionization
— Penning ionization

1.1 Resonant Photon Absorption and Photoionization

1.1.1 Absorption and Photoionization Cross-Sections; Ion Yields; Mass Spectrum

When a molecule absorbs photons of increasing energy from a continuous or from a discrete light source above its first ionization energy, it is possible to measure a photoabsorption curve that expresses the variation of the total absorption cross-section σ_a, as a function of the photon energy. If, simultaneously, the produced total ion current is measured, the variation of the total ionization cross-section σ_i with the photon energy is available. From both curves it is easy to calculate the ionization yield, y_i, as being equal to:

$$y_i = \frac{\sigma_i}{\sigma_a}.$$

1.1

This yield will never exceed unity (one monopositive ion produced for each absorbed photon), but will rather frequently be less than unity, at least in a certain energy range. This is clearly shown in Fig. 1.1 where σ_a, σ_i, and y_i are drawn for NH_3, from very recent data [1]. From this figure, it appears that y_i reaches unity only some 8 eV above the first ionization energy, equal to 10.16 eV for NH_3. Below 18 eV, the production of neutral ammonia molecules with an energy content higher than the first ionization energy is therefore possible. Such excited molecules are designated as "superexcited molecules" [2]. The cross-section for the appearance of the superexcited neutral molecules is given in Fig. 1.2, as equal to:

$$\sigma_n = \sigma_a - \sigma_i.$$

1.2

This last statement needs to be completed by the following remarks:
Superexcited molecules are a priori able to evolve in three ways:
1. Some can autoionize by radiationless transition to one of the adjacent ionization continua, giving rise to a positive ion; the cross-section for production of

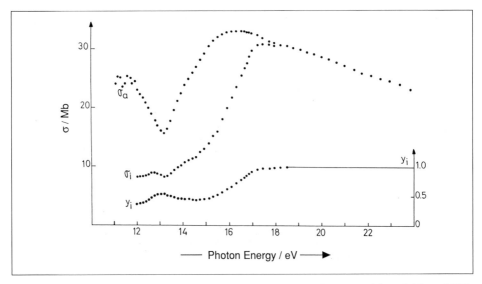

Fig. 1.1. Absorption cross-section σ_a, photoionization cross-section σ_i, and ion yield y_i of NH_3 between 12 and 23 eV. Adapted from [1].

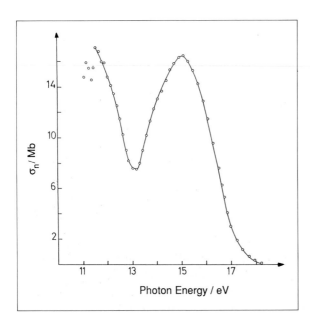

Fig. 1.2. Cross-section for the appearance of superexcited neutrals $\sigma_n = \sigma_a - \sigma_i$ in NH_3. Adapted from [1].

autoionizing neutrals is a part of σ_i and cannot be evaluated from the present data.

2. Some lower their energy by emission of radiation.
3. Some decompose to radicals or to smaller molecules and/or to atoms, by bond dissociation[1].

When the ion yield reaches unity, it does not mean that superexcited molecules are not induced but, on the contrary, that all of them, if present, autoionize.

An intriguing question arises from the concept of total ionization cross-section σ_i: what are the identities of the ions formed in the photoabsorption processes? The best answer to this question lies in the application of mass spectrometry by which a suitable dispersion of the ions in space or in time following their m/e value will give an image (eventually deformed) of the ionic composition hidden under the σ_i values. As an example, we give in Table 1.1 an idea of the mass spectrum of NH_3, observed with photons of increased energy [1].

Such a table illustrates that, beside the molecular ion NH_3^+ present at all the quoted energies, all the possible fragmentary ions appear, each of which being clearly characterized by the occurrence of energy thresholds.

Table 1.1. Relative abundance of ions in photoionization of NH_3 at different energies.

m/e	Ions	15.49 eV	17.71 eV	24.79 eV	31 eV
17	NH_3^+	100	36.68	37.23	36.03
16	NH_2^+		63.32	62.05	55.10
15	NH^+			0.44	3.39
1	H^+			0.27	4.95
14	N^+				0.35
2	H_2				0.15

1.1.2 A Model for the Appearance of the Mass Spectrum of Diatomic Molecules

A pertinent question arises now: does a simple model exist to understand the progressive building up of the mass spectrum at progressively increased energies? This model does exist and an example of it will now be detailed. In Figure 1.3, the potential energy curves of a diatomic AB molecule in its ground state and of a possible set of electronic states (X, A, B, a, C) of the molecular ion AB^+ are drawn. Following the

---

[1] In the case of NH_3, it was shown that an intense fluorescence occurs from about 12 eV up to 18 eV; the fluorescence cross-section varies exactly as the $\sigma_n(E)$ function. The analysis of the fluorescence is ascribed to the $A^3\Pi_i \to X^3\Sigma$ transition of the NH radical. This observation supports the following dissociation process. $NH_3 \to NH(A^3\Pi_i) + 2H$, which has a calculated threshold at 12.11 eV.

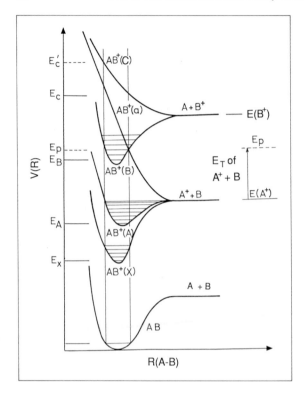

Fig. 1.3. Set of possible potential energy curves for some successive electronic states of a diatomic AB^+ molecular ion.

Franck-Condon principle, the photon absorption from AB to the X, A, B, a, C states of AB^+, will take place by vertical transitions only.

When the energy E_X is absorbed from the ground state of the molecule, the ground state X of the molecular ion appears in its lowest vibrational energy level, at some higher energy the vibrational population of the X state goes up to $v = 4$.

The absorption of photon energies ranging from E_A to $E(A^+)$ induces the appearance of the state AB^+ with a complete manifold of vibrational energy; for an energy $E(A^+)$ the A^+ ion appears by the process: $AB^+(A) \rightarrow A^+ + B$.

This dissociation process does not involve the production of translational energy for the $(A^+ + B)$ pair.

When the energy level E_B is reached the B state of the AB^+ appears in its $v = 0$ level; for somewhat more energy the $v=1$ and $v=2$ levels are populated. For the $v=2$ level, however, the B state becomes unstable by electronic predissociation[2] and the

---

[2] This phenomenon is rather frequə .t; the AB^+(a) state is one of the states which arises from the combination of the electronic state of A^+ and of B; it does not necessarily correspond to one of the singly excited configurations of the molecular ion; it will possibly be correlated with a doubly excited configuration of AB^+.

probability for the system to switch to the repulsive a state of AB^+ is appreciable: the $(A^+ + B)$ pair appears with a rather high translational energy range of $E_p - E(A^+)$. At higher energy the $v = 3,4$ levels will also give rise to $(A^+ + B)$ with successive increment of translational energy E_T, but also with decreasing probabilities.

If an electronic state like a, decomposes during the time needed for a vibrational period ($\approx 10^{-13}$ s), the fragmentary ions produced through a predissociation are able to see their appearance weakly or strongly delayed with respect to the direct dissociation. The lifetime of the electronic state that is predissociated can range from about a microsecond to a nanosecond or shorter, owing to the fact that the predissociation is forbidden or allowed by the appropriate selection rules [3, 4].

The transition to the C state of AB^+ starts when an energy E_C is absorbed.

As the C state is repulsive, no new contribution from the C state appears in the ionic current of AB^+. From E_C to E'_C the $A + B^+$ pair is produced with an excess of translational energy ranging continuously from $E_C - E(B^+)$ to $E'_C - E(B^+)$.

It is relatively easy to summarize the above description: the mass spectrum is due to the successive occurrence of stable, partially unstable, or completely unstable electronic states of AB^+. The stable part of the electronic states contributes to the AB^+ ion current only; the completely or partially unstable part of the electronic states contributes to the appearance of A^+ and B^+ without or with translational energy.

In the case of diatomic molecules, the dissociation processes are generally fast (10^{-13} s); this rate can be strongly diminished when the appearance of fragment ions is due to any kind of predissociation process.

1.1.3 Threshold Laws and Ionization Yield Curves for Diatomic Molecules

Before extending the proposed model to polyatomic molecules, we need to know how the progressive building up of the mass spectrum is experimentally demonstrated; it will be found in the study of the mass selected ion yield curves. Such a study being related to the concept of threshold laws for ionization, it is now appropriate to ask if a theoretical approach to these laws in the case of atoms could be of some help to us.

Approaches made by Wigner [5] and Geltman [6] show that, under certain restrictive assumptions, the cross-section behavior in the threshold region is given by a power law:

$$\sigma(E) = C(E - E_0)^{n-1}, \qquad\qquad 1.3$$

where E is the ionizing electron or photon energy, E_0 is the threshold energy, C is a constant, and n the total number of outgoing electrons in the ionization process.

This law leads to the situations described in Fig. 1.4. As shown by Morrison [7], these laws, if valid over a reasonable energy range, would allow to conclude that for any ionization process:

$$\frac{d\sigma\,(E)_{el}}{dE} = \sigma(E)_{ph},$$ 1.4

where $\sigma\,(E)_{el}$ and $\sigma\,(E)_{ph}$ is the cross-section for electroionization and photoionization, respectively.

If successive ionization thresholds $E_1, E_2 \dots E_i$ are present within the energy limit for the validity of such laws, one will see in the photoionization curve, successive steps at $E_1, E_2 \dots E_i$. Similarly, one could expect a sudden change of slopes for the same energy values in the electroionization yield curve. Unfortunately, in the case of atoms, such a behavior is not verified, except for the sudden increases in the $\sigma(E)$ at photoionization thresholds. As shown in Fig. 1.5, where the single photoionization cross-section of rare gases has been collected, the step function (if any) is restricted to the first few fractions of an electron-volt (eV) above threshold. The shape of the cross-section curves are characterized by a decrease of $\sigma(E)$ with photon energy, (a remarkable exception being observed for neon). Let us, however, point out that the double photoionization behavior, as observed for the appearance of He^{++}, CO_2^{++}, SF_6^{++} is correctly expressed by the linear law with a validity range of about 5 eV above threshold [8, 9, 10, 11]. One could be strongly disappointed about the results observed in the case of atoms and rapidly encouraged by the observations made in the photoionization of molecules.

In this case, the validity of the step function law for photoionization for an energy range wider than the successive vibrational energy steps has been commonly verified.

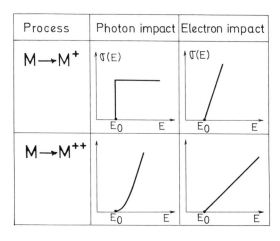

Fig. 1.4. Examples of idealized threshold laws for photoionization and electron impact ionization.

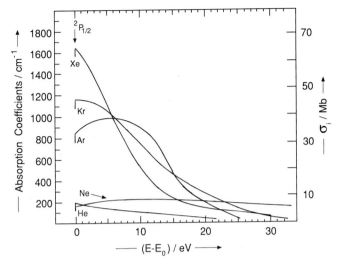

Fig. 1.5. Absorption coefficients (cm^{-1}) and photoionization cross-sections (megabarns) for rare gases [6–8].

Accepting the validity of this behavior enables us to express the phenomena described in Fig. 1.3, by the set of photoionization yield curves drawn in Fig. 1.5. The shape of the molecular ion yield curves of Fig. 1.6 being idealized, it now appears necessary to describe how they will be experimentally observed.

Four main discrepancies can occur:

1. A smearing out of the sharp staircase structure into a series of sigmoidal onsets, due to the triangular slit function of the light-selection devices (monochromator): the points of inflection of sigmoids represent the points to be taken as successive onset energies (Fig. 1.7);

2. Some additional tailing of the onsets and rounding of the steps are introduced as an effect of rotational levels on the direct ionization curve shape;

3. The occurrence, as superimposed on the phenomenon of direct ionization, of autoionization effects. These effects, sometimes hidden in the total ionization cross-section curves, are evidenced here as "resonance phenomena" occurring simultaneously with the direct photoionization. If their intensity is not too high, the staircase structure shall remain detectable, as in the case of NO$^+$ (Fig. 1.8); very often, unfortunately, interactions between the superexcited states and the ionization continuum are so strong that the staircase function becomes undetectable. This is namely the case for H$_2^+$ (Fig. 1.9).

In such unfavorable cases, the discussion of the origin of superexcited states is easier than any hazardous identification of a vibrational progression due to the direct ionization of a molecule.

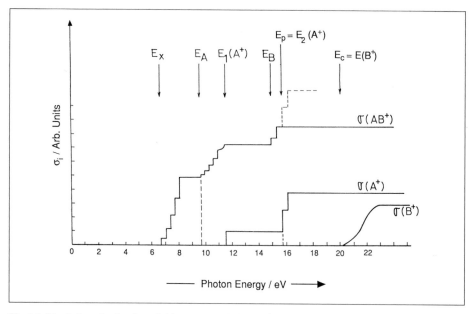

Fig. 1.6. Ideal photoionization yield curves as deduced from the set of potential energy curves for the diatomic AB molecules (Fig. 1.3).

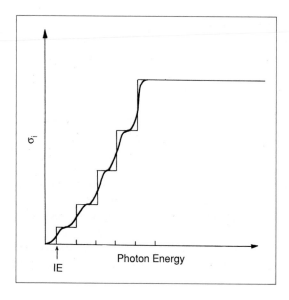

Fig. 1.7. Example of the smearing out of a staircase function through the apparatus function of a VUV mono-chromator.

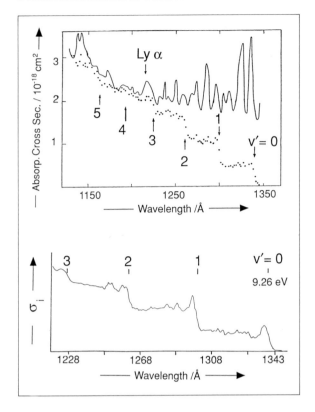

Fig. 1.8. Experimental absorption and photoionization yield curves of NO [12].

4. The cross-section observed at the end of the staircase possibly could not remain constant for high energy values, and could start a regular decrease as observed for rare gases. This decrease could be either similar for or specific to each electronic state involved in the building up of the ionic abundances of the parent and of fragmentary ions.

1.1.4 Time Windows for the Appearance of the Mass Spectrum

The phenomena described in Fig. 1.3 giving rise to the idealized photoionization curves of Fig. 1.6 are not only characterized by the occurrence of successive energy thresholds and by threshold laws, but also by their evolution time scale.

Is this time scale of such a nature as to modify, in certain circumstances, the idealized behavior of Fig. 1.6. Let us first recall that the mass spectrum is an image of the ionization processes, recorded after ion transit times through the mass spectrometer of the order of about several units to several tens of microseconds, depending on the electric potential used to achieve a good mass resolution.

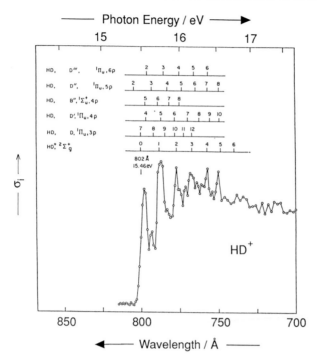

Fig. 1.9. Experimental photoionization yield curve of H_2 [13].

Concerning the time scale for ionization and subsequent dissociation processes, different time ranges are distinguishable:

a) the time for direct ionization is comparable with a Bohr period, about 10^{-16} s;

b) the time required for a direct dissociation as observed from the a or C states of AB^+ is comparable with the time of a vibrational period, about 10^{-13} to 10^{-14} s;

c) the time required for an indirect dissociation or "predissociation" ranges from 10^{-13} to 10^{-3} s, depending on the process responsible for indirect dissociation. Some of these processes will be discussed later.

The last sentence implies that for indirect dissociation, times of the order of the transit times through the mass spectrometer can occur. As a consequence the heights of the steps corresponding to the appearance of parent and fragment ions will include a time-dependent factor. In order to understand this kind of situation, let us consider the case of an AB^+ ion for which one vibrational level is predissociated with a characteristic rate constant k. The corresponding lifetime of AB^+ is within the limits of the transit time of AB^+ through the mass spectrometer.

The transit time through the successive regions of a single focusing mass spectrometer are respectively:

t_1, the time for the extraction of AB^+ from the ionization region;

t_2, the time spent by AB^+ in the acceleration region;

t_3, the time spent by AB^+ between the exit from the acceleration region and its entry into the magnetic sector of the mass spectrometer;

t_4, the time elapsed in crossing the magnetic sector;

t_5, the time spent by AB^+ between the exit of the magnetic sector and the ion collector of the instrument.

If t_i is one of these times and if $N_0(AB^+)$ ions are initially formed and able to decay with a unique rate constant k, the number of these ions collected as AB^+, is:

$$N(AB^+) = N_0(AB^+) \exp\left(- k \sum_{i=1}^{5} t_i\right). \qquad 1.5$$

The number of AB^+ ions that will be collected as A^+ is:

$$N(A^+) = N_0(AB^+) \exp\left(- kt_1\right). \qquad 1.6$$

What happens with AB^+ ions which dissociate during their flight between full acceleration and their entry into the magnetic sector of the mass spectrometers. For these ions only, it is possible to show that their dissociation in flight gives rise to diffuse peaks at an apparent mass m^*, given by [14]:

$$m^* = \frac{m_A^2}{m_{AB}}. \qquad 1.7$$

The number of these ions is

$$N^*(AB^+) = N_0(AB^+)\{1 - \exp\left(- k(t_1 + t_2)\right)\} \exp\left(- kt_3\right). \qquad 1.8$$

These ions, which appear as diffuse peaks at generally nonintegral masses are called "metastable ions"; the fragment A^+ is know as the "metastable fragment" and the peak observed at m^* is the "metastable peak". As recognized very early [15] and as will be extensively discussed later, the shape of such peaks is directly associated with the total translational kinetic energy distribution of the fragments involved in the dissociation process. In diatomic and sometimes in more complex ions the observation of such metastable ions is associated with the occurrence of slow electronic or rotational predissociation. In complex polyatomic ions they can also result from kinetic situations that express the frequent occurrence of vibrational predissociation. All these processes will be described and discussed later.

1.1.5 Origin and Decay of Superexcited States (SE)

1.1.5.1 Origin of SE in Atoms

The presence above the first ionization energy of molecules of superexcited states has been emphasized in Sections 1.1.1 and 1.1.3. As was shown, such SE states can dissociate into neutral atoms or radicals; their study largely concerns photochemistry above the first ionization energy. The result of such investigations helps to understand the dissociative ionization, because some of the observed dissociation processes cause highly excited states of the molecular fragments to appear, the energy levels of which converge to the dissociative ionization limits. A non-negligible part of SE states causes radiationless transitions to one of the adjoining ionization or dissociative ionization continua, giving rise either to molecular or fragmentary ions.

Such radiationless transitions are known as indirect ionization processes; they are able to populate resonantly potential energy curves of molecular ions outside their Franck-Condon region, being thus responsible for an appreciable intricacy of direct and indirect ionization phenomena.

A brief survey of the situations prevailing in atoms shall help us to better understand the situations encountered in molecules. In the helium atom, single excited configurations such as $(1s^1 np^1)$; $(1s^1 nd^1)$, etc., make up the set of energy levels converging to $He^+ (1s^1)$ at 24.5 eV.

One could, however, consider the possibility of the simultaneous excitation of the two electrons in electronic configurations such as $(2s^1 np^1)$, $(3s^1 np^1)$, etc. converging to electronic configurations of He^+ such as $2s^1, 3s^1$, etc. Such neutral levels normally lie between the lowest ionization energy at 24.5 eV and the energy for double ionization at 78 eV. When the excitation of such energy levels is allowed by selection rules, they will appear and eventually decay to one of the continua joining to the $He^+ (2s^1)$, $(3s^1)$ configurations. Such levels have been detected in photoabsorption [16] and one of them, at least, corresponds to the $2s^1 2p^1$ configuration at 58.35 eV [17].

If one considers a more complex atom, Be for instance, which has $(1s^2 2s^2)$ configuration, one can expect for Be two successive ionization energies corresponding to $Be^+ (1s^2 2s^1)$ and $(1s^1 2s^2)$.

The Rydberg series corresponding to the second configuration will be $(1s^1 2s^2 ns)$, $(1s^1 2s^2 np)$, etc. realizing single excited configurations of Be energetically lying far above the first ionization energy of Be. Simultaneously, double excited configurations of the type $(1s^2 2p^1 ns)$ or $(1s^2 2p^1)$ or (nd), converging to excited states of Be^+ such as $(1s^2 2p^1)$ are possibly excited immediately above the first ionization energy.

Both types of excited levels corresponding either to single excited Rydberg series or to double excited configurations, if excited, shall decay to one of the adjacent ionization continua, or will be branched to two or more of these.

Atomic autoionization is an electronic phenomenon. In molecules it will additionally be a vibrational and/or rotational phenomenon.

1.1.5.2 Origin of Superexcited States in Molecules

Electronic structure of molecules is characterized by the filling up with electrons of their successive molecular orbitals. Let us, for instance, consider a linear triatomic molecule N_2O. Its electronic configuration is given by the set of orbitals

$$\dots (6\sigma)^2 (1\pi)^4 (7\sigma)^2 (2\pi)^4,$$

giving rise to the ground state of the molecule: $\tilde{X}^1\Sigma^+$.

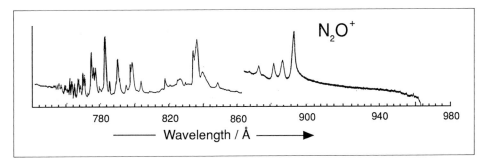

Fig. 1.10. Experimental photoionization yield curve of N_2O [24].

From this structure one can expect four successive ionization energies corresponding to the following ionic configurations and electronic states:

$$(6\sigma)^2 \, (1\pi)^4 \, (7\sigma)^2 \, (2\pi)^3 \, \tilde{X}^2\Pi$$

$$(6\sigma)^2 \, (1\pi)^4 \, (7\sigma)^1 \, (2\pi)^4 \, \tilde{A}^2\Sigma^+$$

$$(6\sigma)^2 \, (1\pi)^3 \, (7\sigma)^2 \, (2\pi)^4 \, \tilde{B}^2\Pi$$

$$(6\sigma)^1 \, (1\pi)^4 \, (7\sigma)^2 \, (2\pi)^4 \, \tilde{C}^2\Sigma^+ .$$

All these ionic states have been observed by photoelectron spectroscopy and their adiabatic ionization energies are, respectively [18], 12.89 eV, 16.4 eV, 17.6 eV and 20.1 eV. Consequently, Rydberg series converging to $\tilde{A}$, $\tilde{B}$, $\tilde{C}$ electronic states of N_2O^+ shall extend from 12.89 up to 20.1 eV at least. Converging to the $\tilde{A}$ state, Rydberg series shall correspond to electronic configurations given by:

$$(7\sigma)^1 \, (2\pi)^4 \, (ns\sigma)^1, \quad (7\sigma)^1 \, (2\pi)^4 \, (np\sigma)^1, \quad \cdots (7\sigma)^1 \, (2\pi)^4 \, (np\pi)^1, \text{ etc.}$$

Each term of these series will be characterized by a vibrational progression and simultaneous rotational sequences. They can autoionize in the $\tilde{X}^2\Pi$ continuum of N_2O^+, respecting some propensity rules still under study.

The rich autoionization structure observed in the photoionization yield curve of N_2O between 12.9 and 16.4 eV shown in Fig. 1.10, corresponds to the autoionized Rydberg series.

1.1.5.3 Non Franck-Condon Populations of Electronic States of Ions through the Decay of Superexcited States

As pointed out above, the decay of SE states through autoionization is able to populate regions of potential energy curves or surfaces of molecular ions lying outside the Franck-Condon zone for direct ionization. If one represents by potential energy curves the situation observed for the X̃ and Ã states of N_2O^+, including the extension of their Franck-Condon zone, it is clear that the autoionization of all the Rydberg levels converging to the Ã state, will give molecular ions in the X̃ state (Fig. 1.11). The energy balance of the autoionization processes will be:

$$N_2O + h\nu \rightarrow N_2O(R) \rightarrow N_2O^+ \ (\tilde{X}^2\Pi \ (v', v'', v''')) + e^- \ (\varepsilon).$$

The kinetic energy of the emitted electron, ε, ranges between zero and $h\nu - TE$ ($\tilde{X}^2\Pi$), where TE is the threshold energy for the ionization of N_2O in the X̃ state. This kinetic energy shall depend on the vibrational population of the N_2O^+ ions resulting from the autoionization process. As it will be shown later, experimental methods for the measurement of the kinetic energy of such photoelectrons shall allow a deeper insight into these processes.

Such indirect population mechanisms are so often responsible for the appearance of fragmentary ions that we think it useful to treat one of these cases immediately.

The first dissociation threshold of N_2O^+ giving rise to $NO^+ (X^2\Pi) + N \ (^4S)$ is measured by photoionization at 15.04 eV [19]. The corresponding thermochemically calculated threshold lies at 14.19 eV. A metastable N_2O^+ ion supporting the appearance of the NO^+ metastable fragment decaying in $5 \cdot 10^{-7}$ s with a total kinetic energy release of about 1 eV, has been shown to appear at the observed fragmentation threshold [20]. Figure 1.11 collects this information in a potential energy diagram which along the N–NO coordinate is an extension to a triatomic problem of the one used in the case of diatomic molecules. From this figure it is clear that the ground state of the molecular ions can only be correlated with the next dissociation threshold at 16.57 eV corresponding to $NO^+ (X^2\Pi) + N(^2D)$. When both dissociation fragments

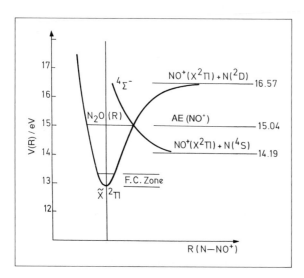

Fig. 1.11. Potential energy curves showing the appearance of the lowest energy dissociation process of N_2O^+, i. e., $N_2O^+ \rightarrow NO^+ + O$.

are formed in their ground state they give a $^4\Sigma^-$ state of N_2O^+. To summarize all the reported observations, it has been accepted that the $^4\Sigma^-$ state predissociates the $X^2\Pi$ state at 15.04 eV, giving rise to NO^+ $(X^2\Pi) + N$ (^4S) with a kinetic energy of about 1 eV. This predissociation is spin and symmetry forbidden and a long lived N_2O^+ ion is observed.

Moreover, the crossing point at 15.04 eV, unattainable by Franck-Condon transitions, is resonantly populated by the decay of a 3 $d\pi$ $(1,0,0)$ Rydberg state converging to the $\tilde{A}^2\Sigma^+$ state, as is shown by both the photoionization curves and the photoelectron energy analysis of zero kinetic energy photoelectrons [21–25].

This rather long discussion of a typical case of indirect population of a long lifetime predissociating state of a molecular ion can be considered as an introduction to a topic which will be examined more extensively in another part of this book (see Chapter 5, Section 2).

1.1.6 Ionization Yield Curves for Polyatomic Molecules

In the case of direct photoionization of diatomic molecules, one has only to consider the effect of a single internuclear distance modification on the vibrational population of electronic states of molecular ions. In polyatomic molecules, however, changes in bond distances, bond angles and even symmetry are expected. The possibilities for vibrational excitations are correspondingly more complex. In many cases, the photoionization threshold curve still has a reasonably sharp onset which probably corresponds to the adiabatic ionization energy. The step function threshold law makes staircase structures appear, including the excitation of few vibrational frequencies. This is particularly true for processes involving the removal of nonbonding electrons, where little change of geometry is expected.

In many cases, significant change of geometry occurs. In these cases the photoionization curve has a gradual onset as observed, for instance, for the appearance of the ground state of NH_3^+ which is planar, whereas the neutral molecule is pyramidal. Such a change in geometry induces a long progression of the umbrella vibrational mode of NH_3^+.

Rather complicated vibrational progressions are excited when an electron is removed from a degenerate orbital. The removal of the electronic degeneracy is expected by vibronic coupling which results in complex vibrational patterns (Jahn-Teller and Renner-Teller effects).

In some cases, the change in geometry does not allow to reach the stable part of the ground state of the molecular ion within the Franck-Condon zone. The molecular ion is not observed as in CF_4^+ where the CF_3^+ fragment ion dominates the mass spectrum.

When the complexity of molecules increases, as for instance in alkanes, the small spacing between vibrational levels and the autoionization contributions are such that the photoionization curves show a very gradual and continuous increase, without any interpretable structure. In these cases, the determination of thresholds either for the successive electronic states or for dissociation processes, is particularly difficult.

1.1.7 Concluding Remarks

Their complexity, the miscellaneous geometries realized by the successive electronic states of their molecular ions, the removal of electronic degeneracies through vibronic couplings, and the interferences between direct and indirect ionization stand very often in the way of an obvious interpretation of the aspects of the photoionization yield curves of polyatomic molecules.

There, therefore, is a need for investigative methods able to clarify the situation encountered in resonant photoionization. Such methods and their results, based on the achievement of high resolution studies of the photoelectron energies, will be described in the Section 1.2.

1.2 Refined Details about Photoionization Processes. Photoelectron Spectroscopies

1.2.1 Photoelectrons as Footprints of the Ionization Processes

The photoionization yield curves quite often appear to be a difficult way to characterize ionization processes. An important partner in the birth of ions has been completely neglected up to now in this work i. e., the photoelectron. It will reveal its particular importance in the case of the studied problem.

When ionization processes such as (1) to (4) are written, a question arises as to what fraction of the energy absorbed by the molecule AB is carried away as translational energy of the photoelectron?

Ionization processes are:

$$AB + h\nu \rightarrow AB^+ + e^- \qquad (1)$$

$$AB + h\nu \rightarrow AB^+ \rightarrow A^+ + B + e^- \qquad (2)$$

$$AB + h\nu \rightarrow (SE) AB \rightarrow AB^+ + e^- \qquad (3)$$

$$AB + h\nu \rightarrow (SE) AB \rightarrow A^+ + B + e^- \qquad (4)$$

Answers to this question will be at variance with the conditions prevailing in the photoelectron emission. They will be detailed in the successive paragraphs of this section.

As we have shown in Section 1.1.6, indirectly populated processes (3) and (4) so often complicate the detection of direct ionization processes (1) and (2) that a convenient way to disentangle direct and indirect ionization processes is necessary. As indirect ionization represents an important aspect of the whole ionization phenomenon, an appropriate way to investigate it will also be needed. Two ways based on the study of photoelectron energies will be now described successively.

1.2.2 Non-Resonant Photoelectron Spectroscopy

Many sophisticated methods used for the determination of translational energies of charged particles are described in the literature; some of them specially adapted for the determination of photoelectron energy distributions are described in detail in [18, 26], and in Part I of this volume.

Coming back to Fig. 1.3 and considering the ideal step function behavior as valid up to high photon energies it is possible to provide the energy distribution of photoelectrons removed from AB through photons of an energy $h\nu$ far in excess with respect to the phenomena detailed in the figure. At this energy the total ionization cross-section σ is given by the sum of the successive cross-sections that characterize the appearance of the X, A, B states of the molecular ion. Each of these $\sigma_X, \sigma_A, \ldots$ are the sum of the successive vibrational step functions. The photoelectrons emitted, for instance, at each adiabatic transition shall carry translational energies given by:

$$\varepsilon_1 = h\nu - E_X; \quad \varepsilon_2 = h\nu - E_A; \ldots \text{etc.,}$$

with $\varepsilon_1 > \varepsilon_2 > \varepsilon_3 \ldots$.

When a progressively decreasing retarding electric potential V_R is opposed to the photoelectron flux before its collection, the result on the photoelectron current is such that for $V_R > \varepsilon_1$ photoelectrons are not observed. For $V_R \geq \varepsilon_1$ a step function appears at the adiabatic transition to the X state with the same height as observed in the photoionization yield curve. The progressive lowering of V_R allows all the successive steps, visible in principle in the photoionization curve, to appear. In this way, a photoelectron spectrum arises entirely equivalent to the sum of parent and fragment photoionization yield curves. As the ionization process is about 10^3 times faster than the fastest dissociation process, the dissociating or predissociating electronic states will show either a vibrational structure or a continuous shape (C state). The photoelectron spectrum corresponding to Fig. 1.3 is shown in Fig. 1.12.

Obviously, the photoelectron spectrum can be scaled by ionization values instead of being scaled by photoelectron kinetic energies: as $\varepsilon_1, \varepsilon_2, \ldots$ are measured E_X, $E_A, \ldots$ are equal to $h\nu - \varepsilon_1, h\nu - \varepsilon_2, \ldots$ The photon energies generally used are those of the resonance lines of HeI (585 Å $\cong$ 21.21 eV) and HeII (304 Å $\cong$ 40.8 eV) easily produced by discharge lamps without any further light dispersion (see Part I, Section 3.1.1).

The photoelectron spectrum resulting from the retarding potential method can also be displayed under the form of its first derivative. In this case, step functions will be replaced by peaks for structured states X, A, B and of a continuous band for the C

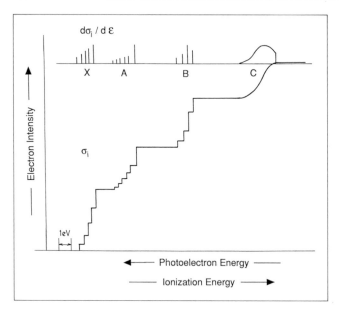

Fig. 1.12. Photoelectron spectrum corresponding to the ionization processes ideally described in Fig. 1.3 for the molecule AB. In the lower part of the figure, the photoelectron spectrum is obtained by the retarding potential technique; in the upper part, through the use of an electrostatic energy selector.

state. If electron energy analyzers were used, the peaks would be directly displayed as they appear in the first derivative of retarding potential curves. In both methods a resolution of about 5 meV (width at half height of the peaks) is easily realized.

One of the most evident advantage of this kind of spectroscopy lies in the fact that, by using a rather high photon energy, superexcited states are not induced. One gets an image of the direct photoionization processes completely free of any autoionization contribution.

A question remains, evoked earlier: if the cross-section diminishes for progressively increasing photon energies, the photoelectron spectrum will no longer give threshold cross-sections, but instead will give the cross-sections for each process at 21.21 eV or 40.8 eV, for instance. This effect does not fundamentally alter the image of the direct photoionization process. As an example of the important progress realized by the photoelectron spectroscopic technique with respect to the photoionization, Fig. 1.13 shows the photoelectron spectrum of H_2 [27]. In this spectrum, the successive vibrational excitations of H_2^+ are easily detected, whereas they are totally hidden by autoionization processes, in the direct photoionization curve (Fig. 1.9).

1.2.3 Photoelectron Spectrum and Molecular Orbitals

The photoelectron spectrum of a hypothetical diatomic molecule AB shows that the photoelectron spectroscopy is evidently an easy way to detect the successive electronic states of a molecular ion. As is well known, the molecular orbital theory describes the electronic configuration of any molecule for a chosen symmetry, in

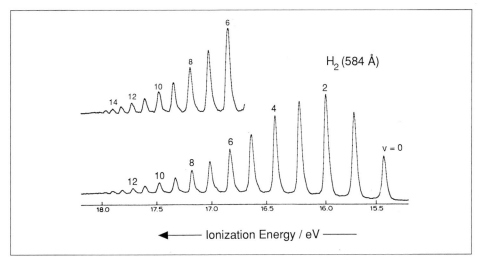

Fig. 1.13. Photoelectron spectrum of H_2 [27].

terms of filling the appropriate molecular orbitals with electrons in the order of their successively increasing energy. Let us first consider a diatomic molecule as N_2. The electronic configuration of its ground state $X^1\Sigma_g^+$ is represented by:

$$KK \ (\sigma_g 2s)^2 \ (\sigma_u 2s)^2 \ (\pi_u 2p)^4 \ (\sigma_g 2p)^2$$

The photoelectron spectrum of N_2 (Fig. 1.14) shows three bands, with their own vibrational spacing. They correspond to the three ionic states of the molecular ion obtained by the removal of one electron from one of the three highest occupied molecular orbitals (Table 1.2).

In this case, as N_2 is a closed-shell molecule, the removal of one electron from each orbital gives rise to one electronic state only.

Proceeding in the same way in order to interpret the photoelectron spectrum of an open-shell molecule, like O_2, will make more complex results appear. The electronic configuration of the $X^3\Sigma_g^-$ ground state of O_2 is written as:

$$KK \ \dots \ (\sigma_g 2s)^2 \ (\sigma_u 2s)^2 \ (\sigma_g 2p)^2 \ (\pi_u 2p)^4 \ (\pi_g 2p)^2 \ .$$

Table 1.2. Electronic configurations of N_2^+.

Configuration	State	Energy (eV)	Frequency (cm^{-1})
$KK \dots (\sigma_g 2p)^1$	$X^2\Sigma_g^+$	15.571	2207.2
$KK \dots (\pi_u 2p)^3 (\sigma_g 2p)^2$	$A^2\Pi_u$	16.693	1875.0
$KK \dots (\sigma_u 2s)^1 (\pi_u 2p)^4 (\sigma_g 2p)^2$	$B^2\Sigma_u^+$	18.757	2419.8

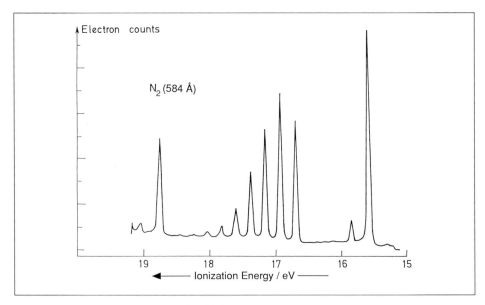

Fig. 1.14. Photoelectron spectrum of N_2 [28].

Table 1.3 collects the electronic states corresponding to the electronic configurations of O_2 obtained by the removal of one electron from one of the three highest orbitals.

The photoelectron spectrum of O_2, as taken with HeI line, is given in Fig. 1.15, and Table 1.4 collects the experimental ionization energies observed compared to quantum mechanically calculated values [30].

The photoelectron spectra of more complex molecules are easily treated in terms of molecular orbitals as it was done for diatomics.

Let us discuss briefly the series of monohalogenated methanes. In all these compounds, the sequence of the outer orbitals, written in C_{3v} symmetry, is:

$$(1e)^4 \, (na_1)^2 \, (2e)^4 .$$

Three successive ionization energies corresponding to $\tilde{X}^2E$, $\tilde{A}^2A_1$, and $\tilde{A}^2E$ states have been observed below 21.21 eV [32]. In Fig. 1.16, we give the threshold energies observed for the four molecules. If, in CH_3F, both the $\tilde{A}$ and $\tilde{B}$ states are undistin-

Table 1.3. Electronic configurations of O_2^+.

Configuration	State
$KK \ldots (\pi_g 2p)^1$	$^2\Pi_g$
$KK \ldots (\pi_u 2p)^3 \, (\pi_g 2p)^2$	$^4\Pi_u, \, ^2\Pi_u(3), \, ^2\phi_u$
$KK \ldots (\sigma_g 2s)^1 \, (\pi_u 2p)^4 \, (\pi_g 2p)^2$	$^4\Sigma_g^-, \, ^2\Sigma_g^-, \, ^2\Delta_g, \, ^2\Sigma_g^+$

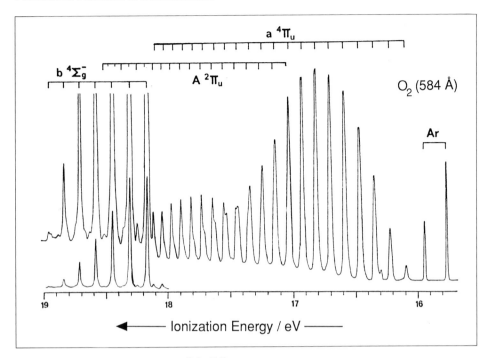

Fig. 1.15. Photoelectron spectrum of O_2 [31].

Table 1.4. Experimentally observed and calculated ionization energies in O_2.

Ionic state	Symbol	Observed IE (eV)	Calculated IE (eV)
$^2\Pi_g$	X	12.31	12.31[a]
$^4\Pi_u$	a	16.82	16.38
$^2\Pi_u(I)$	A	17.73	16.76
$^4\Sigma_g^-$	b	18.17	17.34
$^2\phi_u$		19.10[b]	18.12
$^2\Delta_g$		19.90[b]	19.07
$^2\Sigma_g^-$	B	20.43	19.63
$^2\Pi_u(II)$			20.09
$^2\Pi_u(III)$		24.00[c]	21.71
$^2\Sigma_g^+$		...[d]	(?)

[a]: Set equal to the experimental value; [b]: observed in the photoelectron spectrum of $O_2(^1\Delta_g)$ only [29]; [c]: from HeII line spectrum; [d]: strongly forbidden by selection rules.

guishable, this uncertainty disappears for CH_3Cl, CH_3Br, and CH_3I. Regular trends are observed for the ionization energies as a function of the position of the substituent in the periodic table. For CH_3Br and CH_3I, the spin-orbit interaction removes the degeneracy of the ground electronic state.

If the substitution is made via an OH group (CH_3OH), as the symmetry of the molecule is now C_s, the degenerate orbitals e are split in $a' + a''$ orbitals, the electrionic structure of CH_3OH, being:

$$(3a')^2 \ (1a'')^2 \ (4a')^2 \ (5a')^2 \ (2a'')^2 \, .$$

Figure 1.16 gives the ionization energies observed in CH_3OH [32].

As a conclusion to this brief survey, we describe the photoelectron spectrum of the benzene molecule. Benzene pertains to the D_{6h} symmetry and from ab initio quantum mechanical calculations, the following electronic structure is expected for the nine highest orbitals [33]:

$$(2e_{2g})^4 \ (3a_{1g})^2 \ (2b_{1u})^2 \ (1b_{2u})^2 \ (3e_{1u})^4 \ [1a_{2u}(\pi)]^2 \ (3e_{2g})^4 \ [1e_{1g}(\pi)]^4 \, .$$

A photoelectron spectrum recorded with the HeII line, has been interpreted on the basis that the ratios of the areas below observed peaks to the number of electrons occupying the orbital is constant. Under this assumption the authors accept the above orbital sequence as correct, for the 8 highest orbitals. An inversion is probable between the 10th and the 9th orbitals [34]. Table 1.5 gives the observed values for the successive ionization energies.

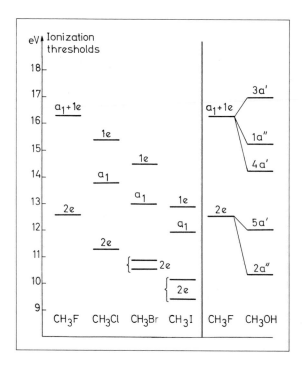

Fig. 1.16. Orbital energy levels in CH_3X (X=F, Cl, Br, I, OH).

Table 1.5. Molecular orbitals and electronic states involved in photoionization of benzene.

Ionized orbital	State	Energy (eV)
$[1e_{1g}(\pi)]^3$	$\tilde{X}^2E_{1g}$	9.247
$[3e_{2g}(\sigma)]^3$	$\tilde{A}^2E_{2g}$	11.49
$[1a_{2u}(\pi)]^1$	$\tilde{B}^2A_{2u}$	12–12.3
$(3e_{1u})^3$	$\tilde{C}^2E_{1u}$	13.8
$(1b_{2u})^1$	$\tilde{D}^2B_{2u}$	14.5–14.7
$(2b_{1u})^1$	$\tilde{E}^2B_{1u}$	15.2–15.4
$(3a_{1g})^1$	$\tilde{F}^2A_{1g}$	16.84
$(2e_{2g})^3$	$\tilde{G}^2E_{2g}$	18.7–19.2

This discussion, outside of its interest concerning the ionization processes in C_6H_6, shows that elementary (Hückel-type) calculations predict the order σ^4, π^2, π^4 for the outermost MOs while ab initio calculations predict the order π^2, σ^4, π^4. It would, however, be better to describe the σ and next π orbitals as being mixed together.

1.2.4 Constant Photoelectron Energy Spectroscopy

Nonresonant photoelectron spectroscopy is not the only way to obtain information about ionization processes through an energy analysis of photoelectrons. Another fruitful way is the use of "constant energy photoelectron spectroscopy" (CEP) obtained by using the electron energy analyzer to transmit photoelectrons of one energy while the photon energy is continuously scanned.

The choice of ε (the photoelectron kinetic energy) is very often $\varepsilon = 0$, in which case one speaks about "zero kinetic energy photoelectron spectroscopy" (ZKE) or "threshold photoelectron spectroscopy" (TPES).

In Section 1.2.1, the four main ionization processes were listed. The possible occurrence of ZKE photoelectrons in each of the processes will now be discussed.

Process 1 is expected to give only ZKE electrons when $h\nu$ is continuously scanned up from the ionization threshold. This ZKE spectrum will reproduce the nonresonant photoelectron spectrum; a possible modification of the relative abundances of the photoelectron bands can occur due to the possible declining regimes of the ionization cross-section with increasing energy.

Process 2 shall certainly give rise to the appearance of ZKE electrons if AB represents a diatomic molecule. When AB is written for a polyatomic molecule, A^+ and B are, respectively, either polyatomic or monoatomic fragments appearing at some energy $h\nu$ above their thermochemical threshold ΔH_0. An incomplete sharing of $h\nu$-ΔH_0 between internal and kinetic energies of $A^+ + B$, could result in the emission of non zero kinetic energy photoelectrons.

However, when ZKE spectra are measured for many molecules [35], extra peaks are observed in the photoelectron spectrum which are in coincidence with all or some of the autoionization structures observed in the direct ionization yield curves. Their origin is obviously due to the important channels 3 and 4, ascribed to the decay of superexcited states. If the superexcited state is excited for an energy $h\nu$ in the ionization continuum of a molecular ion, the only way to explain its appearance in the ZKE photoelectron spectrum is to assume the entire conversion of the difference between photon energy and ionization energy into internal energy of the molecular ion. As most of these SE states lie in the Franck-Condon gap, this phenomenon results in a non Franck-Condon population of the molecular ion (Process 3). If this non Franck-Condon population leads to the appearance of a dissociation, as for the appearance of $NO^+ + O$ from N_2O^+ between $\tilde{X}$ and $\tilde{A}$ state, we are dealing with process 4 (see Fig. 1.11). This indirect population of stable or dissociative region of the potential energy surfaces of electronic states of molecular ions is a frequently occurring phenomenon.

As pointed out before, processes 3 and 4 being good candidates for the appearance of molecular or fragmentary ions, it is necessary to describe now the energy balance realized when they occur in coincidence with photoelectrons ejected with some kinetic energy. A complete answer to this question requires the complete determination of the photoelectron spectrum for each SE state resonance. This has seldom be done except for diatomic molecules where such detailed studies give rise to the statement of "propensity rules" for autoionization mechanisms [36]. Such studies allowed their authors to conclude that for H_2, the vibrational autoionization lifetimes are very short for minimum values of Δv; in the case of N_2, an additional mechanism was detected where the autoionization probabilities are dominated by Franck-Condon factors for the transitions occurring between the SE state and the ionic state. This last process explains the anomalous population of vibrational levels observed in some molecular ions when the ionizing radiation coincides with a SE state [37, 38]. As an example of this phenomenon, a high resolution photoelectron spectrum of O_2 is given in Fig. 1.17, excited by the 736 Å, NeI resonance line. It is clear that if the nonresonant photoelectron spectrum excited by the HeI line (Fig. 1.14) exhibits four vibrational levels of the $X^2\Pi_g$ state of O_2, the resonant excitation at 736 Å of a SE state allows, by Franck-Condon transitions between this state and the $X^2\Pi_g$ state, appearance of some 16 supplementary vibrational levels [39] (Fig. 1.15).

Let us add that, in principle, the occurence of process 4 in polyatomic molecules could (as with process 2) give rise to an energy balance between internal energies of the dissociating fragments and the photoelectron energy.

A final question about the emission of ZKE photoelectrons in resonance with SE states concerns their relative abundance with respect to photoelectrons ejected with some kinetic energy. An example of an answer to this question was given for N_2O [24]: the fraction of the SE states lying between the $\tilde{X}$ and the $\tilde{A}$ state autoionizing by emitting ZKE photoelectrons has been measured as 10 %; 90 % of the same SE states, in this case, autoionizes to be $v=0$ level of $\tilde{X}$ (N_2O). It would certainly not be unusual to see this proportion reversed in more complex molecules.

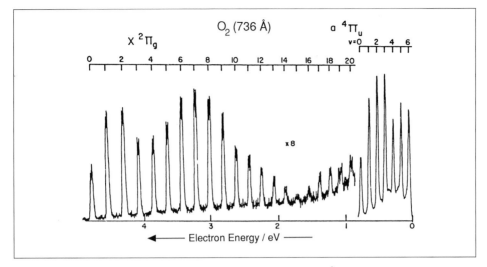

Fig. 1.17. Photoelectron spectrum of O_2 taken with the Ne I (736 Å) resonance line [39].

1.3 Photoion-Photoelectron Coincidence Methods

1.3.1 Generalities about Photoion-Photoelectron Coincidence Methods

We have seen how important is the information to be gained from photoionization yield curves and from the different ways of using photoelectron spectroscopy in a comprehensive study of photoionization in gases. It is now necessary to point out how both these aspects are complementary.

A general complementary relationship is an experimental design in which the photoion abundances are measured in coincidence with photoelectrons of a preselected energy. If this experiment uses a unique photon energy, far in excess above the energies of the successive electronic states of a molecular ion (as is commonly made in non-resonant photoelectron spectroscopy), it will be possible to determine which ions are specifically (either molecular or fragmentary) produced in coincidence with each of the selected electronic states of the molecular ion, without any interference with autoionizing SE states, but with the restrictive assumption that the phenomena are characterized by their cross-section at the unique photon energy used in the experiment. This will be called "non-resonant photoion-photoelectron spectroscopy" or non-resonant PIPECO.

If, otherwise, the coincidences are measured between photoions and ZKE photoelectrons by a scanning of photon energy from the lowest threshold energy up to higher energies, it will be possible to observe PIPECO ionization curves of the direct ionization phenomena with their respective cross-sections. This is valid under the restrictive assumption that the successive autoionization phenomena have a negligible contribution to ZKE photoelectrons. Specific aspects of both these PIPECO methods will now be analyzed.

1.3.2 Non-Resonant PIPECO Mass Spectra

The experimental build up of coincidence mass spectra is abundantly described in the literature [4] and we shall discuss here how the results of these experiments can be interpreted in two main cases: in the first case photoelectron energies are measured by a retarding potential method, in the second, by the use of energy analyzers. Photoions are mass selected by their dispersion in space (magnetic or quadrupole mass spectrometers) or in time (time-of-flight mass spectrometers).

In the first case, coincidence ionization curves are similar to the photoelectrons retarding field curves. They are drawn for increasing ionization energies like the curves of Fig. 1.12. As an example, Fig. 1.18 gives together with the photoelectron spectrum of CH_3OH [37], its non-resonant PIPECO mass spectrum [41]. It is clear from this figure that the $\tilde{X}^2A''$ ground state of CH_3OH^+ gives stable methyl alcohol ions

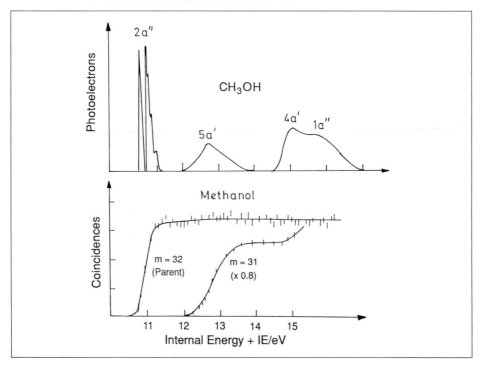

Fig. 1.18. PIPECO mass spectrum of CH_3OH by photoelectron retarding field energy analysis [41].

only. The next electronic state $\tilde{A}^2A'$ is entirely dissociative and gives rise to the $CH_2OH^+ + H$ dissociating couple; a second step in the CH_2OH^+ ion yield curve appears with the excitation of a third electronic state $\tilde{B}^2A'$.

In the second case, before any presentation and discussion of experimental results, it is necessary to discuss the best way to use coincidence data. When the photoelectron energy distributions are determined by energy analyzers, the photoions being transmitted through a mass spectrometer after a correct delay from the ionization event, the following considerations (fully developped in [42]) need to be introduced.

These considerations result essentially from an examination of the quality for the detection of photoions.

This will be appreciated through the definition of a "transmission coefficient" f_i which expresses the relative probability that a particular ion of mass m_k, velocity $\vec{v}$, and position of formation $\vec{r}$ is detected.

The determination of f_i is very difficult except for molecular ions, their velocity distribution being a Maxwell-Boltzmann one. Typical f_i values for thermal ions with $m_k \leq 300$ detected through a quadrupole mass spectrometer range between 0.25 to 0.60. It is clear that for fragmentary ions, $\vec{v}$ can vary due to the possible translational energy gained in the decay process, as also $\vec{r}$, the ion being not necessarily decomposed when its precursor has been ionized.

The result of such coincidence experiments are expressed in terms of a quantity called the "branching ratio": $b[E^*, m_k]_t$.

This is the probability that a molecular ion originally prepared with an internal energy E^* decays within the sampling time t of the experiment into an ion characterized solely by its mass m_k.

With respect of this ideal branching ratio, the experimentally observed one will be equal to:

$$b[E^*, m_k]_t \cdot f_i. \tag{1.9}$$

If $C[E^*, m_k]$ is the true coincidence count rate obtained when ions of mass m_k and electrons of the energy ε^* given by

$$E^* = h\nu - \varepsilon - IE\,(1) \tag{1.10}$$
(*IE* (1) is the first adiabatic ionization energy)

are selected, and if $E\,(E^*)$ is the true photoelectron count rate related to the formation of molecular ions with internal energy E^*, the experimentally observed branching ratio is given by

$$b[E^*, m_k]_t \cdot f_i\,(m_k, \vec{v}, \vec{r}) = \frac{C[E^*, m_k]}{E\,(E^*)}. \tag{1.11}$$

In an ideal experiment the sum of all branching ratios would normally be unity. As it will be shown later, this sum is frequently less than unity owing to the behavior of f_i.

Table 1.6. Decomposition of a molecular ion $ABCD^+$ with increasing excess energy E^* (Fig. 1.19).

E^* (eV)	Electronic state	Ions
0–1.75	$\tilde{X}$	$ABCD^+$
2.5–3.0	$\tilde{A}$	$ABCD^+$
3.0–3.9	$\tilde{A}$	$ABCD^+ \rightarrow ABC^+ + D$
		$ABCD^+ \rightarrow BCD^+ + A$
4.5–6.75	$\tilde{B}$	$ABC^+ \rightarrow AB^+ + C$
		$ABC^+ \rightarrow BC^+ + A$

Ideal examples being, in fact, never realized, we shall once more consider how a photoion-photoelectron coincidence mass spectrum could appear if the experiment was an ideal one. In Fig. 1.19 the non-resonant PIPECO spectrum of an ABCD molecules is given as it would appear if f_i was equal to unity for the entire internal energy range and for all values of m_k. In this case the branching ratio is equal to $C[E^*, m_k]/E(E^*)$.

In the upper part of Fig. 1.19, the hypothetical photoelectron spectrum, as measured with the resolution used in the coincidence experiment, is drawn. In the middle part of the figure, the branching ratios observed for the main ions are shown. The sum curve of all the branching ratios is given below. As the experiment is ideal, the sum curve is equal to unity in the three energy regions where ions are observed, i. e., in coincidence with the three observed electronic states $\tilde{X}, \tilde{A}$, and $\tilde{B}$. This result is called a "breakdown diagram" and is easily described. The ground electronic state of $ABCD^+$ is entirely stable within the range of the first electronic state. Some stable $ABCD^+$ ions appear, up to $E^* = 3$ eV when the second electronic state $\tilde{A}$ is excited. From $E^* = 3$ eV, $ABCD^+$ fragments into ABC^+ ions and from $E^* = 3.45$ eV into BCD^+ ions.

At $E^* = 4.5$ eV when the $\tilde{B}$ electronic state is created, three ions appear: rapidly decreasing ABC^+ ions and correlatively increasing abundances of AB^+ and BC^+.

Such behavior is schematically collected in Table 1.6.

As is evident from Fig. 1.19, the ionic yields are in each case a smooth function of the excitation energy; they are not influenced by the specific electronic state in which the molecular ions is initially created. It follows from this that the rate constants and the characteristic of the reactions as the appearance of translational, vibrational and rotational energies of the fragments, are assumed to depend of only one variable: the internal energy of the molecular ion.

Therefore, the results of this idealized PIPECO experiment lead to the proposal that the mass spectrum of polyatomic molecules is the result of concurrent and consecutive monomolecular reactions of the internally excited molecular ion.

If it was now possible to show, through the play of appropriate correlation rules or through the calculation of potential energy hypersurfaces, that ABC^+ and BCD^+ ions are only able to be connected with the $\tilde{X}$ state of ABC^+ and that AB^+ ions are cor-

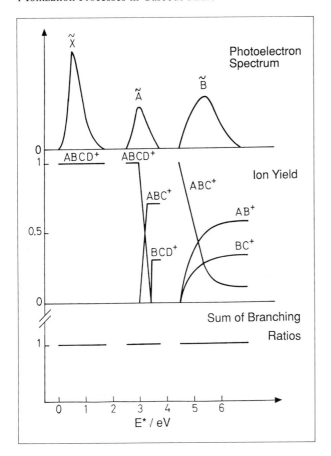

Fig. 1.19. Nonresonant idealized PIPECO breakdown pattern of a complex ABCD molecule (the letters are written here for atoms or groups of atoms).

related to ABC⁺ only, a final question remains: how is the electronic excitation that appears in Ã and B̃ states converted into internal energy of the ground state?

Such electronic energy conversions will be considered later as the result of crossings between the energy hypersurfaces of the successive electronic states, responsible for numerous radiationless transitions between them (see Chapter 2).

From these considerations it appears that any theory which is able to express k as a function of E^* will be suitable to calculate the breakdown pattern of a molecular ion, if the conditions of complete randomization of excitation energy in the molecular ion ground state is really fulfilled[3].

Developments of such theories applied to ionization phenomena in complex molecules will be given in Chapter 4 of this work.

---

[3] One could imagine a situation in which the energy conversion operates completely to one electronic excited state, completely disconnected of the ground state: in this case the validity conditions for the theory would also be filled.

1.3.3 Resonant PIPECO Mass Spectra

As pointed out in Section 1.2.4 on the use of constant energy photoelectron spectroscopy, the observation of ZKE photoelectrons in coincidence with photoions through the continuous scanning of photon energy immediately gives the breakdown diagram for a known energy range of the molecular ion. At each photon energy hv, this internal energy is evidently equal to $hv\text{-}IE$ when IE is the first adiabatic ionization energy. The full validity of the methods is naturally conditioned by the fact that ZKE photoelectrons appearing from the decay through autoionization of SE states can be considered as negligible.

This type of resonant photoion-photoelectron coincidence data has been accumulated over the last 15 years by different laboratories. As an example of the results obtained in this way, we give the measured breakdown diagram for ethane [43], shown in Fig. 1.20.

Without an elaborate discussion, it is evident from this breakdown diagram that, at increasing internal energies of the ground state of $C_2H_6^+$, the following dissociation mechanisms are observed:

$$C_2H_6^+ \begin{cases} C_2H_5^+ + H \rightarrow C_2H_3^+ + H_2 \\ C_2H_4^+ + H_2 \rightarrow C_2H_2^+ + H_2 \\ CH_3^+ + CH_3 \end{cases}$$

It has to be added here that the occurrence of such dissociation mechanisms is strongly supported by the detection of the corresponding metastable transitions in magnetically dispersed mass spectra.

It would not be expedient to present other breakdown diagrams here; more examples will be given later when required to tests the validity of statistical calculations of the rate constant as a function of internal energy. Sophistications allowed by experimental methods will then also be introduced.

1.3.4 From Photoionization Yield Curves to the Breakdown Diagram: an Old-Fashioned Way

In the early days of photoionization mass spectrometry, it was rapidly realized that, on the condition of occasionally using a mild doctoring in order to avoid pronounced occurrence of autoionization, a rather easy progression from photoionization yield curves to the breakdown diagram was possible. If the threshold law for the production by photoionization of a single quantum state of the ion is a step function, the set of normalized first derivatives of the photoion yield curves is just the breakdown diagram for the parent molecular ions. Under the same assumption the deriva-

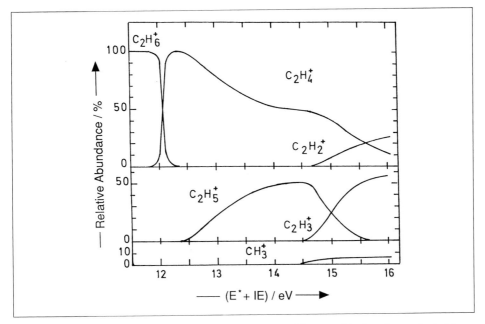

Fig. 1.20. Breakdown diagram for $C_2H_6^+$ from the ZKE coincidences with photoions, adapted from [43].

tive of total ionization with respect to photon energy represents the internal energy distribution produced by photons. This last derivative has to be similar to the photoelectron spectrum at fixed energy and, more accurately, to the ZKE photoelectron spectrum.

The construction of the breakdown diagram is made as follows.

The first derivative of the properly smoothed photoionization yield curves is taken as the differences between ordinates of adjacent points. These first derivatives are normalized by dividing by the sum of all derivatives at each energy, thus forming the desired quantities:

$$\frac{dI_j/dE}{\sum_j (dI_j/dE)}.$$ 1.12

Examples of such a procedure are shown in the literature [44]. The results obtained in this way for C_2H_6 are presented in Fig. 1.21. Compared to the results of the ZKE photoelectron-photoion coincidence from Fig. 1.20, it is clear that some differences appear in the abundance curves, but the general dissociation scheme is reproduced by both methods. Ionic abundance variations are probably due to the occurrence of population through autoionization in the photoionization yield curves.

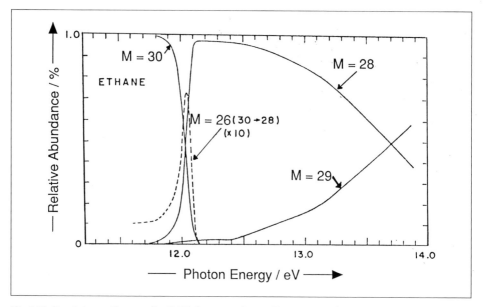

Fig. 1.21. Breakdown diagram for $C_2H_6^+$ from the first differentials of photoionization yield curves [44].

1.4 Electroionization

1.4.1 Resonant Electroionization

A long time before the development of photoionization techniques, electron beams of controlled energy, thermally emitted from metal filaments, were commonly used to investigate ionization phenomena in gases, under collision-free conditions. The use of electron-impact ionization yield curves as a way to study all of the earlier mentioned aspects of ionization phenomena is problematic.

A first problem lies in the rather wide kinetic energy distributions observed for electrons emitted by a hot cathode (≈ 2000 K). Those distributions serve as a convolution function of the theoretically expected ionization yield curves.

A second problem resides in the unfavorable form of the threshold law for electron impact (ideally linear), which for monoenergetic electrons would lead to the appearance of ionization yield curves as a sequence of linear segments, the observed breaks then correspond to the successive thresholds for the ionization processes.

Another problem is the very common occurrence of autoionizing SE states, superimposed to the linear segments described above. Their excitation cross-section no longer has the resonance character observed in photoionization, but varies continuously with electron energies.

Obviously, despite the possible difficulties due to the two last problems, a better approach of electroionization would be possible if the first problem was solved. Solutions to the problem of the rather wide kinetic energy distributions of thermally emitted electrons have been proposed, either by mathematical techniques (deconvolution methods) or by the generation of monochromatized electron beams.

One could attempt to describe the results of all the improvements made up to now in both of these directions; however, this would be rather tedious and not very useful. Except for two experiments in which monochromatized electrons were mainly used to study autoionization processes or to determine the first ionization potential of molecules and of radicals [45], it is still very difficult to give a self-sustaining and unequivocal interpretation of the direct ionization yield curves, even of those obtained with monochromatized electrons. Concerning the use of mathematical methods [46] to remove the convolution of the energy spread of currently used electron beams, it has been pointed out that they can only be applied to electroionization yield curves characterized by a very high signal-to-noise ratio. It is, however, a fundamental endeavor that will be useful in many respects.

Recalling that the first derivative of electron-impact ionization yield curves reproduces the photoionization yield curve, one can electronically produce the first derivative of the electroionization yield curve, measured with a not too dispersed electron beam: such a first derivative will look like a photoionization yield curve convoluted through the electron energy distribution. This results in a poorly resolved photoionization curve; however, it is often an excellent way to extract some information accumulated in a direct ionization curve.

Let us emphasize that when the photoionization yield curve is available, the comparison between the "bad" first derivative and the photoionization results frequently gives an excellent agreement between the two. An example of this is given in Fig. 1.22, from the recently published case of CH_3^+ ions formed at the expense of CH_3F [47, 48]. We will later show that the use of the first derivative of electroionization yield curves is particularly suitable when the determination of the successive dissociation limits of an ion are investigated by the study of the kinetic energy distribution of an ion as a function of the electron energy (see Chapter 3).

1.4.2 High-Energy Electron Impact: Photon Simulation

As early as 1930, Bethe [49] pointed out that high-energy electron impact experiments could provide physical information complementary to that obtained by photon impact. This theory shows that, for the scattering of electrons having velocities far in excess of orbital velocities of the target electrons, the differential scattering cross-section is given by:

$$\frac{\mathrm{d}^2\sigma}{\mathrm{d}\theta\,\mathrm{d}E} = \frac{2}{E} \cdot \frac{k_n}{k_0} \cdot \frac{1}{K^2} \cdot f(K, E), \qquad\qquad 1.13$$

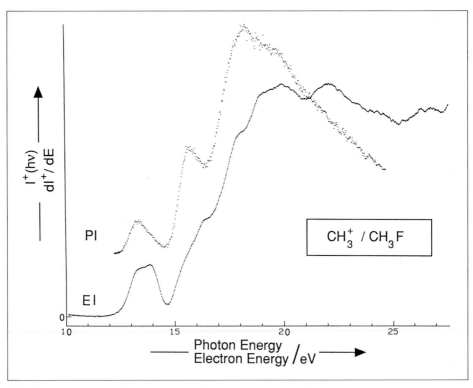

Fig. 1.22. Comparison between the first derivative of electron-impact ionization yield curves and the photoionization yield curves for CH_3^+ ions from CH_3F [47, 48]. (PI = photon impact; EI = electron impact).

where E is the energy transferred to the target, k_n and k_0 are the magnitudes of the incident and scattering momenta, and K is that of the monumentum transfer; $f(K, E)$ is the "generalized oscillator strength" from which it is easy to show that it can be written as a power series:

$$f(K, E) = f(0, E) + a\,K^2 + b\,K^4 + \dots \qquad 1.14$$

It is quite obvious that, for near zero values of K, this series tends to the only dipole term $f(0, E)$.

Two ways are open for the determination of the "dipole oscillator strength"; the first will require the measurement of $f(K, E)$ over a range of K, and extrapolated to $K = 0$ [50]. The second alternative uses the assumption that K is certainly very low for very small values of θ. If contribution of higher terms in the expansion of $f(K, E)$

is negligible, $f(K, E)$ is independent of K, and the angular behavior of the cross-section takes the following form:

$$\frac{d^2\sigma}{d\theta\, dE} = \frac{1}{2\, E_0 \left[\dfrac{E^2}{4\, E_0} + \theta^2 \right]} \qquad\qquad 1.15$$

It was effectively checked that when using 10 keV electrons at $\theta \approx 10^{-2}$ radian, no deviation from this equation is experimentally observed [51]. The measurements of the differential collision cross-sections made in coincidence with one of the collision products (ejected electrons or ions, for instance) will be the electron-impact analogue of photoelectron and photoionization experiments.

Extensive reviews of such experiments are available in the literature [52] and we will only make a short comment about the mimicing of photoelectron spectra and of photoionization yield curves.

An excellent example of photoelectron spectrum is given in Fig. 1.23, in the case of NH_3. In this case the energy of impinging electrons is 3.5 keV; the ejected elec-

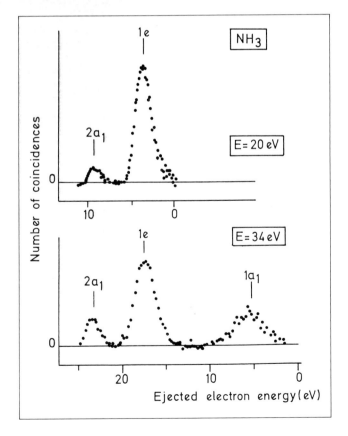

Fig. 1.23. High-energy electron impact simulation of the photoelectron spectrum of NH_3 [51, 52].

trons have been measured for energy losses of the 3.5 keV, i. e., 20 and 34 eV. Let us recall that in C_{3v} symmetry, the electronic configuration of NH_3 is

$$k\,(1\;a_1)^2\;(2\;a_1)^2\;(1\;e)^4\,.$$

From Fig. 1.23 one sees that the simulated photoelectron spectrum obtained for electrons having lost 34 eV reveals the position in energy of the third orbital $1\;a_1$, accessible with the HeII line only. The measurement of electron energy-losses in coincidence with the mass spectrum provides the analog of a resonance photoioniza-tion experiment extended over a wider energy range. Results of this kind are shown in Fig. 1.24 for CH_4 [53]. In this case the mass dispersion is obtained by the time-of-flight technique. An interesting fact must be pointed out about these ionization yield curves. Below 35 eV they are dominated by the results of single ionization processes. In the range of 35 to 40 eV all ion abundances suddenly increase. This energy region is normally dominated by double ionization. As no CH_4^{++} ions are observed in the mass spectrum nor any doubly charged ions of lower mass, it must be concluded that CH_4^{++} dissociates into couples of singly charged ions involving H^+ or H_2^+ such as: $CH_3^+ + H^+, CH_2^+ + H_2^+ \ldots$, in a time shorter than that needed for the extraction of the ions from the collision region. This explains the substantial increase in ionic abundances observed in the 40 eV region. This result is emphasized here in order to make

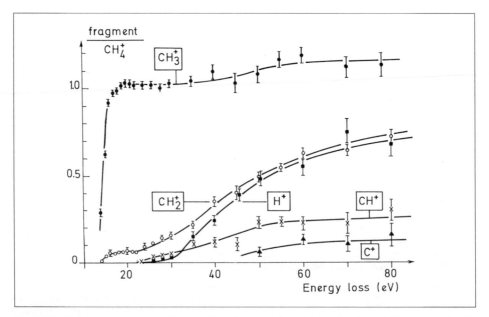

Fig. 1.24. High-energy electron impact simulation of the photoionization yield curves of CH_4 [51, 52].

the reader conscious that when rather high energies are observed by the molecules, the role of double ionization processes has to be taken into account. Such dissociation channels generally involve a large transfer of kinetic energy to the fragments.

The problem of the determination of the kinetic energy distributions as a function of the energy content of the ions will be discussed in Chapter 3. It will be shown that these measurements will open new insights into the dissociation processes of cations.

1.5 Charge Exchange Ionization

1.5.1 Introduction to Charge Exchange Physics and to Charge Exchange Mass Spectra

As was discussed above, the behavior of photons or of fast electron beams with respect to the target molecule results in ionization phenomena entirely dominated by Franck-Condon transitions. This has to be ascribed to the extremely small interaction time between the ionizing particle and the target molecule. As expected, many other ways to prepare electronically excited molecular ions are possible, such as the charge exchange (CE) processes used since 1954 [54, 55] for the study of mass spectra.

The CE ionization takes place when an ion A^+ of known translational energy is allowed to make short-range interactions with another atom or molecule. One distinguishes three main CE processes:

a) the symmetric CE, such as:
$A^+ + A \rightarrow A^+ + A$;
b) the asymmetric CE, such as:
$A^+ + B \rightarrow B^+ + A$;
c) the dissociative CE, such as:
$A^+ + BC \rightarrow B^+ + C + A$.

The symmetric CE being a resonant process, its cross-section is expected to be very high for very low ion impact energies, decreasing monotonically when impact energy increases. It has to be pointed out here that, at thermal energies, the cross-section for such reaction usually reaches considerably higher values than gas kinetic values, which are of the order of $5 \cdot 10^{-16}$ cm^2: for instance, at low kinetic energy the symmetric CE between Ne^+ and Ne has a cross-section of about $30 \cdot 10^{-16}$ cm^2. What has just been stated about symmetric CE emphasizes the most important parameter of any other CE process; the kinetic energy of the projectile ion. If we consider the asymmetric or dissociative CE, another important factor appears: the energy defect ΔE, defined as the internal energy difference between the initial and final states of the system at infinite molecular separation.

Let us illustrate the meaning of ΔE by an example. Suppose that it is needed to induce the population of the $A^2\Sigma(v=0)$ state of HCl^+ at 16.254 eV by CE with Ar^+ ions. The exact threshold energy for the appearance of $Ar^+(^2P_{3/2})$ being 15.759 eV, it is possible to calculate that, for the reaction

$$Ar^+(^2P_{3/2}) + HCl \rightarrow HCl^+(A^2\Sigma, v=0) + Ar(^1S_0),$$

ΔE is equal to $(15.759 - 16.254) = 0.495$ eV.

This energy defect, with respect to the resonance conditions, could be supplied to HCl^+ by a kinetic energy conversion of Ar^+ into internal energy of HCl^+. The question arising now is: what are the conditions to be fulfilled so that this energy transfer occurs adiabatically?

On the basis of simple theoretical arguments, it was shown by Massey [56] that, if l is the distance of closest approach of the projectile ion, v is its relative velocity and ΔE the energy defect if the condition

$$\frac{l \cdot \Delta E}{hv} \gg 1$$

is fulfilled, the collision will be adiabatic. The probability of the electronic transition in the target will be small. If v is reduced to a value v^* such as:

$$v^* \approx \frac{l \cdot \Delta E}{h},$$

the collision will no longer be adiabatic, since the collision time becomes comparable to the transition time $\Delta E/h$. In this case, the cross-section is no longer expected to be small. It will, in fact, reach a peak value for an energy E^*, expressed by:

$$E^* \text{ (eV)} = 36(\Delta E)^2 \, ml^2,$$

where ΔE is expressed in eV, the mass m of the incident particle in atomic mass units and the interaction range l in units of the Bohr radius ($a_0 = 0.53 \cdot 10^{-8}$ cm); l is ordinarily taken as $8 \cdot 10^{-8}$ cm = 8Å.

Indeed, the cross-section must eventually decrease with increasing kinetic energy since the interaction time ultimately becomes too short for the transition to be likely.

With all these remarks in mind, the way in which the mass spectrum can be studied by CE will now be described. In this case, beams of a great variety of ions of such an energy that they can be considered as working in the adiabatic regime and for which ΔE is generally positive, interact in the ion source of a mass spectrometer with the molecules under study. The results of asymmetric and dissociative CE are analyzed with a mass spectrometer and finally transformed in a breakdown diagram.

The underlying hypothesis in these works is that all the recombination energy of the projectile is entirely converted in ionization and excitation of the target molecule. This hypothesis makes the CE technique the only one to be able to give the breakdown diagram without using a coincidence method. It is also to be pointed out that in a charge exchange reaction the molecular ion is formed in a one-step process: autoionization is not allowed. One can, therefore, expect differences in threshold energies for some fragmentary ions, as in non-resonant PIPECO spectroscopy.

As an example of this last statement, let us point out under others the appearance of NO^+ from N_2O. Going back to Fig. 1.11, one calls that the predissociation of the $\tilde{X}^2\Pi$ state of N_2O^+ is necessary to observe the threshold energy of NO^+ at 15.04 eV, concomitantly with the corresponding metastable transition. For a number of projectile ions with recombination energies below 15.04 eV, only N_2O^+ ($\tilde{X}^2\Pi$), ions are observed. The $Ar^+(^2P_{3/2})$ (15.759 eV) is expected to populate the vibrational levels of $\tilde{X}, N_2O^+$ in the predissociation region. In fact, practically no NO^+ ions are produced, implying that in the adiabatic regime the Franck-Condon factor is the major factor for electronic transitions.

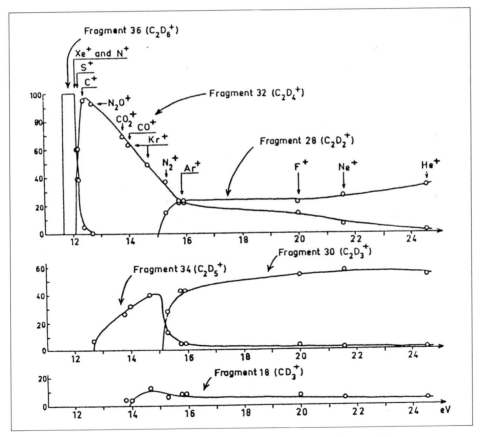

Fig. 1.25. Breakdown diagram of $C_2H_6^+$ from charge-exchange experiments [58].

The NO^+ ions will only be observed through the production of the $\tilde{A}$ state of N_2O^+ by CE studies around 16.5 eV. In this case, the NO^+ ion is produced through the predissociation of the $\tilde{X}$ state through the $^4\Sigma^-$ repulsive curve [57].

For more complex molecules, an as complete as possible breakdown diagram is obtained, shown in Fig. 1.25 (reproduced from [58] on the CE mass spectrum of C_2D_6). It is clear that the general shape of this breakdown diagram is similar to the one obtained either by photoionization (Fig. 1.21) or by ZKE, photoelectron-photoion coincidence (Fig. 1.20) giving strong support to the mechanism and to the statistical behavior of the decay of the $C_2H_6^+$ excited molecular ion.

1.5.2 A first Approach to the Measurement of the Rate Constant for Dissociative Ionization as a Function of Internal Energy

If the correlations about the appearance of nonresonant PIPECO mass spectra (Section 1.3.2), such as those from ZKE, PIPECO mass spectra (Section 1.3.3) are such that, in the case of complex molecules, the breakdown pattern is the result of monomolecular concurrent and consecutive dissociation reactions of the ground state parent ion, with the supplementary assumption that the rate constants of all these reactions are only dependent on the total excitation energy of the parent ion; it is then appropriate to ask if an experimental investigation of the way in which $k(E)$ varies with E is accessible. If we recall that time windows available in ordinary mass spectrometric techniques are in the range of 10^{-6} s (Section 1.1.4), any technique able to extend this time interval over several orders of magnitude will naturally be welcome. Such a technique was developed by Ottinger [59] and it allowed study of the dissociation rate constants of polyatomic ions spanning from 10^5 to 10^9 s^{-1}. By simultaneously using this technique and CE ionization, these authors [60] were able to give the first examples of experimental measurements of the rate constant k for some dissociation mechanisms at increasing internal energies E of the parent ions. Two examples were treated. The first one refers to the process:

$$C_6H_5CN \rightarrow C_6H_5CN^+ \rightarrow C_6H_4^+ + HCN.$$

As is visible from Fig. 1.26, a monotonic increase of the rate constant is observed at increasing excitation energy, as expected from any statistical theory applied to the problem.

The second example refers to two concurrent processes of decay of C_6H_6:

$$C_6H_6 \rightarrow C_6H_6^+ \begin{cases} \nearrow C_6H_5^+ + H & \text{(a)} \\ \searrow C_4H_4^+ + C_2H_2 & \text{(b)} \end{cases}$$

The striking feature of this second example, as shown in Fig. 1.26, is the exceedingly slow variation of $k_a(E)$ compared to $k_b(E)$. This strong difference led the authors to conclude that mechanisms a and b could presumably be not competitive,

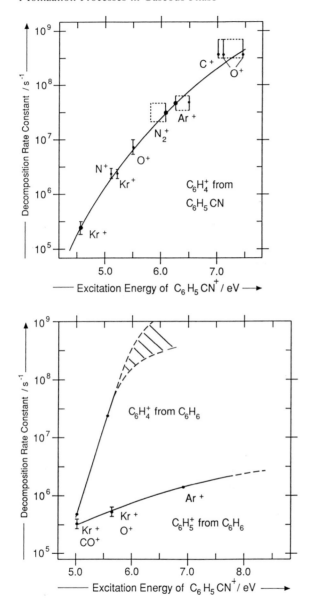

Fig. 1.26. $k(E)$ for $C_6H_5CN^+ \rightarrow C_6H_4^+ + HCN$, for $C_6H_6^+ \rightarrow C_6H_5^+ + H$ and $C_6H_6^+ \rightarrow C_4H_4^+ + C_2H_2$, as obtained from CE experiments with the special technique of Ottinger (see text) [60].

introducing eventual failure of the complete randomization in one electronic state of the molecular ion of the excitation energy E. In other words, process (a) would occur from an electronic state of $C_6H_6^+$ different of the one which gives rise to process (b).

Such situations will be discussed in depth later on, in conjunction with the results of PIPECO in photoionization experiments. It will then be shown how the complete description of the behavior under ionization of molecules as complex as C_6H_6 is difficult to make.

It may now be pointed out that such determinations of $k(E)$ for one or two decomposition processes of a single molecular ion are, with respect to the checking of theoretical predictions, more fruitful than the simultaneous treatment of all the dissociation processes that give rise to a mass spectrum. This manner of testing theoretical speculations will often be followed in the remainder of this work.

1.6 Field Ionization

1.6.1 Description of the Process

Field ionization, developed in the mid-1950s [61], is a technique by which ground state molecules are submitted to such a high field strength Stark effect that the probability for penetration of the wave representing the electron through the barrier surrounding an atom or a molecule becomes appreciable, leading to the appearance of a molecular ion. Such high fields (10^7 to 10^8 Vcm^{-1}) are produced at extremely fine metal tips held at a positive potential of about 10 to 20 kV with respect to a cathode that is the first extracting electrode of the ion source of a mass spectrometer.

As a result of this process, molecular ions appear and remain in the very high field, resulting in the appearance of very few fragmentation processes. The complete mass spectrum is extremely simplified in respect to the electron or photon impact mass spectra.

A first rough interpretation of such mass spectra is based on the migration of the positive charge density from its customary position in an hydrocarbon chain within the high ionizing field. By using theoretical arguments it is possible to show, for instance, that in ionized normal alkanes the maximum positive charge is localized mainly at the β position, giving rise to the preferential breaking of the $\beta(C-C)$ bond in the ion [62], as shown in Table 1.7, where the relative abundances of $C_2H_5^+$ from alkanes is given with respect to the molecular ion intensity.

Such a reasoning, based on positive charge distribution in an ion as a probe of the most probable dissociation mechanism, appears as a typically non-statistical approach of the "dissociation process". It does not take into account the fact that (as will be shown in Section 1.6.2) the rate constants for field dissociation of molecular ions are widely distributed between the highest ($\approx 10^{13}$ s^{-1}) and the lowest ($\approx 10^5$ s^{-1}) values. The particular way in which the ions are formed in field ionization probably

Table 1.7. Relative abundance of $C_2H_5^+$ in field ionization of alkanes.

Substance	nC_3H_8	nC_4H_{10}	nC_5H_{12}	nC_6H_{14}	nC_7H_{16}	nC_8H_{18}	nC_9H_{20}
$C_2H_5^+$	7	18	90	155	240	300	440

leaves no possibility for the formation of electronically excited states of the molecular ion. Its dissociative decay is entirely due to internal excitation of its ground state. This allows the conclusion that the field ionization technique fits particularly well the conditions demanded by a statistical theory of unimolecular reaction, however, in presence of a very high electric field.

The main challenge of such a theory lies in the determination of the energy content of the molecular ion in order to calculate $k(E)$'s.

1.6.2 Lifetimes of Field Ionized Molecular Ions

When the results displayed in Fig. 1.25 are considered, it is easy to conclude that the special technique of Ottinger provides measurements of the rate constant from about 10^5 s^{-1} to 10^9 s^{-1}. With field ionization, measurements of the dissociation rate constant of excited ions range from 10^{13} s^{-1} down to 10^6 s^{-1}. The principle of the method used for the study of the lifetimes of ions will be given here [63].

As we have pointed out, the field ionization takes place between a tip brought at very high potential with respect to a cathode which constitutes the entry slit to the mass spectrometer. In Fig. 1.27, this arrangement is schematically shown with a probable distribution of equipotential lines between the tip and the cathode. If a potential of 10 kV is applied to the tip, the field strength can be of about 0.5 V/Å, the potential drop in front of the tip being of the order of 50 V at a distance of 100 Å.

Ions decomposing at this point lead to fragmented ions with smaller kinetic energy than those fragments found at the tip surface. This energy change can be measured as a shift in the mass number. As the ion needs only a time interval of 10^{-12} s for traveling on 100 Å, decomposition times down to several 10^{-12} s can be resolved. It is possible to observe ions dissociating from about 10^{-6} s at their correct mass number, and all others at mass numbers progressively shifted up with respect to the mass number of the parent ion.

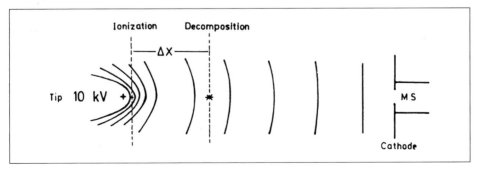

Fig. 1.27. Schematic distribution of equipotential lines between a field ionization tip and a cathode [63].

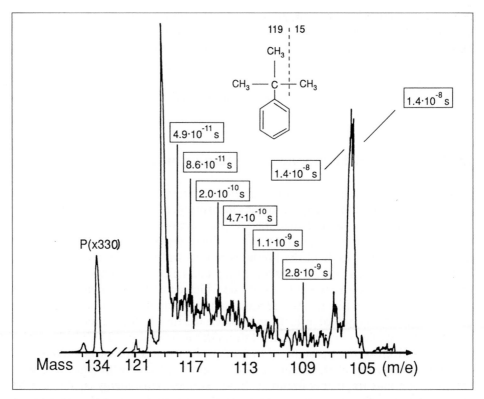

Fig. 1.28. Part of the field ionization mass spectrum of tert-butylbenzene. The calculated decomposition times are indicated. The normal metastable peak is shown on the righthand side [63].

This mass shift can be converted to time shifts by means of an appropriate mathematical treatment [64]. An example of such measurement is shown in Fig. 1.28 in the case of the following process:

$$C_6H_5 - C(CH_3)_3^+ \rightarrow C_6H_5 - C(CH_3)_2^+ + CH_3$$
$$m/e = 134 \qquad\qquad m/e = 119$$

1.7 Penning Ionization

1.7.1 General Description and Considerations

Penning ionization takes place when electronically excited atoms in metastable states for which the energy is higher than the successive ionization energies (IEs) of an atom or a molecule collide with this atom or molecule. As an example of such metastable states, Table 1.8 gives their positions in energy in three rare gases.

Table 1.8 Energies of metastable states in He, Ne, and Ar.

Atom	Term	Energy (eV)
He	^3S	19.818
	^1S	20.614
Ne	$^3P_2$	16.619
	$^3P_0$	16.715
Ar	$^3P_2$	11.548
	$^3P_0$	11.722

With rare gases in these electronic states (generally prepared by electron impact) it is possible to induce ionization processes, such as:

$$M^* + AB \rightarrow AB^+ + M + e^- .$$

Only one electron is released in this reaction, its energy is given by

$$\varepsilon = E(M^*) - [IE\,(AB) + E_{jv}(AB^+)] + \Delta E \,,$$

where $E(M^*)$ is the energy of the metastable particle released in the ionization of AB, $E_{jv}(AB^+)$ is the rotational and vibrational energies of AB^+, and ΔE is the term (positive or negative) that accounts for specific phenomena accompanying Penning ionization, such as:
1. conversion of a part of internal energy of the reactants into kinetic energy of the products;
2. conversion of the kinetic energy of the reactants into the energy of the released electrons;
3. change of electron energy resulting from an intermediate associative ionization, such as:

$$M^* + AB \rightarrow MAB^+ + e^- .$$

For observing electron energy spectra, PIES (Penning Ionization Electron Spectroscopy, realized exactly like PES) is an easy way to measure IEs and $E_{jv}(AB^+)$ distributions. Comparison with the results obtained by PES will allow measurement of the ΔE values, characterizing small but specific interactions of excited or ionized particles with neutrals. As it is not possible to discuss here all the physical aspects of Penning ionization, we will just enumerate the main conclusions obtained by some authors who compared the results obtained by this ionization technique and those obtained from the methods exposed before. From the appropriate literature [65-67] one is able to conclude that:
- ΔE is generally finite, measurable but small ($\pm$ few tens of meV);

● the vibrational distributions for photo and Penning ionization are at least similar in all cases, with small but real differences;

● the electronic transition has Franck-Condon character, as in photoionization, but vertical transitions take place between slightly modified potential energy curves of the molecule, modifications of internuclear distances of the order of 1/100 Å arising in the molecule from their interaction with the collision partner;

● in the case of rather complex molecules, such as C_2H_6, C_3H_8 and C_4H_{10}, the excitation energy distribution curves of the ions differ substantially from those obtained by charge transfer, electron or photon impact.

That is, in Penning ionization more excitation energy is accomodated in the molecular ions than with other methods of ionization. One could, therefore, expect significant differences between the mass spectra as obtained by Penning ionization and electron, photon, or ion impact. Such differences could be a way to check the validity of statistical rate theories applied to ionization phenomena in complex molecules.

References

1. Samson JAR, Haddad GN, Kilcoyne LD (1987) Absorption and Dissociative Photoionization Cross Sections of NH_3 from 80 to 1120 Å. J Chem Phys 87:6416–6422
2. Platzmann RL (1963) Some Remarks on the Nature of Ionization, Ionization Yields, and Isotope Effects in the Ionization of Molecules by Various Agencies. J Chem Phys 38:2775–2776
3. Herzberg G (1950) Molecular Spectra and Molecular Structure: I. Spectra of Diatomic Molecules. D van Nostrand Company Inc, Princeton
4. Lorquet AJ, Lorquet JC, Wankenne H, Momigny J (1971) Excited States of Gaseous Ions: II. Metastable Decomposition and Predissociation of the CH^+ Ion. J Chem Phys 55:4053–4061
5. Wigner EP (1948) On the Behavior of Cross Sections near Thresholds. Phys Rev 73:1002–1009
6. Geltman S (1956) Theory of Ionization Probability near Threshold. Phys Rev 102:171–179
7. Morrison JD (1953) Studies of Ionization Efficiency: Part III The Detection and Interpretation of Fine Structure. J Chem Phys 21:1767–1772
8. Van der Wiel MJ (1972) Threshold Behavior of Double Photoionization in He. Phys Lett 41A:389–390
9. Samson JAR, Kemeny PC, Haddad GN (1977) Double Ionization of CO_2 by Photon Impact. Chem Phys Lett 51:75–76
10. Hitchcock AP, Brion CE, Van der Wiel MJ (1980) Absolute Oscillator Strengths for Valence-Shell Ionic Photofragmentation of N_2O and CO_2 (8–75 eV). Chem Phys 45:461–478
11. Hitchcock AP, Van der Wiel MJ (1979) Absolute Oscillator Strengths (5–63 eV) for Photoabsorption and Ionic Fragmentation of SF_6. J Phys B12:2153–2169
12. Rosenstock HM (1976) The Measurement of Ionization and Appearance Potentials. Int J Mass Spectrom Ion Proc 20:139–190
13. Dibeler VH, Reese RM, Krauss M (1965) Mass-Spectrometric Study of Photoionization: II. H_2, HD, and D_2. J Chem Phys 42:2045–2048
14. Hipple JA, Condon EU (1945) Detection of Metastable Ions with the Mass Spectrometer. Phys Rev 68:54–55
15. Hipple JA, Fox RE, Condon EU (1946) Metastable Ions formed by Electron Impact in Hydrocarbon Gases. Phys Rev 69:347–356
16. Madden RP, Codling K (1965) Two-Electron Excitation States in Helium. Astrophys J 141:364–375

17. Knopp FWE, Kistemaker J (1973) Doubly Excited States of Helium Induced by Low-Energy Electron Impact. Phys Lett 46A:27-28
18. Turner DW, Baker L, Baker AD, Brundle CR (1970) Molecular Photoelectron Spectroscopy. Wiley Interscience, London
19. Dibeler VH, Walker TA, Liston SK (1967) Photoionization of Small Molecules. J Res Nat Bur Standards Sect A 71:371
20. Begun GM, Landau L (1961) Mass Spectra and Metastable Transitions in Isotopic Nitrous Oxides. J Chem Phys 35:547-551
21. Coleman RJ, Delderfield JS, Reuben BG (1969) The Gas-Phase Decomposition of the Nitrous Oxide Ion. Int J Mass Spectrom Ion Proc 2:25-33
22. Newton AS, Sciamanna AF (1970) Metastable Peaks in the Mass Spectra of N_2O and NO_2 II. J Chem Phys 52:327-336
23. Eland JHD, Berkowitz J (1977) Formation and Predissociation of CO_2^+ ($\tilde{C}^2$ Σ_g^+). J Chem Phys 67:2782-2787
24. Baer T, Guyon P-M, Nenner I, Tabché-Fouhaillé A, Botter R, Ferreira LFA, Govers TR (1979) Non-Franck-Condon Transitions in Resonant Autoionization of N_2O. J Chem Phys 70:1585-1592
25. Nenner I, Guyon P-M, Baer T, Govers TR (1980) A Threshold Photoelectron-Photoion Coincidence Study of the N_2O^+ Dissociation between 15 and 20.5 eV. J Chem Phys 72:6587-6592
26. Eland JHD (1983) Photoelectron Spectroscopy. Butterworth, London
27. Gardner JL, Samson JAR (1976) Vibrational Intensity Distributions in the Photoelectron Spectrum of Hydrogen. J Electron Spectrosc Relat Phenom 8:123-127
28. Unpublished Spectrum from our Laboratory
29. Jonathan N, Morris A, Okuda M, Ross KJ, Smith DJ (1974) Vacuum Ultraviolet Photoelectron Spectroscopy of Transient Species. J Chem Soc Faraday Trans 2, 70:1810-1817
30. Dyke JM, Golob L, Jonathan N, Morris A (1975) Vacuum Ultraviolet Photoelectron Spectroscopy of Transient Species. J Chem Soc Faraday Trans 2, 71:1026-1036
31. Edqvist O, Lindholm E, Selin LE, Åsbrink L (1970) On the Photoelectron Spectrum of O_2. Phys Scr 1:25-30
32. Karlsson L, Jadrny R, Mattsson L, Chan FT, Siegbahn K (1977) Vibrational and Vibronic Structure in the Valence Electron Spectra of CH_3X Molecules (X = F, Cl, Br, I, OH). Phys Scr 16:225-234
33. Schulman JM, Moskowitz JW (1967) Benzene and its Ionized States. J Chem Phys 47:3491-3495
34. Åsbrink L, Edqvist O, Lindholm E, Selin LE (1970) The Electronic Structure of Benzene. Chem Phys Lett 5:192-194
35. Stockbauer R (1979) The Formation of High Vibrational States of Ions by Photoionization. Adv in Mass Spectrom 8A:78-79
36. Berkowitz J, Chupka WA (1969) Photoelectron Spectroscopy of Autoionization Peaks. J Chem Phys 51:2341-2354
37. Natalis P, Collin JE (1968) The First Ionization Potential of Nitrogen Dioxide. Chem Phys Lett 2:79-82
38. Natalis P, Collin JE (1968) Experimental Evidence for High Vibrational Excitation in O_2^+ Ground State by Photoelectron Spectroscopy. Chem Phys Lett 2:414-416
    Collin JE, Natalis P (1968) Ionization, Preionization and Internal Energy Conversion in CO_2, COS and CS_2 by Photoelectron Spectroscopy. Int J Mass Spectrom Ion Proc 1:121-132
39. Samson JAR, Gardner JL (1977) The Vibrational Energy Levels and Dissociation Energy of O_2^+ (X^2 Π_g). J Chem Phys 67:755-758
40. Brehm B, Eland JHD, Frey R, Küstler A (1973) Predissociation of SO_2^+ Ions Studied by Photoelectron-Photoion Coincidence Spectroscopy. Int J Mass Spectrom Ion Proc 12:197-211
41. Brehm B, Fuchs V, Kebarle P (1971) Autoionization and Fragmentation Processes in Methanol and Ethanol. Int J Mass Spectrom Ion Proc 6:279-289

42. Dannacher J (1984) The Study of Ionic Fragmentation by Photoelectron-Photoion Coincidence Spectroscopy. Org Mass Spectrom 19:253-275
43. Stockbauer R (1973) Threshold Electron-Photoion Coincidence Mass Spectrometric Study of CH_4, CD_4, C_2H_6 and C_2D_6. J Chem Phys 58:3800-3815
44. Chupka WA, Berkowitz J (1967) Photoionization of Ethane, Propane, and n-Butane with Mass Analysis. J Chem Phys 47:2921-2933
45. Maeda K, Semeluk GP, Lossing FP (1968) A Two Stage Double-Hemispherical Electron Energy Selector. Int J Mass Spectrom Ion Proc 1:395-407
46. Morrison JD (1963) On the Optimum Use of Ionization-Efficiency Data. J Chem Phys 39:200-207
47. Locht R, Momigny J (1986) The Dissociative Ionization of Methyl Fluoride. The Formation of CH_2^+ and CH_3^+. Int J Mass Spectrom Ion Proc 71:141-157
48. Locht R, Momigny J, Rühl E, Baumgärtel H (1987) A Mass Spectrometric Photoionization Study of CH_3F. The CH_2^+, CH_3^+ and CH_2F^+ Ion Formation. Chem Phys 117:305-313
49. Bethe H (1930) Zur Theorie des Durchgangs Schneller Korpuskularer Strahlen durch Materie. Ann Phys 5:325-400
    Inokuti M (1971) Inelastic Conisions of Fast Charged Particles with Atoms and Molecules — The Bethe Theory Revisited. Rev Mod Phys 43:297-347
50. Skerbele A, Lassettre EN (1966) Generalized Oscillator Strengths at Small Momentum Changes for the 1's-2's Transition in Helium. J Chem Phys 45:1077-1078
51. Backx C, Van der Wiel MJ (1974) Coincidence Measurements with Electron Impact Excitation in Vacuum Ultraviolet Physics. Pergamon Vieweg p 137
52. Van der Wiel MJ, Brion CE (1972) Partial Oscillator Strengths for Ionization of the Three Valence Orbitals of NH_3. J Electron Spectrosc Relat Phenom 1:443-455
53. Backx C (1975) Dipole Excitation and Fragmentation of H_2, HD, D_2 and CH_4. PhD Thesis, University of Leiden
54. Lindholm E (1954) Ionisierung und Zerfall von Molekülen durch Stöße mit Ionen. Z Naturforsch 9a:535-546
55. Lindholm E (1966) Charge Exchange and Ion Molecule Reactions Observed in Double Mass Spectrometers. Advances in Chemistry Ser 58:1-19
56. Massey HSW, Burhop EHS (1952) Collision of Electrons with Atoms. In: Fowler R, Kapiza P, Mott NF, Bullard EC (eds), Electronic and Ionic Phenomena vol I. Oxford University Press, London
57. Sunner J, Szabo I (1979) Charge Transfer Mass Spectra of N_2O and Ion-Molecule Reactions in N_2O. Int J Mass Spectrom Ion Proc 31:193-211
58. von Koch H (1965) Dissociation of Ethane Molecule Ions Formed in Charge Exchange Collisions with Positive Ions: Ion-Molecule Reactions of Ethane. Arkiv för Fysik 28:559-574
59. Ottinger Ch (1967) Messungen der Zerfallszeiten von Molekülionen. Z Naturforsch 22a:20-40
    Hertel I, Ottinger Ch (1967) Zerfallszeiten von Molekülionen II. Z Naturforsch 22a:1141-1156
60. Andlauer B, Ottinger Ch (1971) Unimolecular Ion Decompositions: Dependence of Rate Constants on Energy from Charge Exchange Experiments. J Chem Phys 55:1471-1472
61. Inghram MG, Gomer R (1954) Mass Spectrometric Analysis of Ions from the Field Microscope. J Chem Phys 22:1279-1280
    Gomer R, Inghram MG (1955) Applications of Field Ionization to Mass Spectrometry. J Am Chem Soc 77:500
    Inghram MG, Gomer R (1955) Massenspektrometrische Untersuchungen der Feldemission Positiver Ionen. Z Naturforsch 10a:863-872
    Beckey HD (1963) Field Ionization Mass Spectroscopy. Adv Mass Spectrom 2:1-24
62. Beckey HD (1964) Production of the Ionized State of Molecules by High Electric Fields. Bull Soc Chim Belges 30:26-56
63. Beckey HD (1970) Recent Studies on FI Mass Spectrometry. In: Ogata K, Hayakawa (eds) Recent Developments in Mass Spectroscopy. University Park Press, Baltimore

64. Beckey HD (1961) Messung Extrem Kurzer Zerfallszeiten Organischer Ionen mit dem Feldion-
    isations-Massenspektrometer. Z Naturforsch 16a:505–510
65. Cermák V (1966) Retarding-Potential Measurements of the Kinetic Energy Released in Pen-
    ning Ionization. J Chem Phys 44:3781–3786
66. Cermák V, Ozenne JB (1971) Penning Ionization Electron Spectroscopy Ionization of Noble
    Gases H_2, NO, C_2H_4, C_3H_6 and C_6H_6. Int J Mass Spectrom Ion Proc 7:399–413
67. Hotop H, Niehaus A (1970) Reaction of Excited Atoms and Molecules with Atoms and Mole-
    cules. Z Phys 238:452–465
68. Samson JAR (1966) The Measurement of the Photoionization Cross Sections of the Atomic
    Gases. Adv Atom Mol Phys 2:178–261

2 The Occurrence of Transitions between the Electronic States of Molecular Ions

Introduction

It was repeatedly shown in Chapter 1 that the absorption of energy by molecular systems above their first ionization energy results in the creation of all possible electronic states of the molecular ion allowed by the appropriate selection rules. The decay of such excited electronic states of molecular ions can take place either by emission of radiation or by successive non radiative processes. Allowed emission of radiation requires times ranging from 10^{-9} to 10^{-7} s. Other decay processes are generally faster than radiative desexcitation, e. g. direct dissociation (10^{-14} to 10^{-13} s) and non-radiative transitions, precisely invoked when the radiative emissions are completely quenched.

In some cases electronic predissociations can compete with radiative emission, the lifetime of a predissociated state lying between 10^{-13} s and 10^{-5} s. As will be seen later, vibrational and rotational predissociations can result in a molecular ion lifetime eventually longer than 10^{-7} s up to some 10^{-3} s.

As was shown in Chapter 1, PIPECO spectra of complex molecules are very often free of any memory of the many observed electronically excited states. This was obvious in many breakdown diagrams that are interpreted as resulting from a complete redistribution of electronic energy of molecular ions into internal energy of the ground state molecular ion, which is then allowed to dissociate.

Such a situation leads to the question: what are the energy redistribution mechanisms and on what time scale are they generally operating? The following addresses this problem.

2.1 Non-Radiative Transitions between Electronic States

2.1.1 Non-Adiabatic Interactions between Electronic States

As expected, any interpretation of the evolution of electronically excited molecular systems is possible only in the frame of potential energy curves or surfaces representations. Such curves or surfaces can sometimes be determined within their Franck-Condon zone from spectroscopic information. Outside this rather limited region they are known by "ab initio" quantum calculations. Potential energy curves or surfaces are "adiabatic" within the validity of the Born-Oppenheimer approximation. This means that, if the variation of the electronic-wave function of a given molecule is sufficiently slow. the electronic- and nuclear-wave functions can be calculated separately. It is, however, clear that any nonradiative evolution of electronic energy results from interactions between potential energy curves or surfaces, thereby coupling electronic and nuclear motions in the molecular ion to a certain extent. In such a case, the validity of the Born-Oppenheimer approximation breaks down, giving rise to the concept of "nonadiabatic" interaction between sets of interacting electronic states.

As with increasing energies, the density of electronic states of polyatomic molecules increases and the frequent occurrence of such nonadiabatic interactions results from nonadiabatic couplings of potential energy surfaces, either of the same spin and symmetry or of different spins and symmetries. Such couplings result in the occurrence of either allowed or nonallowed crossings between surfaces.

2.1.2 Allowed and Avoided Crossings between Potential Energy Curves

Let us consider two electronic states of a diatomic molecule which cross in the way indicated in Fig. 2.1 for an interatomic distance R_0. If $U_1(R)$ and $U_2(R)$ are the analytical expression of their potential energy curves, the energies $E_1 = U_1(R_0)$ are given by

$$H(R_0) \, \psi_1 = E_1 \, \psi_1 \qquad\qquad 2.1$$

$$H(R_0) \, \psi_2 = E_2 \, \psi_2, \qquad\qquad 2.2$$

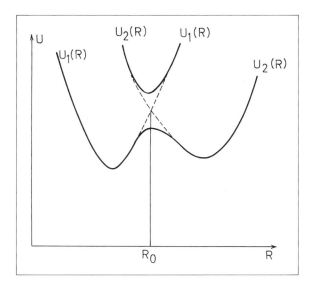

Fig. 2.1. A typical case of avoided crossing between two potential energy surfaces $U_1(R)$ and $U_2(R)$.

ψ_1 and ψ_2 being the electronic wave functions of the two states and $H(R)$ the Hamiltonian of each system.

In fact, it is more useful to consider that E_1 differs from E_2 by a small amount $(E_1 \neq E_2)$. Such a situation can be expressed by a small modification of $H(R_0)$, written as

$$H(R) = H(R_0) + \left(\frac{\delta H}{\delta R}\right)_{R_0} \delta R = H_0 + V, \qquad\qquad 2.3$$

where V is now a perturbation term.

In other words, one looks for a ψ function such as

$$\psi = C_1 \psi_1 + C_2 \psi_2, \qquad\qquad 2.4$$

V, C_1 and C_2 being functions for R.

This ψ function will normally result from the solution of

$$(H_0 + V)(C_1 \psi_1 + C_2 \psi_2) = E(C_1 \psi_1 + C_2 \psi_2), \qquad\qquad 2.5$$

or (recalling Eqs. 2.1 and 2.2) of

$$C_1(E_1 + V - E)\,\psi_1 + C_2(E_2 + V - E)\,\psi_1 = 0. \qquad\qquad 2.6$$

If (2.5) is now multiplied by ψ_1^* and integrated, one gets

$$C_1 E_1 \int \psi_1 \psi_1^* \, d\tau + \underline{C_2 E_2 \int \psi_1^* \psi_2 \, d\tau} + C_1 \int \psi_1^* V \,\psi_1 \, d\tau + C_2 \int \psi_1^* V \,\psi_2 \, d\tau$$
$$= C_1 E \int \psi_1^* \psi_1 \, d\tau + \underline{C_2 E \int \psi_1^* \psi_2 \, d\tau}. \qquad\qquad 2.7$$

The underlined integrals vanish as including orthogonal wave functions.

Therefore, Eq. 2.6 is equivalently written as

$$C_1(E_1 + V_{11} - E) + C_2 V_{12} = 0, \qquad\qquad 2.8$$

recalling that V_{11} is written for $\int \psi_1^* V \,\psi_1 \, d\tau$ and V_{12} for $\int \psi_1^* V \,\psi_2 \, d\tau$.

By multiplying Eq. 2.5 by ψ_2^* and by integration, one equivalently gets

$$C_2(E_2 + V_{22} - E) + C_1 V_{21} = 0. \qquad\qquad 2.9$$

The compatibility condition for Eqs. 2.8 and 2.9 is written:

$$\begin{vmatrix} E_1 + V_1 - E & V_{12} \\ V_{21} & E_2 + V_{22} + E \end{vmatrix} = 0. \qquad\qquad 2.10$$

Solving for E, one gets

$$E = (E_1 + E_2 + V_{11} + V_{12}) \pm \sqrt{\frac{(E_1 - E_2 + V_{11} + V_{22})^2 + V_{12}}{4}}. \qquad 2.11$$

Owing to the $\pm$ signs, this represents *two* potential energy curves. In order to observe a crossing between both, the discriminant needs to be zero. This implies two conditions:

$$(E_1 - E_2 + V_{11} + V_{12})^2 = 0 \qquad 2.12$$

$$V_{12} = 0. \qquad 2.13$$

The size of δR being the only arbitrary parameter that determines the perturbation term V, the two conditions (2.12) and (2.13) are not able to be filled simultaneously; the crossing is therefore forbidden or "avoided".

It can be, however, that V_{12} is identically zero. It is possible to show that in a diatomic molecule, this arises only when the symmetry of the interacting electronic states are different: different Λ values, parities or multiplicities [1].

2.1.3 Crossings between Energy Hypersurfaces — Conical Intersections

It has been recognized that the non-crossing rule has to be somewhat modified in polyatomic molecules [2]. In these cases, since several internuclear distances exist, the two crossing conditions can be fulfilled for certain R_i values, even when the two states have the same species, because the electronic-wave functions depend on R_s internuclear distances as parameters.

If s is the number of independent nuclear distances, the number of nuclei in the molecule being N, s is given by $3N - 6$. Each term $U_n(r_1 \ldots r_s)$ represents a $(s+1)$ dimensional surface. Those n surfaces can cross; the geometrical locus of their interaction can vary from zero (the crossing being reduced to one point) up to $(s-1)$ dimensions.

The discussion made in Section 2.1.2 for the crossing remains valid with the difference that the perturbation V is now determined, not just by one, but by s parameters, these being the elongations $\delta R_1, \delta R_2 \ldots \delta R_j$. As two parameters are enough to determine both Eqs. (2.12) and (2.13) this leads to an important result: in polyatomic molecules, two electronic states are always able to cross. If they belong to the same symmetry, the crossing is governed by the two conditions (2.12) and (2.13) and the number of dimensions of the intersection locus is $(s-2)$. If the states belong to different symmetries, one condition survives and the interactions locus is a $(s-1)$ dimensional hypersurface. If the problem of crossing between two potential hypersurfaces is reduced to the consideration of a three-dimensional space $(U = f(R_1, R_2))$ or

$(U = f(R_1, \alpha))$, the electronic states represented by tridimentional surfaces will cross along curves $(s-1)=1$ for terms of different symmetries and at a point for terms of the same symmetry $(s-2)=0$.

Such an equation as (2.11) represents an elliptical cone. In the neighbourhood of the intersection points the electronic states are represented by an elliptical cone with two arbitrarily oriented sheets. This particular case of crossing is called "a conical intersection". An example of conical intersection is shown in a molecule like XYZ.

In $D_{\infty h}$ symmetry two states such as Π and Σ can cross freely (Fig. 2.2 a). For any fixed value of $\alpha = 180°$, the molecules belong to the C_s symmetry and the Σ and Π states are now, respectively, A' and A' + A" states. In this case the crossing of the two A' states is avoided (Fig. 2.2 b). When the system is now described by $U = f[\alpha, R(X-Y), R(Y-Z)]$, being fixed, one will assist to the conical intersection described in Fig. 2.2 c.

It should be noted that the statement that two potential energy surfaces are always allowed to cross does not give any information about the energy regions where the crossings could eventually occur. The only way to know if conical intersections or situations analogous to avoided crossings arise is with careful ab initio quantum calculation of potential energy surfaces.

An example of such a situation is given by a two $^2A_1$ electronic state of CH_2^+, for which a situation analogous to an avoided crossing is shown to occur in the adiabatic potential energy surfaces [3].

The constant energy gap contours calculated between these two states are given in Fig. 2.3 [4] as a function of two coordinates defining the problem.

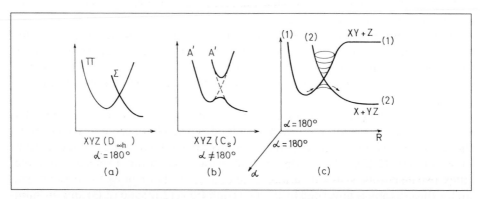

Fig. 2.2. a) Allowed crossing between states of different symmetry (Σ and Π states) of linear molecule ($D_{\infty h}$). **b)** Avoided crossing between the two A' states of the same molecule when distorted to the C_s symmetry. **c)** Conical intersection between the same states for a continuous variation of X and of $R(Y-Z)$, $R(X-Y)$. The arrows indicate the lowest energy path from surface (1) to surface (2), with XYZ → X + YZ dissociation.

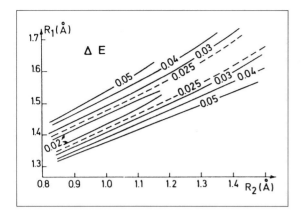

Fig. 2.3. The quantum mechanically calculated energy gap between two $^2A_1$ states of CH_2^+ in function of the R_2(H–H) distance (C_{2v} symmetry) [4].

2.1.4 Non-Adiabatic Interactions and the Time Scale for Energy Redistribution

The occurrence of numerous possibilities of nonadiabatic interactions between potential energy surfaces characterizing the dense manifold of electronically excited states realized in polyatomic ions, raises a question about the time scale for the processes responsible for the redistribution of electronic energy in internal energy of the molecular ion ground state. Such characteristic time windows can be calculated quantum mechanically [4, 5]. Without describing in detail such calculations, it is possible to report a lower limit of the nonradiative transitions rates. As was pointed out in the introduction, radiative transitions between electronic states have lifetimes of about 10^{-9} to 10^{-7} s. The very frequent lack of radiative transitions in electronically excited polyatomic molecular ions, means that nonradiative transitions take place faster than the radiative one.

This means that nonradiative transitions can occur in 10^{-10} s or less.

Without detailing the calculations [4–5] it is possible to indicate that, owing to the kind of interaction, the interaction time is very fast for conical intersections ($\approx 10^{-14}$ s) and slower for avoided crossings. The authors of the calculations indicate, however, that if the time to turn around the cone (the lower energy path) is very short, it can be enlarged by curvatures in the transitions path. In the case of avoided crossings like the one drawn in Fig. 2.4, it was shown that transitions between both states need the excitation of a vibrational mode perpendicular to the "seam" between nonadiabatic surfaces, this being a limiting factor of the transition rate.

Let us recall that, whatever the case, all these nonadiabatic transition rates remain equal or higher than the some 10^{10} s^{-1}, needed for the complete quenching of all radiative emission.

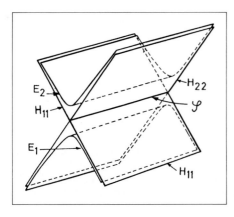

Fig. 2.4. A drawing of diabatic E_1, E_2 energy surfaces and adiabatic H_{11}, H_{22}, illustrating an avoided surface crossing. The line φ is the "seam" between both surfaces [5].

2.2 Radiative Transitions between Electronic States of Molecular Ions

Until the 1970s, very few ions had been characterized by their emission spectra, except triatomic molecular ions and the diacetylene cation [6]. This situation changed rapidly afterwards and presently the emission spectra of more than 50 complex organic cations have been observed, generally due to transitions from their first electronic excited state to their ground state. Some reviews about the developments of this spectroscopy have been published [7–9]. Interestingly, the main aspect of these works concerns the discovery, in certain cases, of a competition between radiative and nonradiative transitions, including the simultaneous occurrence of dissociative phenomena of the cation. Some examples of such a behavior have been described in recent years [10]; as an example, the decay of *cis*- and *trans*-difluoroethylene cations will be discussed here [11].

In the case of the *cis*-$C_2H_2F_2$, after its preparation in the $\tilde{A}^2A_1$ state, either a fluorescence photon ($\tilde{A} \rightarrow \tilde{X}$) or a C_2HF^+ ion is produced. The quantum yield for fluorescence and the associated rate constant have been obtained from photoelectron-photon coincidence measurements as well as the parent-ion to fragment-ion branching ratio, from the photoelectron-photoion coincidence mass spectrometry. All these results can be schematically expressed as follows:

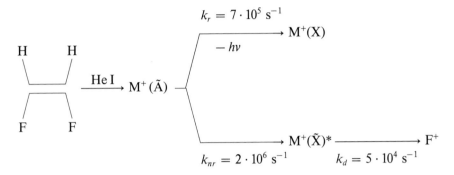

The respective rates k_r and k_{nr} are obviously such that one can consider any direct dissociation of the $\tilde{A}$ state as highly improbable. The most surprising information about the behavior of cis- and trans-$C_2H_2F_2$ isomers, comes from the trans-isomer for which the fluorescence yield is just at the detection limit. The conclusion is that in the trans-isomer k_{nr} is by far more important than k_r.

This is summarized in the following decay scheme for the trans-isomer:

$$
\begin{array}{c}
\text{H} \quad\quad \text{F} \\
\diagdown \underline{} \diagup \\
\diagup \quad\quad \diagdown \\
\text{F} \quad\quad \text{H}
\end{array}
\xrightarrow{\text{He I}} M^+(\tilde{A}) \xrightarrow{k_{nr} \geq 10^{10}\ \text{s}^{-1}} M^+(\tilde{X})^* \xrightarrow{k_d = 5 \cdot 10^4\ \text{s}^{-1}} F^+
$$

In both cases the unimolecular fragmentation rate constants have been extracted from the variation of the parent ion yield as a function of modified total transit times through the quadrupole mass spectrometer used. It is noteworthy that the rather slow rate for internal conversion of the cis-isomer, comparable with normal ionic fragmentation rates, excludes the behavior of the cis-isomer of the application of any statistical theory, for which dissociative rates may be much slower than internal conversion. As will be shown later, other facts place this molecule (and some others) in a category rather hard to reconcile with a full validity of any statistical rate theory.

References

1. Landau L, Lifschitz E, Mecanique Quantique, 2nd edn. Edition MIR. Moscow 1967 p 423
2. Teller E (1937) The Crossing of Potential Surfaces. J Phys Chem 41:109–116
3. Sakei S, Kato S, Morokuma K, Kusunoki I (1981) Potential Energy Surfaces of the Reaction $C^+ + H_2 \rightarrow CH^+ + H$. J Chem Phys 75:5398–5409

4. Desouter-Lecomte M, Dehareng D, Leyh-Nihant B, Praet MT, Lorquet AJ, Lorquet JC (1985) Nonadiabatic Unimolecular Reactions of Polyatomic Molecules. J Phys Chem 89:214–222
5. Desouter-Lecomte M, Leyh-Nihant B, Praet MT, Lorquet JC (1987) Avoided Crossings: A Study of the Nonadiabatic Transition Probabilities. J Chem Phys 86:7025–7034
6. Herzberg G (1966) Molecular Spectra and Molecular Structure III. Electronic Spectra and Electronic Structure of Polyatomic Molecules. Van Nostrand Reinhold Company, New York
7. Maier JP (1979) Decay Processes of the Lowest Excited Electronic States of Polyatomic Radical Cations. In: Ausloos P (ed) Kinetics of Ion-Molecule Reactions, Vol B40. Plenum Press, New York
8. Leach S, Dujardin G, Taieb G (1980) Radiationless Transitions in Gas Phase Molecular Ions. J de Chim Physique et de Physico-Chimie Biologique 77:705–718
9. Maier JP (1982) Open-Shell Organic Cations: Spectroscopic Studies by Means of their Radiative Decay in the Gas Phase. Acc Chem Res 15:18–23
10. Dannacher J (1980) Photoelectron-Photoion Coincidence Study of Polyatomic Radical Cations for which Competition between Radiative and Dissociative Decay is Observed. Adv Mass Spectrom 8A:37–46
11. Stadelmann JP, Vogt J (1980) Isolated State Dissociations from Electronically Excited Radical Cations of Fluoroethenes Studied by Photoelectron-Photoion Coincidence Spectroscopy. Adv Mass Spectrom 8A:47–55

3 Energy Balance in the Dissociation Processes of Molecular Ions

3.1 Experimental Approach to the Thermochemistry of Dissociation Processes

3.1.1 Threshold Energies for Dissociation. Appearance Energies

Any dissociation process of a molecular ion can be written as

$$AB^+ \rightarrow A^+ + B,$$

where A^+ and B are written for atoms if AB^+ is diatomic, or for more complicated species if AB^+ is polyatomic.

Such processes can be characterized by many dissociation limits if one takes into account the possibility of electronic excitation of A^+ and B when dealing with atoms, and of rotational, vibrational, and electronic excitation when A^+ and B are polyatomic species.

First, considering AB^+ as a diatomic species, it is easy to calculate the possible thresholds for dissociation as equal to

$$TE_{0,1\ldots i} = D(A\text{--}B) + IE(A) + E^*_{0,1\ldots i},$$

where $D(A\text{--}B)$ is the dissociation energy of AB, $IE(A)$ is the lowest ionization energy of A and E^* is zero or equal to the electronic energy of A^+ or of B or of $A^+ + B$.

As an example, the three first TE_i of O_2 are given in Table 3.1.

Threshold energies are therefore calculated values of the energy position of the successive possible dissociation limits of a molecular ion. In the case of diatomic as of polyatomic ions the correlation rules allow knowing all the possible electronic sta-

Table 3.1. Threshold energies for the three lowest dissociation channels in O_2^+.

Dissociation process	TE (eV) (1)	E^* (eV)
$O_2 \rightarrow O_2^+ \rightarrow O^+(^4S) + O(^3P)$	18.729	0
$O_2 \rightarrow O_2^+ \rightarrow O^+(^4S) + O(^1D)$	20.697	$1.968 : O\ ^3P \rightarrow {}^1D$
$O_2 \rightarrow O_2^+ \rightarrow O^+(^2D) + O(^3P)$	22.054	$3.325 : O^+\ ^4S \rightarrow {}^2D$

(1) Using $D(O\text{--}O) = 5.115$ eV and $IE(O) = 13.614$ eV

tes of a molecular ion correlated with a given dissociation limit. When such rules are used together with the results of photoelectron spectroscopy and of electronic configurations, a system of potential energy curves like the simplified one presented in Fig. 1.4 can be drawn. Taking once more the fictitious case of Fig. 1.3 into consideration, it is possible to discuss the relation between TE values and what we refer to as *appearance energy* values (AE) for the fragment ions, *defined as the experimentally accessible minimum energy values for the appearance of ionic fragments possibly in their successive combinations of electronic states.*

As it will be pointed out, a complete experimental treatment of this problem will only be possible if the kinetic energy distributions of fragment ions are measured. What the experimental expectations are from the set of potential energy curves of Fig. 1.3 will now be described with the help of Fig. 3.1.

From Fig. 3.1 it is clear that by a direct ionization process, the $A^+ + B$ dissociation is possible by two different mechanisms:

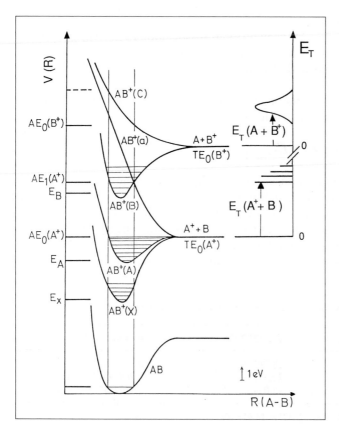

Fig. 3.1. System of potential energy curves of AB^+; on the upper righthand side the kinetic energy distributions observable for the dissociation processes. The AE's are experimentally observable appearance energies, the TE's are the calculated threshold energies for each dissociation process.

1. the Franck-Condon transition takes place to the dissociation limit quoted as $TE_0(A^+)$ in narrow coincidence with this limit. As a consequence, no kinetic energy of $A^+ + B$ appears, and in this case:

$$AE_0(A^+) = TE_0(A^+);$$ 3.1

2. the Franck-Condon transition takes place to the $B(AB^+)$ state, predissociated from the $v=2$ up to $v=5$ vibrational levels. In this case, the lowest possible AE is:

$$AE_1(A^+) = TE_0(A^+) + E_T,$$ 3.2

where E_T is the total kinetic energy of $(A^+ + B)$.

In this circumstance successive values of AE could be observed, such as

$$AE_1 + n\omega_e \ (n = 1, 2, 3),$$

corresponding to the successive values of $TE_1 + n\omega_e$. In other words, AE_1 differs from TE_0 and the kinetic energy distribution of $A^+ + B$ will be modulated through the frequency of the predissociated vibrational levels.

3. When the Franck-Condon zone reaches the entirely repulsive C-state for an $AE_0(B^+)$ value above $TE_0(B^+)$, a kinetic energy distribution of $(A^+ + B)$ starts for a value of $AE_0(B^+) - TE_0(B^+)$, goes through a maximum, and stops at a value corresponding to the end of the Franck-Condon zone. This distribution is continuous and can be considered as resulting from the reflection of the $\psi^2 \ (v=0)$ of the ground neutral state into the potential energy curve of the C-state [1].

From these considerations it clearly appears that for diatomic molecules or ions, AE_i values will often be different from TE_i values, the differences corresponding to kinetic energy distributions of the dissociation products.

It follows that

$$AE_i(A^+) = TE_i(A^+) + E_T.$$ 3.3

The kinetic energy of A^+ is

$$E_T(A^+) = \frac{m_B}{m_A + m_B} E_T.$$ 3.4

In other words, a linear relation exists between the kinetic energy values measured for an ion and the excess above the dissociation limit. In the case of diatomic ions, this relation is linear with a slope just equal to m_B^+/M in the case of A^+ and m_A^+/M in the case of B^+, M being equal to $m_A + m_B$.

If, for transition probabilities reasons, zero kinetic energy A^+ ions are not formed, the observed straight line will extrapolate to $TE_i(A^+)$, giving a way to determine this TE_i value where AE_i differs from TE_i. If the dissociation process is observed in a polyatomic molecular ion, the straight line behavior can also be expected. However, a partition of the internal energy of AB^+ is now possible on the products, not only as translational, but also as vibrational and rotational energy. Therefore, one could observe a slope which is no longer always the calculated one, but can be less by an amount expressing the fraction of internal energy transformed in vibrational and rotational energies. An example of the possible observables in the diatomic case of Fig. 3.1 is given in Fig. 3.2, the assumption being made that $M = 60$, with $m_A = 20$ and $m_B = 40$ mass units.

If AB represents a polyatomic molecule, the available excess energy will generally appear as kinetic and internal (vibrational, rotational) energy of the fragments.

3.1.2 Kinetic-Energy-Distribution Measurements

Determination of kinetic energy distributions of ions is possible within the frame of two limiting cases: in the first one, they are measured irrespective of the molecular ion lifetime; in the second one, they are measured within the classical metastable time window (10^{-5} to 10^{-7} s). This section will exclusively be concerned with the first possibility, Section 3.2.4 will be devoted to the second.

We will not explain here [2] all the methods used for the measurement of kinetic energy distributions of ions, e. g., retarding potential, parallel plate electric condenser, deflection in magnetic fields, and time-of-flight methods. It will just be remembered that after a survey of all methods, some authors [3] pointed out that the retarding potential method was the most resolutive one, provided that the electric field in the ion-formation region is kept at zero or at such a small value that the ther-

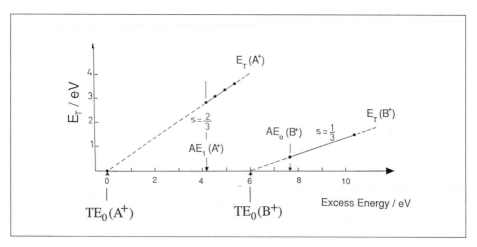

Fig. 3.2. Translational energies of A^+ and B^+ as a function of the excess energy. The excess energy scale refers to the threshold of the lowest dissociation limit.

mal energy distributions of molecular ions is not appreciably disturbed. In their photoionization experiment they observed full width at half maximum (fwhm) of the distribution ions of 0.03 eV for Ar^+, indicating an expected kinetic energy resolution better than this width. This has encouraged another group [4], beginning in 1974, to use a retarding potential device for the study of dissociative electroionization and, more recently, photoionization [5].

The fwhm of a Boltzmann distribution obtained at 473 K (ion source temperature) by electroionization was of about 0.08 eV for Ar^+. This resolution is almost sufficient for resolving a possible modulation of the kinetic energy distribution by vibrational levels of a diatomic predissociated electronic state. This was in fact observed in the case of O_2^+, N_2^+, and CO^+ [4, 6, 7]. By simultaneously using kinetic energy measurements coupled with the observation of successive AEs in the first derivative of the ion yield curves, these authors were able to study with the help of diagrams such as Fig. 3.2 the detailed behavior of about 10 molecules, the more complex being CH_3F for which electro- and photoionization have been used, showing a full agreement between their results. As an example the kinetic energy distribution of CH_3^+ from CH_3F is shown in Fig. 3.3 for both electron and photon impact at the same energy [8]; Figure 3.4 gives the translational energy of CH_3^+ observed as a function of both the photon and electron energies, showing an excellent agreement between both the observed behaviors [5].

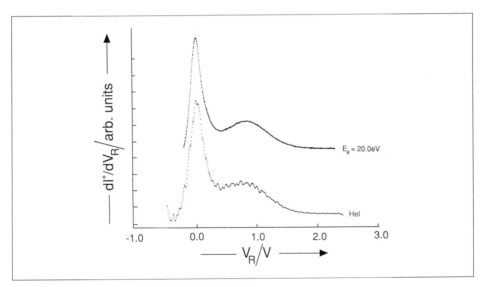

Fig. 3.3. Kinetic energy distributions of CH_3^+ as observed with isoenergetic photons and electrons for the process. $CH_3F^+ \rightarrow CH_3^+ + F$. Two distributions are clearly visible: a quasithermal one and a very wide one, extending from about zero to 2.0 eV [5].

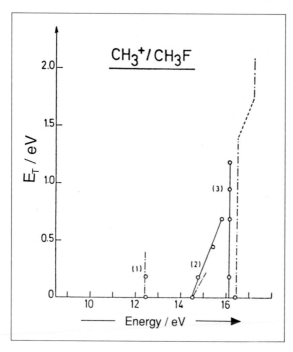

Fig. 3.4. The E_T (CH$_3^+$) diagram as a function of photon (full lines) or electron energy (dotted lines) shows an excellent agreement between the results from both methods for the process: CH$_3$F$^+$ → CH$_3^+$ + F: Line (1) relates to the ion-pair process; CH$_3$F → CH$_3^+$ + F$^-$; line (2) gives an AE_0 of 14.5 eV in agreement with TE_0 for the process CH$_3$F$^+$ → CH$_3^+$ + F; the vertical line (3) expresses the fact that at 16.2 eV the whole wide kinetic energy distribution appears [5].

3.1.3 Influence of Lifetimes of the Molecular Ions on *AE* Measurement. Kinetic Shift

If the reader goes back to Section 1.1.4, he will remember that in the case of the occurrence of predissociation processes, such as the one shown in Fig. 3.1 for the decay of the B state of AB$^+$ into A$^+$ + B, the lifetimes of the successive predissociated vibrational levels of B with respect to dissociation can range from 10^{-3} to 10^{-13} s, depending on the predissociation process.

Quite often, even in the case of diatomic ions, the lifetime of the predissociated levels is in the range of the characteristic transit times through the mass spectrometer. As shown in Section 1.1.4, this circumstance is responsible for a modification of the relative intensities of the parent and the daughter ions. This implies that if the vibrational level coinciding with AE_1(A$^+$) and the next one have lifetimes longer than the transit time of AB$^+$ through the mass spectrometer the only observable AE_1 will be realized for $v=4$ instead of $v=2$; the real $AE_1(v=2)$ will be two vibrational quanta lower in energy and could only be observed for much longer transit time, or by introducing a longer known delay between ionization and extraction of ions.

This difference in *AE*'s values is called the "kinetic shift" for the appearance of fragmentary ions.

This kinetic shift concept will play a considerable role in dissociation processes of complex polyatomic ions, when they are ruled by a statistical kinetic behavior. In such cases, the rate constant for TE will often be too low to become observable with a normal mass spectrometer and methods allowing the introduction of measureable delays between ionization and acceleration of the ions have been developed that allow the AE to be followed down in function of time delays [9]. As an example of such measurements, we quote the study of the AE for the process

$$C_6H_6^+ \rightarrow C_6H_5^+ + H$$

for electron impact. For a delay time of 1 μs one gets $AE\ (C_6H_5^+) = 14.3 \pm 0.1\ eV$; for a delay time of 900 μs the AE value drops to 12.8 ± 0.2 eV.

Such a method has been used in the framework of the measurements of ZKE photoelectron-photoion coincidences, in the time delay range of 1 to 10 μs, showing similar effects in the decay of complex polyatomic molecules [10, 11].

The reality of such a kinetic shift can lead to the expectation of a lower AE for the metastable ion signals than for the corresponding metastable fragments. This was shown in some of the cases collected in Table 3.2 [12].

Table 3.2. Values of AE (fragment) $- AE$ (metastable)

Compound	Transition	Kinetic shift (eV)
n-C$_3$H$_8$	$C_3H_8^+ \rightarrow C_2H_4^+ + CH_4$	0.3 to 0.4
C$_6$H$_5$CN	$C_6H_5CN^+ \rightarrow C_6H_4^+ + HCN$	0.65 to 0.75
1-4 Dioxan	$C_4H_8O_2^+ \rightarrow C_3H_6O^+ + CH_2O$	0.25 to 0.30
Toluene	$C_7H_7^+ \rightarrow C_5H_5^+ + C_2H_2$	0.8 to 1.0
Benzene	$C_6H_6^+ \rightarrow C_4H_4^+ + C_2H_2$	0.9

3.2 The Concept of Ion Structure

3.2.1 General Considerations

Starting with a neutral molecule of a given structure, one may ask what the structure of the corresponding molecular ion will be, either in its ground electronic state or in its excited electronic states.

Such an answer is, in some cases, obtained from the electronic emission or absorption spectra often used in conjunction with results from photoelectron spectroscopy, at least for some rather simple molecular ions. If the conventional equation expressing the lowest ionization energy of a molecule like NH$_3$ writes:

$$\tilde{X}\,NH_3 + h\nu \rightarrow \tilde{X}\,NH_3^+ + e^-$$

it gives no information about the structure of NH_3^+. We are only able to guess that the pyramidal structure of NH_3 survives the ionization phenomenon. If the photoelectron spectrum of NH_3 is considered, it will be seen that the appearance of NH_3^+ is accompanied by a long progression of the umbrella frequency of NH_3^+ (v_2, A_1).

This means that the transition between NH_3 and NH_3^+ is characteristic of a transition between a C_{3v} pyramidal molecule and a D_{3h} planar molecular ion [13].

Such information is of importance in the framework of rate constant calculations for the dissociations of the ground state molecular ion by means of statistical theories. The problem of clarifying ionic structures is of fundamental importance when the description of the products of a molecular ion dissociation is needed.

As far as different structures can be written for the fragment ion and its neutral counterpart, ambiguities remain about the scheme of the considered process. Their removal is of importance in the correct assignment of a correct threshold energy.

Let us choose as an example an observed dissociation process of $C_6H_6^+$. At a certain energy the process

$$C_6H_6^+ \rightarrow C_4H_4^+ + C_2H_2$$

is observed.

In this case, assuming the ethyne structure of C_2H_2, one is faced with the possibility to write 10 different structures for $C_4H_4^+$ [14], under which, for instance:

$$HC \equiv C-CH=CH_2^+ \qquad H_2C=C=C=CH_2^+ \qquad H_2C=C\underset{\diagdown CH}{\overset{\diagup CH}{\diagup}} +$$

It is obvious that methods are needed in order to specify the real signification of the above dissociation process. At this point, the attention of the reader is called to the fact that if 10 isomers of $C_4H_4^+$ can be written, the $C_6H_6^+$ formula is able to give rise to 217 isomers!

This shows how it could be particularly impossible, if $C_6H_6^+$ appeared as a fragment ion of more complex molecules, to decide which of those 217 isomers appears in the dissociation process. When the fragment ions, however, can be characterized by a small number of structures, the assignment of a definite dissociation mechanism will quite often be possible. Many methods proposed and used in order to solve the problem of ion structures, will now be successively described.

3.2.2 Quantum Mechanically Calculated Energies and Geometries of Ions

The availability of powerful computers allows quantum chemists "ab initio" calculations of the total energy of molecules and ions in their possible structures, proposing for the studied species a set of molecular orbitals which will facilitate the determination of the successive electronic states of a given species [15]. At a lower degree of sophistication semiempirical molecular orbital calculations are used [16] as empirical molecular mechanics calculations [17, 18].

As an example, a brief report is given here on ab initio quantum calculations related to five $C_2H_4O^+$ isomeric ions among the 11 isomers which have been studied [19]. The three first are known to be non interconverting observable species. In Table 3.3, the structures of these isomers are indicated, their energy being scaled with respect to the most stable calculated structure.

Table 3.3. Ab initio calculations of the structure and stability of the $C_2H_4O^+$ isomers [19].

Structure	Energy (kJ mol^{-1})	Symmetry	Electronic ground state
$CH_2=CH-OH^+$	0	C_s	$^2A''$
CH_3 $C=O^+$ H	52.3	C_s	$^2A'$
O^+ $H_2C \triangle CH_2$	182.6	C_{2v}	$^2B_1$
$CH_2-O-CH_2^+$	124.9	C_s	$^2A'$
$CH_2-CH_2-O^+$	237.3	C_s	$^2A''$

3.2.3 Determination of Ion Structure through ΔH_f (Ions)

If it is verified (or assumed) that for the dissociation process

$$AB \rightarrow AB^+ \rightarrow A^+ + B,$$

the following conditions are fulfilled:

a) the kinetic shift, if present, and the kinetic energy distributions of the fragment ion have been measured allowing corrected AE values to be considered as TE values; and

b) the fragments $A^+ + B$ are unequivocally correlated with the AB^+ ground state only,

it is then correct to write

$$\Delta H_R = TE = \Delta H_f(A^+) + \Delta H_f(B) - \Delta H_f(AB). \qquad 3.5$$

From this relation, $\Delta H_f(A^+)$ can be calculated. In this equation ΔH_f is, unless otherwise stated, written for ΔH_f^{298}.

If IE (A^+) and $\Delta H_f(A)$ are separately available, the fitting of the above hypotheses is easily checked by using the relation

$$\Delta H_f(A^+) = \Delta H_f(A) + IE\,(A^+). \qquad 3.6$$

An example where it has been verified that all the above conditions are fulfilled is the process [8]

$$CH_3F \to CH_3F^+\,(\tilde{X},\,{}^2E) \to CH_3^+\,(\tilde{X},\,{}^1A_1) + F\,({}^2P_u)\,.$$

From photoelectron spectroscopy it is known that the ground state of the molecular ion is filled up to 14 eV by Franck-Condon transitions. An appearance energy of 14.5 ± 0.06 eV has recently been measured with dispersed synchrotron radiation. This energy is the lowest threshold energy for the above-mentioned process, which is populated through autoionization.

Figure 3.4 (Process 2) shows that this value corresponds to the appearance of CH_3^+ with zero kinetic energy. No metastable transition is known for this process; one will assume that the kinetic shift is zero. Therefore, it is allowed to write

$$14.5 \pm 0.06 \text{ eV} = \Delta H_f(CH_3^+) + \Delta H_f(F) - \Delta H_f(CH_3F)\,. \qquad 3.7$$

As $\Delta H_f(F)$ is 0.818 eV and $\Delta H_f(CH_3F) = -2.424$ eV, this relation leads to:

$$\Delta H_f(CH_3^+) = 11.258 \pm 0.06 \text{ eV}\,.$$

The value of $\Delta H_f(CH_3)$ being known as 1.474 eV, the calculated IE of (CH_3) is:

$$IE\,(CH_3^+) = 9.783 \pm 0.06 \text{ eV},$$

which compares favorably with the value 9.842 ± 0.002 eV obtained from Rydberg series.

An extensive compilation of ΔH_f (ions) is given in the literature [20].

As a first example of the use of ΔH_f (ion) estimations from AE's of the same ion from different molecules, Table 3.4 collects values of $\Delta H_f(C_2H_2^+)$ from different origins. The AE's have been measured by photoionization [21].

It is easy to deduce from this table, within the limits of experimental error, that all the reactions lead to the appearance of the ground state acetylenic ion. Two divergent results (reactions 6 and 4) deserve appropriate discussion. For reaction 6, the difference of about 20.9 kJ mol^{-1} has to be ascribed to either the appearance of kinetic energy on the fragments, or to vibrational excitation of C_2H_2. Reaction 4 is more troublesome, because the observed difference is of about 129.7 kJ mol^{-1} (1.34 eV).

It could first be thought that C_2H_2 is not formed from $C_2H_3F^+$ by 1-2, but by 1-1 elimination of HF leading to a $CH_2 = C^+$ (vinylidene) ion. This explanation is ruled out by a quantum mechanically calculated $\Delta H_f(CH_2 = C^+)$ of 1476.9 kJ mol^{-1}, higher than the value of $\Delta H_f(C_2H_2^+)$ from reaction 4.

Table 3.4. Photoreactions leading to $C_2H_2^+$.

Reaction	Observed AEs (eV)	$\Delta H_f(C_2H_2^+)$ (kJ mol^{-1})
1 $C_2H_2 + h\nu \rightarrow C_2H_2^+ + e^-$	11.396 $\pm$ 0.003	1326.3
2 $C_2H_4 + h\nu \rightarrow C_2H_2^+ + H_2 + e^-$	13.13 $\pm$ 0.02	1317.9
3 $C_2H_3Cl + h\nu \rightarrow C_2H_2^+ + HCl + e^-$	12.47 $\pm$ 0.1	1317.9
4 $C_2H_3F + h\nu \rightarrow C_2H_2^+ + HF + e^-$	13.5 $\pm$ 0.02	1456.0
5 $C_2H_3Br + h\nu \rightarrow C_2H_2^+ + HBr + e^-$	12.25	1297.0
6 $1\text{-}1\,C_2H_2F_2 + h\nu \rightarrow C_2H_2^+ + 2F + e^-$	19.08 $\pm$ 0.03	1347.2

From PIPECO measurements with the He I line [23] it was shown that the unimolecular rate constant for HF elimination varies with the ionization energy of the parent ion. This result shows that an AE of 13.5 eV correlates with a k value of about $8 \cdot 10^4$ s^{-1}. At 13.35 eV, k ranges around $5 \cdot 10^3$ s^{-1}. These results are confirmed by a photoionization mass spectrometric measurement with residence times of 30 to 50 μs, which lead to a value of 13.31 eV [24].

In the microsecond range a metastable transition is observed which allows the estimation of a kinetic energy excess on both fragments of about 0.98 $\pm$ 0.05 eV [25].

As the TE is calculated at 12.16 eV, it is probable that at this energy the rate constant is at least of 1 s^{-1}. In the microsecond range a kinetic shift of about 1.34 eV is realized – about 70 % of this energy appears as kinetic energy, and 30 % as vibrational excitation in $C_2H_2^+$ and HF.

A second example deals with the most stable isomers of $C_2H_4O^+$ ions. From the IE of ethanol and oxirane, $\Delta H_f(C_2H_4O^+)$ are respectively evaluated as 820 kJ mol^{-1} and 966.5 kJ mol^{-1}. As the vinylalcohol is not a stable molecule, the $\Delta H_f(CH_2=CH-OH^+)$ can only be evaluated from the AE of an appropriated dissociation process. From the measurements performed with an electron energy selector [26] on the process

$$\begin{array}{ccc} H_2C & \!\!\!\!\!-\!\!\!\!\! & CHOH \\ | & & | \\ H_2C & \!\!\!\!\!-\!\!\!\!\! & CH_2 \end{array} \longrightarrow CH_2=CH-OH^+ + C_2H_4 \,,$$

a value of 761.4 kJ mol^{-1} was deduced for $\Delta H_f(CH_2=CH-OH^+)$.

Table 3.5 gives the values observed for the three most stable isomers of $C_2H_4O^+$ in comparison with the calculated values of Table 3.3 [19] and shows an excellent agreement.

This evaluation of $\Delta H_f(CH_2=CH-OH^+)$ leads to the possibility to give an estimation of the IE of the vinylalcohol. An estimation of $\Delta H_f(CH_2=CH-OH)$ made by a group contribution method (see Section 3.2.4) gives -110.8 ± 8.2 kJ mol^{-1} for this value. From $\Delta H_f(CH_2=CH-OH^+)$ the value of IE of the vinylalcohol is deduced as 9.0 $\pm$ 0.15 eV.

Table 3.5. Observed stability of some $C_2H_4O^+$ isomers.

Isomer	ΔH_f (kJ mol^{-1})	Difference with respect to the most stable	
		Measured	Calculated
$CH_2 = CH-OH^+$	761.4	0	0
$CH_3-C = O^+$ $\quad\vert$ $\quad$ H	820.0	58.6	52.3
$H_2C \triangle^+ CH_2$ (O)	966.5	205.1	182.6

3.2.4 Evaluation of ΔH_f (Molecules) and of ΔH_f (Ions) through Group Contribution Schemes and from a Macroincrementation Approach

Evaluation of ΔH_f (ions) is possible only if ΔH_f (molecules) is known. Such values are not always known experimentally and need to be known from sources other than "ab initio" calculations.

A first method to do so was proposed [27] and is known as "group equivalent method" based on additivity rules and evaluation of corrections for structure peculiarities.

The scheme is as follows:

$\Delta H_f(C_2H_6)$ being equal to -84.68 kJ mol^{-1}, the group equivalent of CH_3 is evaluated to be -42.34 kJ mol^{-1}. From $\Delta H_f(CH_3-CH_2-CH_3)$ equal to 103.84 kJ mol^{-1} and from $\Delta H_f(CH_3)$ the group equivalent of CH_2 is evaluated to be equal to -19.16 kJ mol^{-1}.

Applied to $n-C_6H_{14}$, this method leads to a $\Delta H_f(n-C_6H_{14})$, calculated to be

$$\Delta H_f(n-C_6H_4) = 2 \ \Delta H_f(CH_3) + 4 \ \Delta H_f(CH_2) = -161.32 \text{ kJ mol}^{-1},$$

which is in good agreement with an experimental value of -167.19 kJ mol^{-1} [20].

Such calculations are rather easy to generalize except when electron delocalizations play a role in the molecule.

If the concept of the positive-charge localization in the ion is accepted in a first approximation, as in $s-C_3H_7^+$ or $t-C_4H_9^+$ ions where it is probable that the structures of the ions are

$^{\oplus}H_2C\diagdown\diagup^{CH_3}_{CH_3}$ and $^{H_3C}_{H_3C}\diagdown^{H_3C}\diagup C^{\oplus}$, from $\Delta H_f(s-C_3H_7^+)$ and $\Delta H_f(t-C_4H_9^+)$, respectively measured as equal to 799.1 and 883.6 kJ mol^{-1}, it is possible to deduce group equivalent values for $> CH_2^{\oplus}$ and $> C^+$ of, respectively, 883.6 and 825.5 kJ mol^{-1}. Similarly, with groups like $C_6H_5^+$, CH_2OH^+, $H-C=O^+$, the halogens, and many other ionized groups, this method can be used with fair success in calculating the heat of formation of ions involving them. On the other hand, when the charge cannot be considered as localized, like in conjugated hydrocarbons and derivatives, the method breaks down.

Another more recent approach is called the macroincrementation method [28]. In this approach it is assumed that for each of two sets of molecules, the total number of bonded atoms and structural types is the same. Then, if all but one of the heats of formation are available, the remaining one can be estimated by simple arithmetic. In this method, resonance energy, ring strain, and other interactions are more explicitly included in the calculations [29]. As, generally, more than one set of schemes can be composed, their results, when compared, show eventual self-consistencies.

Two examples of such a process will be shown here: the first one is an estimate of ΔH_f (fulvene). By using the following scheme (all numbers in the scheme are in units of kJ mol^{-1}):

| 310 | 32.6 | x | 121.6 |

one finds $x = \Delta H_f$ (fulvene) $= 221$ kJ mol^{-1}.
Another scheme could be:

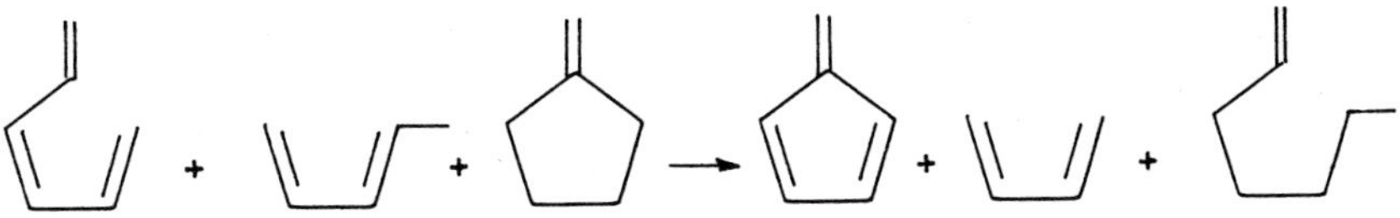

| 170.5 | 76.2 | 16.8 | x | 109.8 | − 62.4 |

which leads to $x = \Delta H_f$ (fulvene) $= 216.1$ kJ mol^{-1}, in good agreement with the first estimated value.

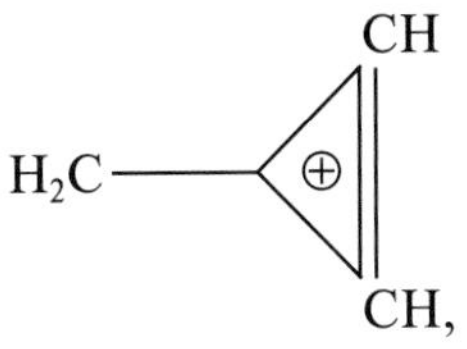

A similar technique is useful for the calculation of the ΔH_f of ions. If we choose, for instance, as a structure for $C_4H_4^+$ the formula a simple macroincrementation scheme can be written as

$$
\underset{\substack{\text{HC}\quad\text{CH}\\[2pt]1003\ \text{kJ mol}^{-1}}}{\overset{\text{CH}_3}{\triangle_{+}}}
+ \underset{107.9\ \text{kJ mol}^{-1}}{\text{CH}_3-\text{CH}_2} \rightarrow
\underset{\substack{\text{HC}\quad\text{CH}\\[2pt]\text{x}}}{\overset{\text{CH}_2}{\triangle_{\oplus}}}
+ \underset{-84\ \text{kJ mol}^{-1}}{\text{CH}_3-\text{CH}_3}
$$

which gives $x = \Delta H_f\ (C_4H_4^+) = 1194.9\ \text{kJ mol}^{-1}$.

The latest reported experimental value for $\Delta H_f\,(C_4H_4^+)$, as measured by photoionization of benzene [30] being $1208.9\ \text{kJ mol}^{-1}$, the methylene cyclopropenium ion appears as a good candidate for the structure of this $C_4H_4^+$ ion, as it is formed from the benzene molecule.

3.2.5 The Metastable Peak Observation

In Part 1 the importance of time evolution in the appearance of fragmentary ions was stressed. Indirect dissociation appears as a consequence of electronic, vibrational, or rotational predissociation, the lifetime of the predissociated state matching the time-of-flight of the ions through the mass spectrometer. This means, in the most frequent case, lifetimes of 10^{-5} to 10^{-6} s. As we reported in Section 1.1.4. Hipple was the first to give a correct interpretation of this phenomenon and showed that in a magnetic sector mass spectrometer (as in double-focussing instruments under normal operation) the occurrence in flight of processes such as

$$m_i^+ \rightarrow m_f^+ + (m_i - m_f) \qquad\qquad\qquad 3.8$$

is characterized by a diffuse "metastable peak" observed for a mass number given by

$$m^* = \frac{m_f^2}{m_i}. \qquad\qquad\qquad 3.9$$

It was simultaneously shown that the shape of the metastable transition was directly related to the kinetic energy released during the dissociation process.

So many efforts and techniques have been developed to use this property that they will not be described in detail; an excellent review of them was published in 1973 [31], followed by another review in 1980 [32].

For a general understanding of this introduction, we will give shortened information. The time scale for evolution of metastable ions being known, three facts are available from observing them:

their appearance energy: as shown earlier, the AE values can be equal to the AE of the metastable fragment or lower, due to the occurrence of a "kinetic shift". In some cases, the observation for the AE of the metastable of a value higher than the one of the AE for the fragment ion implies a complex behavior of the chemical dynamics of the reaction, as in the case of $H_2CO^+ \rightarrow HCO^+ + H$, which shows an m^* appearing at 14 eV, while the TE of the process in 12 eV [33–35];

their shapes: the shape of a metastable transition is generally close to Gaussian, or slightly distorted (Gaussian shape, or complex Gaussian); In some cases the shape realizes what is customarily called a "dished, flat-top metastable". In some cases, composite peaks are observed;

the kinetic energy release: often characterized by the fwhm of the peak. This represents the lowest degree of characterization of a metastable peak.

A better method is the determination of the real kinetic energy distribution by deconvoluting the metastable peak for the apparatus function [36, 37].

One of the most useful ways to record metastable transition with a good energy resolution is the use of a double-focusing mass spectrometer in which the electric sector precedes the magnetic analyzer in the so-called defocusing mode [38]. In these experiments, fragmentations taking place in the first field free region (between ion source and electric sector) are examined. The electric sector voltage E_1 and the ion accelerating voltage V_1 are both reduced from their normal beam transmitting values in the ratio m_f/m_i. The daughter ion m_f is then selected by adjustment of the magnetic-field strength. Increasing the accelerating voltage from $V_1 \cdot \dfrac{m_f}{m_i}$ to V_1 permits observation of all precursor ions from m_f to m_i.

The last m_f derived directly from m_i will be transmitted at V_1.

An example of such a scan is shown in Fig. 3.5, when the CDO^+ ions, formed in CD_3OH, appear from different precursors: $CDHO^+$, CD_2O^+, and CD_2OH^+. This figure also gives a good idea of different kinds of metastable shapes: the first one is Gaussian; the second is a composite one; the third is a "dished, flat-top peak". It is easy to show that from intensity distributions in function of V_1, the associated total kinetic energy release E_T is given by [31]:

$$E_T = \frac{m_f}{4(m_i - m_f)} \frac{(V_1^* - V^*)^2}{V^*}. \qquad 3.10$$

In this relation, V_1^* is the accelerating voltage for which a metastable intensity appears, V^* being the theoretical accelerating voltage for which the metastable peak is observed. From this relation it is easy to conclude that for $V_1^* = V, E_T$ is equal to zero. It is generally accepted that (except for flat-top metastables) the metastable ions have their maximum intensity for $E_T = 0$. From this relation, the following values have been deduced for the processes of Fig. 3.5 (Table 3.6) [39].

It is remarkable that, coming form apparently the same ionic structure, the metastable process 1) is so different from the metastable process 2). As the metastables are related to reactive configuration on the microsecond scale, this observation leads

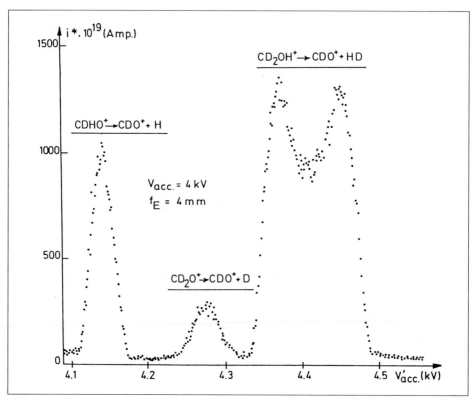

Fig. 3.5. The IKE metastable shape spectrum of the precursors of CDO^+ ions from CD_3OH: the first peak shows a quasi-Maxwellian distribution, the second is a composite peak, and the third a dish-topped peak.

Table 3.6. Kinetic energy release in metastable processes yielding CDO^+

Metastable process	Observed kinetic energy (eV)		
	100 % Peak height	50 % Peak height	Near 0 % peak height
1) $CDHO^+ \rightarrow CDO^+ + H$	0.044 ± 0.005	0.34 ± 0.013	1.173
2) $CD_2O^+ \rightarrow CDO^+ + HD$	—	—	0.854
3) $CD_2OH^+ \rightarrow CDO^+ + HD$	1.288 ± 0.015	—	2.540

to the conclusion that, probably, the precursor ions have different reacting structures. It has been shown [39] that, if D_2CO^+ is the structure involved in 2), the one involved in 1) is $DCOH^+$ (or $HCOD^+$).

Such a method of characterization of ionic configurations realized on the microsecond scale was pioneered in 1973 [31] and immediately further developed [40]. We will just report here an application of the method in assigning reactive configurations to $C_2H_4O^+$ ions on the basis of the shape of metastable transitions corresponding to

$$C_2H_4O^+ \rightarrow C_2H_3O^+ + H.$$

Possible structures for $C_2H_4O^+$ ions have been given in Section 3.2.2 (Table 3.3) under the 11 isomers studied by ab initio calculations. In Fig. 3.6, the metastable peak shapes for the loss of a hydrogen atom from $C_2H_4O^+$ ions formed in a) ethylene oxide, b) acetaldehyde, c) ethylvinylether, and d) dimethylether are given [40]. The argument is now such that if a $C_2H_4O^+$ ion generated from a more complex molecule looses an H atom, giving rise to a metastable transition similar in shape to a), b), c), d), one can reasonably expect that the reactive configuration of this ions has one of the following structures:

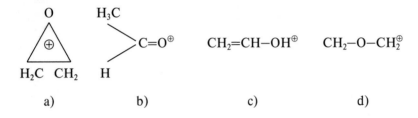

a)   b)   c)   d)

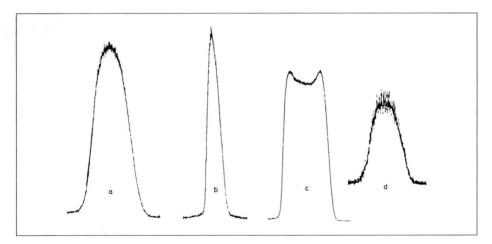

Fig. 3.6. Shape of metastable peaks observed for the decay $C_2H_4O^+ \rightarrow C_2H_3O^+ + H$, the precursor molecule being a) ethylene oxide, b) acetaldehyde, c) ethylvinylether, and d) dimethylether. The E_T values (translational energy measured at half height) are: a) 430 meV; b) 43 meV; c) 320 meV; and d) 580 meV [40].

For instance, when the glycidol is ionized, two structures can be expected from simple cleavage considerations of the parent molecule:

$$\rightarrow CH_2=CH-OH^\oplus + CH_2O$$

The observed metastable shape for the H-atom loss from $C_2H_4O^+$ ions formed in glycidol shows that only the vinylalcohol ions appear. Similarly, cyclobutanol and cyclopentanol also give rise to the appearance of $CH_2=CH-OH^\oplus$, as it is expected from

In numerous cases, one precursor molecular ion can simultaneously give rise to different configurations; in this case a composite metastable is observed which can often give an idea of the relative abundances of the different $C_2H_4O^+$ precursors.

3.2.6 The Collision-Induced Dissociation of Ions

Before giving information about the collision-induced dissociation of ions, it appears useful to introduce for the study of both non-collision-induced metastable shapes and collision-induced fragment ions peak shapes, a modified double-focussing mass spectrometer called a "reversed geometry mass spectrometer" [41, 42]. Such an instrument is schematically shown in Fig. 3.7. From this figure, it is easy to understand that positive ions formed in an ion source, accelerated to a rather high energy (2 to 8 keV), mass analyzed in the magnetic sector B, are allowed to cross a collision region, filled or not by a target gas. The ionized product of any dissociation, either monomolecular or collision-induced, taking place in the second field free region is identified and the kinetic energy analyzed by an electric sector scan. This method is known as MIKE (Mass analyzed Ion Kinetic Energy) spectroscopy. This appears as a nice complement to the IKE method, described in Section 3.2.4. The IKE scan gives a scan of all metastable precursors of an ion; the MIKE provides a scan of all fragments from the same precursor.

An example of IKE was given in Fig. 3.5 where three metastable precursors of CDO^+ ions are given as they appear from CD_3OH; an example of MIKE is given in Fig. 3.8 in the case of the pyridine molecular ion decaying monomolecularly to $C_4H_4^+$ and collisionally to two more fragments: $C_4H_3^+$ and $C_4H_2^+$. It is not possible to dicuss here all aspects of the collision-induced dissociation of ions. They have been abundantly reviewed since the beginning of such experiments [43-45]. The collisional activation spectrum quite simply represents the mass spectrum of an ion which has sufficient energy to decay in the microsecond time frame (the time required to fly from the ion source to the second field free region of the reversed geometry instument).

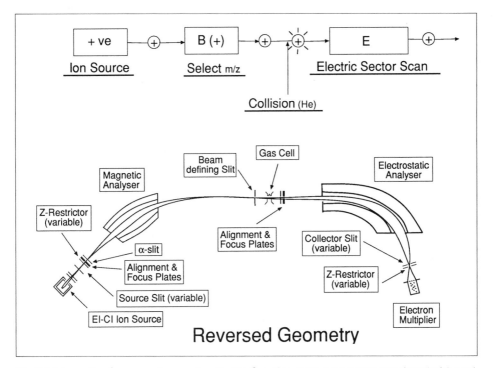

Fig. 3.7. Schematic of a reversed geometry, double-focusing mass spectrometer equipped with a collision cell in the second field free region [52].

One can, consequently, expect that such a mass spectrum provides structural information on the ion, just as a conventional mass spectrum provides the fragmentation pattern which allows structural conclusions about molecules introduced in the ion source. Through its violent interaction with the target gas, it is expected that the collisionally activated ion suffers only vertical transitions to higher electronic states, the Massey criterion (see Section 1.5.1) being fulfilled; the decay modes of these electronic states are entirely controlled by the structural identity of the colliding ion, the hypothesis being that its main energy content is unimportant.

As an example of the structure determinations obtained by this technique, we quote the case of $C_2H_5O^+$ ions for which at least three structures a), b), c) are possible:

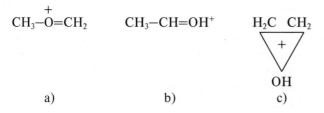

$$\overset{+}{CH_3-O}=CH_2 \qquad CH_3-CH=OH^+ \qquad H_2C \quad CH_2$$

a)   b)   c)

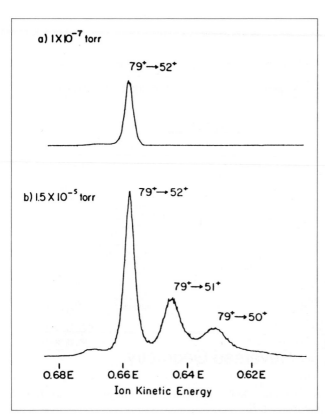

Fig. 3.8. MIKE spectrum of pyridine molecular ion: at low pressure in the collision cell the non-collision-induced metastable transition $79^+ \rightarrow 52^+ + 27$ is the only one observed; at high pressure, additional collision-induced dissociations appear [43].

If the idea is accepted that $C_2H_5O^+$ ions from CH_3-O-CH_2Y precursors have an a) structure, the $C_2H_5O^+$ ions from $CH_3-CHY-OH$ precursors have a b) structure, and those from protonated ethylene oxide a c) structure, one can consider as a footprint of each structure the collision-induced dissociation spectra of each ion. In case a) the CID spectrum is characterized by abundant CH_3^+ and HCO^+ ions; in case b) HCO^+ ions dominate the spectrum, with c) giving rise to abundant CH_2OH^+ ions. Concerning $C_2H_5O^+$ ions of other origin, they are generally characterized by one of the a), b) or c) spectrum, or by a mixture of these which allows speculations about possible isomerizations in the microsecond time frame.

Since the CID mass spectrum is simply the mass spectrum of an ion, it can be used in deducing ion structures without the necessary prior assignment of a given mass spectrum to a given structural type.

As an example, let us mention the fact that $C_2H_3^+$ ions are formed much more readily from $C_2H_5O^+$ ions issued from $CH_3-CHY-OH$ precursors than from those arising from CH_3-O-CH_2Y precursors, thereby confirming structure a) for these $C_2H_5O^+$ ions. If, following the same reasoning, we consider the ions formed by loss of CH_3 from 2-hexanone, they may have structure d) or e), depending on which methyl group is lost:

$$CH_3-CH_2-CH_2-CH_2-CO^+$$

d)

$$\begin{array}{c} \overset{+}{} \\ H_2C \underset{|}{\quad\quad} O \\ \underset{|}{} \quad\quad\quad \diagdown \\ \quad\quad\quad\quad\quad\quad C-CH_3 \\ \underset{}{} \quad\quad\quad \diagup \\ H_2C \underset{}{\quad\quad} CH_2 \end{array}$$

e)

By using deuterium labelling to distinguish these ions, it was found through CID that the one ion fragments by losing CO and $CH_2=CO$ (hence assigned as structure d)), whereas the other fragments by loss of C_2H_4, C_3H_6, and CH_3, consistent with structure e) [45, 46].

As a last example of CID study, we cite [47], which used this method to show that molecular ions of alkanes and alkenes do not isomerize under typical mass spectrometric conditions, whereas their even electron fragment ions do.

3.2.7 The Ion Cyclotron Resonance Method

Ion cyclotron resonance spectrometry (ICR) offers a possibility to identify ions having insufficient energy to fragment in the millisecond or higher time frame, by examining their reactions with various neutral molecules (ion-molecule reactions).

If one remembers that a particle of charge e, mass m and velocity $\vec{v}$ moving in a magnetic field $\vec{H}$ is submitted to a force $\vec{F}$ [48]:

$$\vec{F} = e\left[\vec{v} \times \vec{H}\right] \tag{3.11}$$

and also that this force is always normal to the instantaneous velocity, the particle moves circularly in a plane normal to $\vec{H}$, such that

$$\frac{mv^2}{R} = evH,$$ 3.12

R being the radius of the orbit and v the velocity component normal to $\vec{H}$. From this equation it follows that the angular frequency ω_c (called the cyclotron frequency of the orbital motion is

$$\omega_c = \frac{v}{R} = \frac{eH}{m}.$$ 3.13

The cyclotron frequency is independent of the velocity of the particle, whereas the radius of the orbit is directly proportional to v.

A beam of ions of a given mass will have a distribution of velocities, but a sharp cyclotron frequency ω_c, which depends only on m, e, and H. If now an alternating electric field $E(t)$ is applied perpendiculary to H at a frequency ω, when $\omega = \omega_c$, the ions absorb energy from the alternating field and are accelerated to larger velocities and orbital radii. The absorption of energy from $E(t)$ at $\omega = \omega_c$ by ions is thus detectable if the source of $E(t)$ is an oscillator extremely sensitive to changes in load. A recording of the energy level of the oscillator as a function of H for a fixed oscillator frequency displays the cyclotron resonance spectrum for a mixture of ions of various masses on a linear mass scale. When sample pressures in the range of 10^{-5} mbar are used, the coherent cyclotron motion is occasionally interrupted by collisions between ions and neutral molecules and ion-molecule reaction products can arise. It is well known that ion-molecule reaction rate constants are usually dependent on the relative ion neutral velocities. If the velocity of a reactant ion A^+ is increased by application of a strong radiofrequency electric field $E(t)$, oscillating at a frequency $\omega_a = eH/m_a$, then a substantial change can occur in the concentration of an ion C^+, if A^+ and C^+ are coupled by a reaction such as

$$A^+ + B \rightarrow C^+ + D.$$

The consequent changes in cyclotron resonance intensity and line shape can be observed with a weak radiofrequency field $E(t)$ oscillating at ω_c, ω_c being equal to eH/m_c.

It is also possible to completely eliminate a reactant ion of a specific mass from the cell, using a variety of irradiation and modular options. The effect of the removal of a specific reactant on the product distribution allows the reaction pathways to be identified.

As an example of application of this method to ion structure determination, the case of $C_2H_5O^+$ ions formed by electron impact on dimethylcarbonate will be detailed [49].

The main ions observed in the electron impact mass spectrum of dimethylcarbonate ($m/e = 59$ and 45) are due to the following reactions:

$$\begin{array}{c} \text{CH}_2=\text{O}^+ \\ \hspace{2cm} \diagdown \\ \text{CH}_3-\text{O} \hspace{0.3cm} \diagup \end{array} \text{C}=\text{O} \xrightarrow[m/e=89]{-\text{CO}_2} \text{CH}_3-\overset{+}{\text{O}}=\text{CH}_2 \; (m/e = 45)$$

$$\begin{array}{c} \text{CH}_3-\text{O} \\ \hspace{1cm}\diagdown \\ \text{CH}_3-\text{O}\diagup \end{array}\text{C}=\text{O}^+$$

$$m/e = 90$$

$$-\text{CH}_3\text{O}$$

$$\text{CH}_3-\text{O}-\overset{+}{\text{C}}=\text{O} \quad (m/e = 59).$$

In order to determine if $\text{CH}_3-\overset{+}{\text{O}}=\text{CH}_2$ is a correct structure for $m/e = 45$, both 59 and 45 ions are allowed to react with the parent compound.

In these conditions, ions are observed at $m/e = 91$ and 105.

The probable formulae for these ions are: $(\text{CH}_3\text{O})_2\text{C}=\text{OH}^+$ for the protonated parent compound, and $(\text{CH}_3\text{O})_3\text{C}^+$ for the trimethoxycarbonium ion.

The double-resonance spectrum of $m/e = 105$ product exhibits contributions from $m/e = 45$ and 59, indicating the occurrence of reactions:

$$\text{CH}_3-\overset{+}{\text{O}}=\text{CH}_2 + \begin{array}{c}\text{CH}_3\text{O}\\ \hspace{0.3cm}\diagdown \\ \text{CH}_3\text{O}\diagup \end{array}\text{C}=\text{O} \rightarrow (\text{CH}_3\text{O})_3\text{C}^+ + \text{CH}_2\text{O}$$

$$\text{CH}_3-\overset{+}{\text{O}}-\text{C}=\text{H} + \begin{array}{c}\text{CH}_3\text{O}\\ \hspace{0.3cm}\diagdown \\ \text{CH}_3\text{O}\diagup \end{array}\text{C}=\text{O} \rightarrow (\text{CH}_3\text{O})_3\text{C}^+ + \text{CO}_2,$$

and confirming the structures proposed for 45 and 59 ions.

3.2.8 The Collision-Induced Dissociative Ionization of Neutrals (CIDI)

In some cases, when a dissociation equation is written, some doubt remains about the identity of the neutral molecule or radical appearing in the process. This uncertainty could be removed by threshold measurements and thermodynamic calculations. If some possibility could be found to study the mass spectrum of these neutral particles, more information would be available for the unequivocal disentangling of the ionic dissociation process. With the help of the reversed geometry instrument, this is quite easily possible, as will now be explained.

When a mass selected ion M_1 passes through the second field free region of a reversed geometry mass spectrometer, it can fragment unimolecularly; scanning the electrostatic sector downwards from the value transmitting M_1 produces the well-known "metastable ion mass spectrum" corresponding to

$$M_1^+ \rightarrow M_2^+ + M .$$

If the ion source accelerating voltage is V_A, then the translational energy of M_2 is $V_A\, m_2/m_1$ and that of the neutral fragment is $V_A\, m/m_1$.

In this equation, m_2, m_1, and m are the masses of the ions and neutral involved in the dissociation process.

Both M_2 and M will cross the collision chamber and will be subject to collision-induced dissociations with the target gas introduced in the collision cell. If a positive potential V_C is applied to the collision cell, such that $V_C > V_A$, only the neutral particles can enter the cell and undergo collision. As their energy is in the range of keV, the collision will result in their dissociative ionization.

This CIDI reaction being:

$$M \rightarrow M^+ \rightarrow P^+ + Q ,$$

the translational energy of P^+, when it reaches the exit of the collision cell, is given by

$$E_T = \frac{m_p}{m} \cdot \frac{m}{m_1} V_A + V_C = \frac{m_p}{m_1} V_A + V_C . \qquad 3.14$$

This means that all fragment ions from the ionized neutral M have translational energies in excess of V_C, being simultaneously independent of the mass of the neutral M.

As examples of the help expected from this method two metastable processes will be treated. In the first one the dissociations of pyridine and aniline molecular ions metastably losing a neutral fragment of mass 27 will be considered.

This mass corresponds either to HCN or to HNC. It was generally accepted that the dissociation process in pyridine was

$$C_6H_5N^+ \rightarrow C_4H_4^+ + HCN ,$$

when some authors [50, 51] discovered that the appearance energy of $C_5H_6^+$ from aniline was 0.8 eV too high with respect to the calculated threshold energy of the dissociation process:

They proposed that in aniline ion dissociation CHN was formed instead of HCN, the ΔH_f (CNH) being some (0.6 ± 0.1) eV higher than the ΔH_f (HCN). This hypothesis has been tested [52] by CIDI of both the neutrals from pyridine and aniline. The CIDI mass spectra are shown in Fig. 3.9. The spectrum of $m = 27$ neutral from pyridine is entirely similar to the one obtained by CIDI of HCN [53], showing, as expected, C^+ and CH^+ as prominent fragments. The spectrum of $m = 27$ neutral from aniline is strongly different from the first one, showing NH^+ and C^+ ion only, as expected, for the mass spectrum of hydrogen isocyanide. Both these results corroborate the thermochemical argument.

In the second example the structure of C_2H_3O radical will be considered, as it results from the dissociation reaction:

$$
\begin{array}{c}
O \\
\parallel \\
H_3C-C-O-CH=CH_2^+
\end{array}
\rightarrow H_3C-CO^+ +
\left\{
\begin{array}{ll}
H_2\overset{\circ}{C}-CH=O & \text{b)} \\
CH_3-\overset{\circ}{C}=O & \text{a)}
\end{array}
\right.
$$

In order to solve this problem the CIDI mass spectrum of $CH_3-C=O$ radical produced in the dissociation:

$$H_3C-CO-CO-CH_3^+ \rightarrow H_3C-CO^+ + H_3C-\overset{\circ}{C}=O$$

was produced and compared to the CIDI mass spectrum of a) [54] (Fig. 3.10).

The observed mass spectrum is the mass spectrum b) which shows a C_2H_3O radical giving rise to an unstable CH_2-CHO^+ ion, and ruling out the process a).

The two examples show how useful the CIDI of neutrals can be (recent extensive reviews on this method are available [55, 56]).

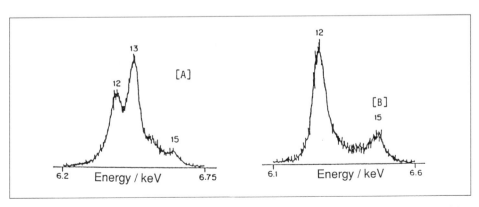

Fig. 3.9. CIDI mass spectrum of HCN issued form pyridine [A] and of CNH from ionized aniline [B], [52].

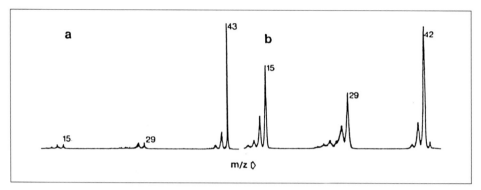

Fig. 3.10. The CIDI mass spectrum of CH_3CO radical issued from the decay of $CH_3COCOCH_3$; the spectrum is practically reduced to the very stable and dominant molecular ion CH_3CO^+; the CIDI mass spectrum b) originating from the decay of vinylacetate shows many fragment ions from the relatively unstable CH_2-CHO^+ ion [56].

3.2.9 Photoabsorption and Photodissociation of Ions

Trapping molecular ions in an ICR cell is a way to observe their absorption spectra or their photodissociation. An excellent review of the possibilities of such a method has been given [57]. The absorption spectrum of molecules represents an excellent tool for their structure determinations, as do also the absorption spectra of molecular ions.

An early and particularly nice example of ion structure determination will be selected here. It has been shown that different molecules of C_7H_8 formula give very similar mass spectra when submitted to electron impact. This suggests that the dissociating structure giving rise to these mass spectra is the same, regardless of the original structure of C_7H_8 compound. It was, on the contrary, shown that the photodissociation spectra of toluene, cycloheptatriene, and norbornadiene cations were all different, showing that in the time frame of their photodissociation the $C_7H_8^+$ retains their indentity: the spectrum of the parent ion of the toluene shows that the neutral structure is definitely retained. The ion from cycloheptatriene give a single absorption band near 500 nm in excellent agreement with the $\pi \to \pi$ transition expected on the basis of the PES of the neutral species. The ion from norbornadiene gives a spectrum unlike that suggested by PES, and the possibility exists that a rearrangement of this ion occurs to an, up to now, unassigned structure.

It was shown more recently [58] that n-butylbenzene and 2-phenyl-ethanol can undergo the Mc Lafferty rearrangement process:

The photodissociation spectrum of the methylene-cyclohexadiene cation shows no conversion of this structure to either cycloheptatriene or toluene ions, a last non interconvertible structure of the $C_7H_8^+$ cation.

It is, however, now the moment to point out that there is no real contradiction in the fact that the mass spectra of all these four ions are similar, because mass spectra are due to a selection of those ions which have a sufficient internal energy to fragment in the microsecond range. The photodissociation experiment, however, selects only those ions which are unable to dissociate spontaneously and shows that these ions are not sturcturally labile, until their internal energy was conveniently increased.

References

1. Herzberg G (1950) Molecular Spectra and Molecular Structure Spectra of Diatomic Molecules. Van Nostrand Company, New York
2. Momigny J (1980) Electronic, Rotational and Vibrational Predissociations in the Decay of Electronically Excited Molecular Ions. J de Chim Physique et de Physico-Chimie Biologique 77:725-738
3. Chupka WA, Berkowitz J (1971) Kinetic Energies of Ions Produced by Photoionization of HF and F_2. J Chem Phys 54:5126-5132
4. Schopman J, Locht R (1974) The Observation of Predissociation in the Oxygen Molecular Ion by Low-Energy Electron Impact, Chem Phys Lett 26:596-600
   Locht R, Schopman J (1974) The Dissociative Ionization in Oxygen. Int J Mass Spectrom Ion Proc 15:361-378
5. Locht R, Momigny J, Rühl E, Baumgärtel H (1987) A Mass Spectrometric Photoionization Study of CH_3F. The CH_2^+, CH_3^+ and CH_2F^+ Ion Formation. Chem Phys 117:305-313
6. Locht R, Schopman J, Wankenne H, Momigny J (1975) The Dissociative Ionization of Nitrogen. Chem Phys 7:393-404
7. Locht R, Dürer JM (1975) New Predissociations of CO and CO^+ Induced by Low Energy Electron Impact. Chem Phys Lett 34:508-512
   Locht R (1977) The Dissociative Ionization of Carbon Monoxide. Chem Phys 22:13-27
8. Locht R, Momigny J (1986) The Dissociative Ionization of Methyl Fluoride. The Formation of CH_2^+ and CH_3^+. Int J Mass Spectrom Ion Proc 71:141-157
9. Lifshitz C, Mackenzie Peers A, Weiss M, Weiss MJ (1974) A Direct Measurement of the Kinetic Shift in Benzene. Adv Mass Spectrom 6:871-875
10. Stockbauer R, Rosenstock HM (1978) Kinetic Shift in Methane and Allene Ion Fragmentation. Int J Mass Spectrom Ion Proc 27:185-195

11. Dannacher J, Rosenstock HM, Bull R, Parr AC, Stockbauer RL, Bombach R, Stadelmann JP (1983) Benchmark Measurement of Iodobenzene Ion Fragmentation Rates. Chem Phys 75:23-35
12. Hickling RD, Jennings KR (1970) Kinetics Shifts and Metastable Transitions. Org Mass Spectrom 3:1499-1503
13. Rabalais JW, Karlsson L, Werme LO, Bergmark T, Siegbahn K (1973) Analysis of Vibrational Structure and Jahn-Teller Effects in the Electron Spectrum of Ammonia. J Chem Phys 58:3370-3380
14. Hehre WJ, Pople JA (1975) Molecular Orbital Theory of the Electronic Structure of Organic Compounds. XXVI. Geometries, Energies, and Polarities of C_4 Hydrocarbons. J Am Chem Soc 97:6941-6955
15. Schaefer HF (1977) Modern Theoretical Chemistry. Methods of Electronic Structure Theory and Applications of Electronic Theory, vols 3 and 4. Plenum Press, New York
16. Segal GA (1977) Modern Theoretical Chemistry. Semiempirical Methods of Electronic Structure Calculating, vols 7 and 8. Plenum Press, New York
17. Engler EM, Andose JD, Schleyer PR (1973) Critical Evaluation of Molecular Mechanics. J Am Chem Soc 95:8005-8025
18. Allinger NL (1976) Calculation of Molecular Structure and Energy by Force-Field Methods. Adv Phys Org Chem 13:2-82
19. Bouma WJ, Macleod JK, Radom L (1979) An Ab Initio Molecular Orbital Study of the Structures and Stabilities of the $C_2H_4O^+$ Isomers. J Am Chem Soc 101:5540-5545
20. Rosenstock HM, Draxl K, Steiner BW, Herron JT (1977) Energetics of Gaseous Ions, J Phys Chem Ref Data, Vol 6 Suppl 1
21. Krässig R, Reinke D, Baumgärtel H (1974) Photoreaktionen Kleiner Organischer Moleküle. II. Die Photoionenspektren der Isomeren Propylen-Cyclopropan and Acetaldehyd-Äthylenoxyd. Ber Bunsenges Phys Chem 78:425-436
22. Davis JH, Goddard III WA, Harding LB (1977) Theoretical Studies of the Low Lying States of Vinylidene. J Am Chem Soc 99:2919-2930
23. Dannacher J, Schmelzer A, Stadelmann JP, Vogt J (1979) A Photoelectron-Photoion Coincidence Study of Vinylfluoride. Int J Mass Spectrom Ion Proc 31:175-186
24. Williamson AD, Beauchamp JL (1976) Ion Molecule Reactions in Vinylfluoride by Photoionization. Effects of Vibrational Excitation on Major Reaction Pathways. J Chem Phys 65:3196-3202
25. Cooks RG, Kim KC, Beynon JH (1974) Hydrogen Fluoride Elimination from Fluoroalkanes and Fluoroalkenes: Energy Partioning and Thermochemistry. Int J Mass Spectrom Ion Proc 15:245-254
26. Holmes JL, Terlouw JK, Lossing FP (1976) The Thermochemistry of $C_2H_4O^+$ Ions. J Phys Chem 80:2860-2862
27. Franklin JL (1953) Calculation of Heats of Formation of Gaseous Free Radicals and Ions. J Chem Phys 21:2029-2033
28. Rosenstock HM, Dannacher J, Liebman JF (1982) The Role of Excited Electronic States in Ion Fragmentation: $C_6H_6^+$. Radiat Phys Chem 20:7-28
29. Greenberg A, Liebman CF (1978) Strained Organic Molecules. Academic Press, New York
30. Rosenstock HM, Stockbauer R, Parr AC (1981) Unimolecular Kinetics of Pyridine Ion Fragmentation. Int J Mass Spectrom Ion Proc 38:323-331
31. Cooks RG, Beynon JH, Caprioli RM, Lester GR (1973) Metastable Ions. Elsevier, Amsterdam
32. Holmes JL, Terlouw JK (1980) The Scope of Metastable Peak Observations. Org Mass Spectrom 15:383-396
33. Bombach R, Dannacher J, Stadelmann JP, Vogt J (1981) Fragmentation of Formaldehyde Molecular Cations. Int J Mass Spectrom Ion Proc 40:275-285
34. Bombach R, Dannacher J, Stadelmann JP, Vogt J (1981) The Fragmentation of Formaldehyde Molecular Cations: The Lifetime of CD_2O^+ ($\tilde{A}^2B_1$) Chem Phys Lett 77:399-402
35. Vaz Pires M, Galloy C, Lorquet JC (1978) Unimolecular Decay Paths of Electronically Excited Species. I. The H_2CO^+ Ion. J Chem Phys 69:3242-3249

36. Holmes JL, Osborne AD (1977) Metastable Ion Studies (VIII): An Analytical Method for Deriving Kinetic Energy Release Distributions from Metastable Peaks. Int J Mass Spectrom Ion Phys 23:189-200
37. Mändli H, Robbiani R, Küster Th, Seibl J (1979) Automatic Aquisition and Shape Analysis of Metastable Peaks. Int J Mass Spectrom Ion Phys 31:57-64
38. Beynon JH, Saunders RA, Williams AE (1964) Metastable Ions in Double-Focusing Mass Spectrometers. Nature 204:67-68
39. Momigny J, Wankenne H, Krier C (1980) Correlation Diagram Approach to the Dissociative Ionization Mechanisms of Methanol. Int J Mass Spectrom Ion Proc 35:151-170
40. Holmes JL, Terlouw JK (1975) Metastable Ion Studies. V. The Identification of $C_2H_4O^+$. Ion Structures from their Characteristic Kinetic Energy Releases. Can J Chem 53:2076-2083
41. Wachs T, Bente III PF, McLafferty FW (1972) Simple Modification of a Commercial Mass Spectrometer for Metastable Data Collection. Int J Mass Spectrom Ion Proc 9:333-341
42. Beynon JH, Cooks RG, Amy JW, Baitinger WE, Ridley TY (1973) Design and Performance of a Mass-Analyzed Ion Kinetic Energy (MIKE) Spectrometer. Anal Chem 45:1023A
43. Cooks RG (ed) (1978) Collision Spectroscopy. Plenum Press, New York
44. Holmes JL (1985) Assigning Structures to Ions in the Gas Phase. Org Mass Spectrom 20:169-183
45. McLafferty FW, Kornfeld R, Haddon NF, Levsen K, Sakai I, Bente III PF, Tsai SC, Schuddemage HDR (1973) Application of Collisional Activation Spectra to the Elucidation of Organic Ion Structures. J Am Chem Soc 95:3886-3892
46. Van de Graaf B, Dymerski PP, McLafferty FW (1975) Detection of Stable Cyclic $C_2H_5O^+$ and $C_2H_5S^+$ Ions by Collisional Activation Spectroscopy. J Chem Soc Chem Comm 978-979
47. Levsen K (1975) Isomerisation of Hydrocarbon Ions: I. Isomeric Octanes: A Collisional Activation Study. Org Mass Spectrom 10:43-53
    Levsen K (1975) Isomerisation of Hydrocarbon Ions: II. Octenes and Isomeric Cycloalkanes: A Collisional Activation Study. Org Mass Spectrom 10:55-63
48. Baldeschwieler JD, Sample Woodgate S (1971) Ion Cyclotron Resonance Spectroscopy. Acc Chem Res 4:114-120
49. Beauchamp JL, Dunbar RC (1970) Identification of $C_2H_5O^+$ Structural Isomers by Ion Cyclotron Resonance Spectroscopy. J Am Chem Soc 92:1477-1485
50. Baer T, Carney TE (1982) The Dissociation of State Selected Metastable Aniline Ions by Single and Multiphoton Ionization. J Chem Phys 76:1304-1308
51. Lifshitz C, Gotchiguian P, Roller R (1983) Time Dependent Mass Spectra and Breakdown Graphs. The Kinetic Shift in Aniline. Chem Phys Lett 95:106-108
52. Mommers AA (1985) Collision Spectrometry of Fast Gaseous Organic Ions and Neutrals. Thesis, University of Utrecht
53. McLafferty FW, McGilvery DC (1980) Gaseous HCN^+, HNC^+ and $HCNH^+$ Ions. J Am Chem Soc 102:6189-6190
54. Holmes JL, Lossing FP (1984) Heats of Formation of Organic Radicals from Appearance Energies. Int J Mass Spectrom Ion Proc 58:113-120
55. Holmes JL (1989) Ion Structures, Neutral Structures; Explorations in Mass Spectrometry. Adv in Mass Spectrom 11A:53-79
56. Terlouw JK (1989) Analysis of Neutral(ized) Species by Collision Induced Dissociative Ionization (CIDI) and Neutralization Reionization Mass Spectrometry (NRMS). Adv in Mass Spectrom 11B:984-1010
57. Dunbar RC (1979) Ion Photodissociation. In: Bowers MT (ed) Gas Phase Ion Chemistry, vol 2. Academic Press, New York
58. Dunbar RC, Klein R (1977) Spectroscopy of Radical Cations. The McLafferty Rearrangement Product in Fragmentation of n-Butylbenzene and 2-Phenylethanol Ions. J Am Chem Soc 99:3744-3746

4 Statistical Theories for Unimolecular Rate Constant Calculation in Isolated Systems

Introduction

The ways by which the electronic energy accumulated in molecular ions by the absorption of some 10 eV of energy excess with respect to the lowest molecular ionization energy is rapidly redistributed in the ground state molecular ion have been described in Chapter 2.

If the dissociation channels of the molecular ion ground state are now considered, it is clear that an additional hypothesis needs to be proposed in order to solve the problem of the a priori calculation of the rate constant for these dissociations. This problem is entirely analoguous to the problem of adiabatic unimolecular dissociation of thermally or chemically activated molecules. The hypothesis made for the decay of molecular ions will therefore be entirely similar to the one made for molecules: we shall accept that the internal energy of molecular ions is completely and statistically randomized within the ground state hypersurface of the molecular ions, in a time shorter than the fastest dissociation processes. This hypothesis allows the a priori calculation of the rate constants for the successive dissociation processes of the molecular ion as the evaluation of their variation with energy excess E above the lowest energy needed for their appearance threshold energy E_0.

4.1 The First Approach of Unimolecular Reaction Rate Theories

4.1.1 Basic Assumptions of RRKM and QET Methods

The probable occurrence of a rapid redistribution of electronic energy in internal energy of the ground state molecular ion has been demonstrated in Chapter 2. The building of theoretical formulas for the rate constant calculation for dissociation processes of this ground state molecular ion was set up simultaneously by Marcus [1] and by Rosenstock et al. [2] around 1952.

Both these theoretical approaches, called, respectively, RRKM (after its four authors, Rice, Ramsperger, Kassel, Marcus) and QET (quasi–equilibrium theory) require the following basic assumptions.

The assumptions of transition state theory can be expressed as:
- validity of the Born-Oppenheimer approximation;
- description of the nuclear motion by classical mechanics;

- occurrence in the system configuration space of a surface, completely separating reactants and products, with the property that all systems crossing this surface in the direction rectants-to-products, cross once and only once, and will terminate as products. This "configuration of no return" is often called the "transition state". Once the system has reached this configuration it is assumed to proceed further and to form the products. That path, followed on the potential energy hypersurface along which the reactants reach the critical configuration and, therefore, proceed to products, is called the "reaction path" and is defined by the successive values of the "reaction coordinate".

The assumption of the quasi-equilibrium expressed as:

- subject to known conservation laws the density in phase space of energized molecules, prior to their reaction, is uniform and independent of the method of initial preparation.

As a consequence of these assumptions, the rate constant for each dissociation process of a molecular ion can be considered as depending only on its total energy E, each dissociation process being characterized by a minimum value E_0 of the energy, for which the rate constant has its minimum non-zero value

$$k(E) = f(E - E_0) \text{ with } k(E_0) \neq 0.$$

There are many ways to deduce statistical expressions of the rate constant $k(E)$ for isolated systems [3, 4] and they will not be detailed here. One of them will be choosen in order to establish the general RRKM formula expressing the statistical rate constant $k(E)$, mainly because it is pedagogically excellent [5].

4.1.2 The RRKM Expression for $k(E)$

4.1.2.1 Dissociation Rate Constant of a Diatomic Molecule

The reasoning is based on the expression of the real potential energy curve of an AB molecule as a function of the interatomic distance R, as represented in Fig. 4.1. In

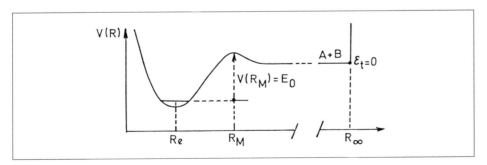

Fig. 4.1. Potential energy curve illustrating the dissociation rate constant of a diatomic molecule.

this model, $V(R)$ becomes infinite for $R = R_\infty$; This means that the definite separation of A and B is considered impossible in such a way that an equilibrium between AB and A + B can be taken into account.

The real value of R_∞ does not need to be quoted and it will consequently be eliminated in the calculations. From the model, it results that both atoms A and B are considered as captured within a one-dimensional potential box; their translational energies ε_t inside this well $(0 < R < R_\infty)$ are given by

$$\varepsilon_t = \frac{h^2}{8\mu \, R_\infty^2} \cdot n^2,$$ 4.1

μ being the reduced mass of AB, h the Planck constant, and n an integer.

From Fig. 4.1, it is clear that the occurrence of a translation of A and B requires that the molecule acquire at least an energy E_0 above its zero-point energy. As the masses A and B are important, the ε_t constitute a dense set of states, namely

$$d \, \varepsilon_t = \frac{h^2}{8\mu \, R_\infty^2} \, 2n \, dn \, .$$ 4.2

From 4.2 the density of translational states $N(\varepsilon_t)$ is given by

$$N(\varepsilon_t) = \frac{dn}{d \, \varepsilon_t} = \frac{4\mu \, R_\infty^2}{h^2} \cdot \frac{1}{n} \, .$$ 4.3

From Eq. 4.1 n is derived:

$$n = \frac{2R_\infty}{h} \, (2\mu \, \varepsilon_t)^{1/2} \, .$$

Therefore Eq. 4.3 can be written

$$N(\varepsilon_t) = \frac{R_\infty}{h} \left(\frac{2\mu}{\varepsilon_t} \right)^{1/2} \, .$$ 4.4

If v is the relative speed of A and B, it follows that

$$\varepsilon_t = \frac{1}{2} \, \mu v^2 \text{ or } v = \left(\frac{2\varepsilon_t}{\mu} \right)^{1/2} \, .$$ 4.5

The dissociating pair A and B shall increase their relative distance from R to R_∞ and back in a time equal to

$$\frac{2(R_\infty - R)}{v} \, ;$$

during this time, the pair A + B increases its interatomic distance only once; it follows that the inverse of this time is the rate constant $k(\varepsilon_t)$ for the dissociation of a molecule AB with an excitation energy $E = E_0 + \varepsilon_t$. This dissociation leaves A and B in a translational energy state ε_t and

$$k(\varepsilon_t) = \frac{v}{2\,R_\infty} = \frac{(2\,\varepsilon_t/\mu)^{1/2}}{2\,R_\infty}. \qquad 4.6$$

The uncertainty relation between time and energy implies that the energy of the decomposing molecule can only be specified within an energy interval δE; this means that the value of the rate-constant $k(\varepsilon_t)$ is also specified in an energy interval δE.

In this energy interval, the number of translational states is given by

$$\frac{dn}{d\,\varepsilon_t} \cdot \delta E\,. \qquad 4.7$$

For each of these states a rate constant $k(\varepsilon_t)$ exists. It follows, therefore, that the contribution to the rate over the range δE, is expressed by

$$k(E) = \frac{dn}{d\,\varepsilon_t} \cdot \delta(E) \cdot k(\varepsilon_t) = \frac{\delta E}{h}, \qquad 4.8$$

when the Eqs. 4.4 and 4.6 are introduced in Eq. 4.7. The Eq. 4.8 is characteristic of one exit channel within the translational quasi-continuum of the fragment A and B.

As A and B states are non degenerate, δE contains one state of AB only. Equation 4.8 expresses, therefore, the rate constant for an "elementary chemical act", schematically described by

Initial state (i) $\xrightarrow{k_{ij}}$ Final state (j),

for which

$$k_{ij} = \frac{\delta E}{h}\,. \qquad 4.9$$

Such a process, correlating a unique state of AB to a unique dissociation channel, is obviously characterized by a rate-constant independent of E, i, and j.

If AB is an oscillator of frequency v in the harmonic approximation, its quantum states are spaced by hv. It follows that one state i is present in the hv interval or, in other words, $\delta E = hv$. Therefore,

$$k(E) = v\,. \qquad 4.10$$

For a harmonic oscillator, the hypothetical rate constant for unimolecular dissociation would be independent of energy

4.1.2.2 Dissociation Rate Constants for a Polyatomic Molecule

If it is now supposed that AB is written for a polyatomic molecule, able to dissociate in fragments A and B, both A and B can exist with vibrational state distributions. They are also able to carry translational and rotational energies. If the discussion is limited to the occurrence of vibrational excitation, in first approximation, Fig. 4.1 again represents the situation. The difference is that, within the interval δE, $N(E)\,\delta E$ quantum states are present, $N(E)$ being the density of AB states at energy E. In this case, Eq. 4.8 becomes, when $E = E_0$,

$$k(E_0) = \frac{\delta E}{hN(E_0)\,\delta E} = \frac{1}{hN(E_0)}. \qquad 4.11$$

Equation 4.11 expresses the rate constant for the dissociation of a polyatomic molecule at its threshold energy, in a unique exit channel, the energy E_0 being insufficient to account for any vibrational exitation of A and B. If the energy E of AB is above E_0, one writes the rate-constant of dissociation of a molecule AB with an exitation energy E in a unique exit channel as

$$k^1(E) = \frac{1}{hN(E)}. \qquad 4.12$$

If the energy of a polyatomic molecule AB exceeds E_0 by a finite difference $E - E_0$, the dissociation products A and B will, in general, appear with some internal energy. The $E - E_0$ energy difference is shared between the available quantum states of A and B and the translational continuum. Therefore, one can consider that an exit channel is realized for each internal energy state of the products. Equation 4.12 gives the rate constant per channel. As a consequence, the rate constant relative to all exit channels will be obtained by multiplying $k^1(E)$ through the number of available channels. This number of channels is given by the number of ways through which the $E - E_0$ energy can be distributed between A and B quantum states and the translational continuum.

The use of this concept is, however, limited by the difficulty to know exactly the number and nature of the energy states of reaction products.

It is possible to remove this problem with a hypothesis: the activated molecules are considered as being in equilibrium with the molecular "transition state". As postulated in Section 4.1.1, this transition state being realized, it decomposes by giving rise to the products. This constitutes a way to replace the difficult enumeration of the A and B energy states by the enumeration of the energy states of the transition state.

If the number of states of the transition state is given by $G^*(E - E_0)$ in the energy range $E - E_0$, there are also equivalent numbers of exit channels, and the rate constant for all the channels will be written as

$$k(E) = \frac{G^*(E - E_0)}{hN(E)}.$$
4.13

This constitutes the rate constant expression known as the RRKM rate-constant. As is easily seen, when $E = E_0$, the number $G^*(E - E_0)$ is reduced to one state and

$$k(E_0) = \frac{1}{hN(E_0)}.$$
4.14

This threshold rate-constant is the lowest possible value for $k(E)$.

This minimum value of the rate-constant can be a precious test of the theory in the case of the dissociation process of a molecular ion, when $k(E_0)$ reaches the range 10^5 to $10^6 \, \text{s}^{-1}$. In this case, metastable ions, when detected in the mass spectrometer, give an immediate confirmation of the validity of RRKM calculations. Some examples of this test will be given later.

However, since it is possible in principle to measure rates as a function of the total angular momentum J, it is useful to express the rate constant as a function of J. Hence,

$$k(E, J) = \frac{G^*(E - E_0, J)}{hN(E, J)},$$
4.15

where k, G^* and N are functions of E, J, and J_z (J_z being the projection of J on a fixed-space axis), $N(E, J)$ being the vibrational-rotational density of the energized molecule.

Some additional information about the use of Eq. 4.13 needs to be introduced in the following two sections.

4.1.2.3 Quantum Effects in the Vicinity of the Potential Barrier

The reaction criterion included in the expression of $k(E)$ is such that all molecules with a total energy $E < E_0$ do not react, and all those with a total energy $E > E_0$ react with a unit probability. The reaction results from surmounting a potential barrier; it is now necessary to recall that, quantum mechanically, the crossing of potential barriers involves two possibilities:

● a non-zero probability to cross the barrier for $E < E_0$ (tunneling), and
● a non-unit probability to cross the barrier for $E > E_0$ (reflexion).

The rate-constant Eq. 4.13 therefore needs to be completed by a transition coefficient $\alpha(E)$ in order to express the reflexions occurring for $E > E_0$.

The occurrence of tunneling has to be taken into account, if necessary, through the building up of a quantum model for the reaction path.

Both these effects will reduce the $k(E)$ values to either below or above E_0. The evaluation of $\alpha(E)$ will depend so strongly on the form of the potential energy barrier that a general solution is not possible [3, 4]. It is fairly obvious that the role played by $\alpha(E)$ is limited to the immediate vicinity of the threshold and only for reactions where μ^* (the effective mass of the particle) and E_0 are both small. Such quantum effects will, therefore, be taken into consideration only exceptionally, and we will admit that, generally, $\alpha(E) = 1$ for $E \geq E_0$.

One special case where calculations of $\alpha(E)$ might have some merit is when relative rates of evolution of isotopically substituted molecules are involved. In such cases, the form of the barrier will be very similar; only μ^* and E_0 will be modified, the E_0 modification being due to the differences in zero-point energies between isotopic molecules.

4.1.2.4 Degeneracy of the Reaction Path

As far as possible the meaning of "transition state" and "reaction path" will be given below, remembering that, frequently, many symmetrical possibilities exist in a molecule, in order to realize a reaction. If the dissociation reaction

$$CH_4^+ \rightarrow CH_3^+ + H$$

is considered, it is clear that the transition state $H_3C^+ \ldots H$ can be realized in four equivalent ways: one can say that the reaction path degeneracy σ is equal to 4.

If the two similar reactions

$$CH_3D^+ \rightarrow CH_2D^+ + H$$

$$CH_3D^+ \rightarrow CH_3^+ + D$$

are now considered, with analoguous transition states, it is clear that $\sigma = 3$ for the first, and $\sigma = 1$ for the second.

In any case, care has to be taken of the geometry attributed to the transition state before any σ value attribution.

4.1.2.5 The Most General $k(E)$ Expression

If one takes into account the occurrence of a transmission coefficient and of a reaction path degeneracy, the most general expression of $k(E)$ becomes

$$k(E) = \alpha(E) \cdot \sigma \cdot \frac{G^*(E - E_0)}{hN(E)}. \qquad\qquad 4.16$$

where $\alpha(E)$ is a transmission coefficient which will be very often considered as unity for $E \geq E_0$, σ is the reaction path degeneracy, $G^*(E - E_0)$ is the number of quantum states of the transition state at energy E, and E_0, is the reaction energy threshold; $N(E)$ is the density of states of the reactant at energy E (number of quantum states per unit energy); and h is the Planck constant.

In principle, this expression allows the calculation of $k(E)$. Its challenge lies in the determination of the transition state molecular properties (geometry, degrees of freedom, vibrational frequencies, and possible occurrence of internal rotations), and also in the more or less exact way by which $G^*(E - E_0)$ and $N(E)$ will be evaluated.

Before treating this important problem, a definition of "transition state" and "reaction path" concepts will be given.

4.1.3 Potential Energy Profile along the Reaction Path

When diatomic molecules dissociate, the potential energy profile along the reaction path coincides exactly with the potential energy curve of the system, expressed as a function of the internuclear distance R.

In this case, R is a rigorously defined reaction coordinate.

If one turns now to the dissociation of non linear polyatomic molecules of N atoms, one deals with a system of $3N - 6$ internal degrees of freedom.

The complete potential energy function of such molecules will be represented by a $3N-6$ dimensional hypersurface in a $3N-5$ dimensional hyperspace. As an example, if the HCO molecule is considered, with an equilibrium configuration given by Fig. 4.2, its complete potential energy function will vary with $R(C-H)$, $R(C-O)$, and θ; in other words V will be written

$$V = f[R(C-H), R(C-O), \theta].$$

If it is admitted that V can be calculated quantum mechanically, the visualization of this relatively simple hypersurface still appears somewhat difficult. Practically, it will

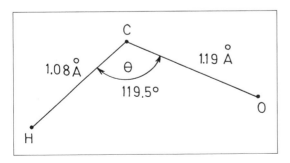

Fig. 4.2. Equilibrium configuration of the HCO radical.

be necessary to draw equipotential curves as a function of either R(C–H) or R(C–O), for constant values of θ. When R(C–H) increases enough, the C–H bond breaks and the system is reduced to H + CO. When the C–O bond distance increases, the C–O bond breaks and the system goes to HC + O. A hypothetic diagram of this type is given in Fig. 4.3 for an ABC linear molecule. In this diagram, equipotentials are given in energy units and the bottom figure shows the potential energy profile along the reaction path in the case of an elongation of the R(A–B) distance leading to A + BC, and also for an elongation of the R(B–C) distance leading to AB + C.

The transition state is considered as localized at the top of the potential energy profile, i. e., around the point M for the former case and somewhere on the asymptote for the latter. It is clear that the second case, rather often encountered in the simple bond cleavage, appears as a challenge, i. e., concerning the exact location of the

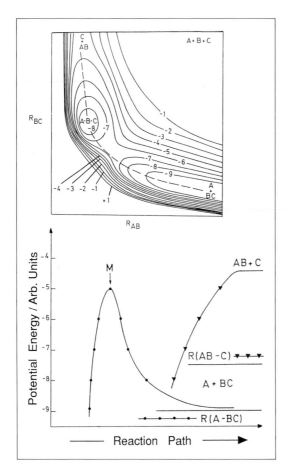

Fig. 4.3. Hypothetical energy surface of an ABC molecule (energies are given on equipotentials in arbitrary units). At the lower part of the figure the reaction paths deduced from the curves are given for the processes A + BC and AB + C.

transition state, in the determination of the exact intermolecular distance for the products seperation.

This difficulty can be overcome if rotational motions of the molecule are taken into consideration. As is well known, molecules have degrees of freedom corresponding to overall rotation that give rise to additional centrifugal and Coriolis forces. A rigorous treatment of the coupling of rotational degrees of freedom with internal degrees of freedom would be so difficult that it is often useful to replace it by adding a rotational potential to the electronic energy for each nuclear configuration. An effective potential is obtained in this way, which, in cases similar to a diatomic molecule AB, has the form shown in Fig. 4.1. The transition state will be localized at the top of this small barrier, transforming the second case of Fig. 4.3 into a case analogous to the first one.

An example of such a treatment will be given later in the case of $CH_4^+ \rightarrow CH_3^+ + H$ dissociation, for which it will be shown that tunneling through the rotational barrier is one of the ways to explain the occurrence of very long lifetimes for excited CH_4^+ ions.

4.1.4 Transition States

A first aspect of what the theory of unimolecular reactions call a transition state, will just be given here. In 1932, when the so-called absolute theory of reaction rates was born, it was believed that it would rapidly be possible to calculate all the details of potential energy surfaces for chemical reactions, consequently locating the position of the "saddle point" on these surfaces as the vibrational and rotational characteristic of the transition state.

Some 50 years later, it is clear that one is very far from such an ideal situation in most cases.

The only way to overcome this difficulty is to postulate the transition state, a reasonable structure related to the type of reaction studied, and to infer by different approximate means the best values for its rotational and vibrational frequencies.

A holistic approach to this problem exists, reported in specialized literature aiding the choice of vibrational and rotational characteristics of transition states [3, 4].

Here, we offer an example of the type of reasoning used. If the isomerization of oxirane into ethanal is considered [6]:

$$
\begin{array}{ccc}
H_2C & & H_3C \\
\diagdown & & \diagdown \\
\triangleright\!\!=\!\!C\!\!=\!\!O & \rightarrow & \diagup\!C\!\!=\!\!O \\
\diagup & & \diagup \\
H_2C & & H
\end{array}
$$

an example of plausible activated complex could be

$$
\begin{array}{c}
H_2C \\
H \diamond C{=}O \\
CH
\end{array}
$$

A reasonable correct rate constant has been obtained, admitting that in the transition state the deformation vibration frequency of the cycle was 650 cm^{-1} instead of 895 cm^{-1} in the oxirane molecule, in such a way as to correspond to the deformation vibration C$-$C$=$O realized in CH$_3$$-$COH. Simultaneously, the respiration frequency of the cycle was taken as 1500 cm^{-1} instead or 1260 cm^{-1}, in order to become similar to a C$=$O valence vibration frequency. It is also to be noted that a reaction path degeneracy factor $\sigma = 4$ is plausible for the chosen transition state, each of the four hydrogen atoms being able to realize the hydrogen bridge.

4.2 The Evaluation of Sums and Densities of Molecular Quantum States

The calculation of $k(E)$ rate constants depends critically on the ability to evaluate $G^*_{v,r}(E - E_0)$ for a postulated model of the transition state and $N_{v,r}(E)$ for the reacting molecule.

For the purpose of the present chapter, superscripts affecting $G(E)$ and $N(E)$ will be ignored, E being the energy where $G_{v,r}(E)$ and $N_{v,r}(E)$ need to be evaluated. One shall precise that the v, r indices are written for vibrational and rotational quantum states. The different levels considered for the calculation of $G(E)$ and $N(E)$ are extensively described in specialized books [3, 4]. What will be presented here will merely consist of the reasoning, the assumptions made to get the formula expressing $G(E)$ and $N(E)$, and the energy ranges for their validity.

4.2.1 Terminology, Definitions, Assumptions

● $N(E)$ is the number of quantum states per unit energy;
  $W(E)$ is the number of quantum states at energy E, in other words, the quantum degeneracy at energy E; it is a dimensionless number;
  $G(E)$ is the total number of states at energy E or the integrated density of states between zero and E; it is also a dimensionless number.

● The quantum mechanical partition function is written:

$$Q = \sum_E W(E)\, e^{-\frac{E}{kT}} .$$ 4.17

In this expression $W(E)$ is the number of states at energy E, this energy being affected by an uncertainty δE. It is convenient to notice that if $N(E)$ expresses the levels density:

$$W(E) = N(E)\, \delta E .$$ 4.18

For high values of $N(E)$, Eq. 4.18 leads to the semiclassical expression of the partition function:

$$Q \text{ (semiclassical)} = \int_0^E N(E)\, e^{-\frac{E}{kT}}\, dE .$$ 4.19

● The evaluation of $N(E)$, $G(E)$, and Q refer exclusively to collections of harmonic oscillators and rigid free rotators, without any kind of internal coupling (lack of any cross term in the potential energy function).
● If the quantum partition function is used, discrete values of E are only taken into consideration. In this case, $W(E)$ is always zero, except for the quantum mechanically allowed values of E; then, $W(E)$ is normally equal to unity except when the quantum energy level is degenerated. It follows from this that $W(E)$ is a succession of δ functions.

In this case, $G(E) = \sum_{E=0}^{E} W(E)$ is a step function as is shown in Fig. 4.4.

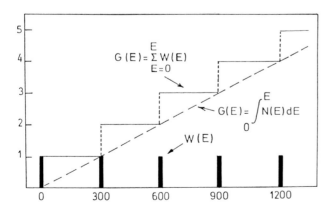

Fig. 4.4. Correlation between $W(E)$ and $N(E)$. The width of levels is proportional to δE (uncertainty on energy E) [4].

In this figure, vertical lines are equal to the quantum degeneracy $W(E)$. Their widths represent the uncertainty δE on the E values, the step function at energy E expresses $G(E)$ and the straight lines represents the continuous approximation of $G(E)$ or, in the other words,

$$G(E) = \int_0^E N(E) \, dE . \qquad\qquad 4.20$$

The correctness of this approximation improves when energy intervals become smaller and smaller, as is the case for complex molecules. Quantum mechanically $G(0) = 1$; in the continuous approximation $G(0) = 0$.

4.2.2 Evaluation of $W(E)$ and $G(E)$ by Direct Count

In simple molecular systems, for not too high energy values E, $W(E)$, and $G(E)$ are able to be evaluated step-by-step, as will be shown, for instance, in the case of CO_2.

This molecule of linear symmetry has two non-degenerate vibrations and a doubly degenerate one, the frequencies being, respectively, 2349, 1388, and 667 cm^{-1}.

The energy of the system counted up from the zero-point energy is

$$E = 2349 \, v_1 + 1388 \, v_2 + 667 \, v_3 . \qquad\qquad 4.21$$

The direct count of the lowest vibrational levels of CO_2 is given in Appendix I.

It is quite clear from this Appendix that for a bigger number of vibrational degrees of freedom or for much higher energies, the direct count will be so tedious that faster methods are needed in evaluating $N(E)$ and $G(E)$. It will now be shown how such methods can be found in the inversion of partition functions.

4.2.3 $N(E)$ and $G(E)$ from the Inversion of Partition Functions

4.2.3.1 Inversion of the Semiclassical Partition Function

The semiclassical partition function 4.19 in which s is written for $(kT)^{-1}$, becomes

$$Q = \int_0^\infty N(E) \, e^{-sE} \, dE . \qquad\qquad 4.22$$

Q appears as the Laplace transform of $N(E)$, or

$$Q = l(s) = \lambda\{N(E)\} . \qquad\qquad 4.23$$

In this expression $\lambda\{...\}$ denotes the operation to be performed on $N(E)$ to get Q; s is called the transform parameter.

Standard methods are described [7, 8] in order to reverse the process. By using tables of "inverse Laplace transform", $N(E)$ can be calculated

$$N(E) = \lambda^{-1}\{l(s)\};$$ 4.24

λ^{-1} denotes symbolically the operation of inversion of the Laplace transform.

A theorem of operational calculus, called the integration theorem, asserts that

$$\frac{l(s)}{s} = \lambda\left\{\int_0^E N(x)\,dx\right\} = \lambda\{G(E)\}.$$ 4.25

The inversion of $G(E)$ is written

$$G(E) = \lambda^{-1}\left\{\frac{l(s)}{s}\right\}.$$ 4.26

Equations 4.24 and 4.26 can be combined to

$$\lambda^{-1}\left\{\frac{l(s)}{s^k}\right\} = \begin{cases} N(E) & \text{when } k = 0 \\ G(E) & \text{when } k = 1 \end{cases}$$ 4.27

4.2.3.2 Approximations to $N(E)$ and $G(E)$ for Rotational States

It is easy to establish a good approximation to $N_r(E)$ and $G_r(E)$ if one assumes the "separability" of rotations. Two cases of rotator will only be taken into account: the one-dimensional and the two-dimensional rotator. The one-dimensional rotator is characterized by one axis of rotation and one moment of inertia; this is typically the case of an internal rotator. Its partition function will be written as $Q_{r=1}$.

The two dimensional rotator will be characterized by two axes for rotation and two moments of inertia; this is typically the case of a diatomic or of a polyatomic linear molecule. Its partition function will be written as $Q_{r=2}$.

If a "very prolate symmetric top" molecule is considered, with $I_1 \neq I_2 = I_3$ it shall be assumed that the lowest moment of inertia I_1 is separable from both others and can be treated as a one-dimensional rotator; therefore,

$$Q_{r=3} = Q_{r=1}\ Q_{r=2}.$$

If an asymmetric top rotator is considered with $I_1 \neq I_2 \neq I_3$, it shall be assumed that

$$Q_{r=3} = (Q_{r=1})_1\,(Q_{r=1})_2\,(Q_{r=1})_3.$$

It follows from this that only one of two partition functions will be used, either $Q_{r=1}$ or $Q_{r=2}$.

From considerations of statistical mechanics, it is easy to show that in the general case for n_1 one-dimensional rotators and n_2 two-dimensional rotators, the total partition function Q_r is written

$$Q_r = Q'_r\, s^{-\frac{r}{2}}, \qquad\qquad 4.28$$

with $s = (kT)^{-1}$ and $r = n_1 + 2n_2$.

In this general case, Q'_r is defined as

$$Q'_r = \left(\frac{8\pi^2}{h^2}\right)^{\frac{r}{2}} \cdot \pi^{\frac{n_1}{2}} \cdot \prod_{j=1}^{n_2} \left(\frac{I_i}{\sigma_i^2}\right)^{1/2} \prod_{j=1}^{n_1} \left(\frac{I_j}{\sigma_j}\right), \qquad\qquad 4.29$$

when σ_i and σ_j are the symmetry numbers for rotation of the i one-dimensional and j two-dimensional rotators.

From tables of Laplace transforms, it is easy to find that

$$\lambda^{-1}\left\{s^{-\frac{r}{2}}\right\} = \frac{E^{\frac{r}{2}-1}}{\Gamma\left(\frac{r}{2}\right)}, \qquad \lambda^{-1}\left\{\frac{s^{-\frac{r}{2}}}{s}\right\} = \lambda^{-1}\left\{s^{-\left(1+\frac{r}{2}\right)}\right\} = \frac{E^{\frac{r}{2}}}{\Gamma\left(1+\frac{r}{2}\right)}, \qquad 4.30$$

therefore,

$$N_r(E) = \frac{Q'_r\, E^{\frac{r}{2}-1}}{\Gamma\left(\frac{r}{2}\right)} \qquad\qquad 4.31\,a$$

$$G_r(E) = \frac{Q'_r}{\Gamma\left(1+\frac{r}{2}\right)}. \qquad\qquad 4.31\,b$$

These approximate $N_r(E)$ and $G_r(E)$, when compared to the results of exact counting techniques, are found to be excellent above $500\ \mathrm{cm}^{-1}$, a rather unusual energy for a dissociation process. This approximation is, therefore, suitable when used at energies ordinarily encountered for excited molecules concerning $N(E)$. It will certainly also be acceptable for $G^*(E - E_0)$ near the threshold, owing to the hypotheses generally made about the geometry of transition states.

4.2.3.3 Inversion of the Partition Function for Vibrational States

For one classical harmonic oscillator of energy $h\nu$, the value of $N(E)$ is $1/h\nu$. Therefore, Q_v in its semiclassical form will be written

$$Q_v = \frac{1}{h\nu} \int_0^\infty e^{-E/kT}\, \mathrm{d}E = \frac{kT}{h\nu}. \qquad\qquad 4.32$$

For v independent harmonic oscillators, it is written

$$Q_v = \frac{(kT)^v}{\prod\limits_{i=1}^{v} h\nu_i}.$$ (4.33)

If one writes, $s = (kT)^{-1}$

$$Q_v = \frac{1}{s^v \prod\limits_{i=1}^{v} h\nu_i}.$$ (4.34)

It follows from 4.27 that

$$N(E) = \lambda^{-1} \left\{ \frac{1}{s^v \prod\limits_{i=1}^{v} h\nu_i} \right\} \quad G(E) = \lambda^{-1} \left\{ \frac{1}{s^{v+1} \prod\limits_{i=1}^{v} h\nu_i} \right\}.$$

From tables of Laplace transforms, one sees immediately that

$$N_v(E) = \frac{E^{v-1}}{\Gamma(v) \prod\limits_{i=1}^{v} h\nu_i}$$ (4.35)

$$G_v(E) = \frac{E^v}{\Gamma(v+1) \prod\limits_{i=1}^{v} h\nu_i}$$ (4.36)

4.2.3.4 Vibrorotational States Density and Number

The joint vibrational and rotational semiclassical partition function is written

$$Q_{v,r} = \frac{Q'_r s^{-\frac{r}{2}}}{s^v \prod\limits_{i=1}^{v} h\nu_i} = \frac{Q'_r}{s^{v+\frac{r}{2}} \prod\limits_{i=1}^{v} h\nu_i}.$$ (4.37)

Therefore,

$$N_{v,r} = \lambda^{-1} \left\{ \frac{1}{s^{v+\frac{r}{2}}} \right\} \frac{Q'_r}{\prod\limits_{i=1}^{v} h\nu_i} = \frac{Q'_r}{\Gamma\left(v+\frac{r}{2}\right) \prod\limits_{i=1}^{v} h\nu_i} E^{v-1+\frac{r}{2}}$$ (4.38)

$$G_{v,r} = \lambda^{-1} \left\{ \frac{1}{s^{v+\frac{r}{2}+1}} \right\} \frac{Q'_r}{\prod\limits_{i=1}^{v} h\nu_i} = \frac{Q'_r}{\Gamma\left(v+1+\frac{r}{2}\right) \prod\limits_{i=1}^{v} h\nu_i} E^{v+\frac{r}{2}}$$ (4.39)

4.3 Expressions of $k(E)$ in Classical Approximations

4.3.1 The Energized Molecule is Considered as a Collection of Harmonic Vibrators

If the reactant molecule and the activated complex are described as a collection of independent harmonic oscillators only, recalling that the reactant molecule has v oscillators and the activated complex has $(v-1)$ oscillators, the one released loosed being transformed in the reaction coordinate by using Eqs. 4.13, 4.35, 4.36, the rate constant $k(E)$ can be written as

$$k(E) = \frac{G^*(E - E_0)}{hN(E)} = \frac{(E - E_0)^{v-1}}{\Gamma(v) \prod_{i=1}^{v-1} hv_1^*} \cdot \frac{\Gamma(v) \prod_{i=1}^{v} hv_i}{hE^{v-1}}, \qquad 4.40$$

or

$$k(E) = \frac{h^v}{h^{v-1} \cdot h} \cdot \frac{\Gamma(v)}{\Gamma(v)} \cdot \frac{\prod_{i=1}^{v} v_i}{\prod_{i=1}^{v-1} v_i^*} \left(\frac{E - E_0}{E}\right)^{v-1} = \frac{\prod_{i=1}^{v} v_i}{\prod_{i=1}^{v-1} v_i^*} \left(\frac{E - E_0}{E}\right)^{v-1}. \qquad 4.41$$

4.3.2 The Energized Molecule is a Collection of Uncoupled Harmonic Oscillators and Independent Internal Rotators

If the molecule is considered as composed of r internal rotators and $(v - r)$ oscillators, the transition state being composed of r^* internal rotors and $(v - 1 - r^*)$ oscillators, one can write

$$N_{v,r} = \frac{Q_r' E^{v-r-1+\frac{r}{2}}}{\Gamma\left(v - r + \frac{r}{2}\right) \prod_{j=1}^{v-r} hv_j} = \frac{Q_r'}{\Gamma\left(v - \frac{r}{2}\right)} = \frac{E^{v-1-\frac{r}{2}}}{h^{v-r} \prod_{j=1}^{v-r} v_j} \qquad 4.42$$

$$G_{v,r} = \frac{Q_{r^*}' (E - E_0)^{v-1-r^*+\frac{r^*}{2}}}{\Gamma\left(v - 1 - r^* + 1 + \frac{r^*}{2}\right) \prod_{j=1}^{(v-1-r^*)} hv_j^*} = \frac{Q_{r^*}'}{\Gamma\left(v - \frac{r^*}{2}\right)} \qquad 4.43$$

$$G_{v,r} = \frac{(E - E_0)^{v-1-\frac{r^*}{2}}}{h^{v-1-r^*} \prod_{j=1}^{(v-1-r^*)} v_j^*} \qquad 4.44$$

$$k(E) = \frac{Q'_{r^*}}{Q'_r} \cdot \frac{\Gamma\left(v - \dfrac{r}{2}\right)}{\Gamma\left(v - \dfrac{r^*}{2}\right)} \cdot \frac{h^{v-1-r}}{h^{r-1-r^*}} \cdot \frac{\displaystyle\prod_{j=1}^{v-r} v_j}{\displaystyle\prod_{j=1}^{(v-1-r^*)} v_j^*} \cdot \frac{(E - E_0)^{v-1-\frac{r^*}{2}}}{E^{v-1-\frac{r}{2}}} \qquad 4.45$$

$$k(E) = A \left(\frac{E - E_0}{E}\right)^{v-1-\frac{r}{2}} (E - E_0)^{\frac{r-r^*}{2}} \qquad 4.46$$

$$\text{with } A = \frac{Q'_{r^*}}{Q'_r} \cdot \frac{\Gamma\left(v - \dfrac{r}{2}\right)}{\Gamma\left(v - \dfrac{r^*}{2}\right)} \cdot \frac{\displaystyle\prod_{j=1}^{v-r} v_j}{\displaystyle\prod_{j=1}^{v-1-r^*} v_j^*} \, h^{r^*-r}.$$

Both Eqs. 4.42 and 4.46 were first deduced by Rosenstock in his fundamental thesis on QET theory for unimolecular reactions in ionized molecules [2]. The first of these equations was proposed much earlier by RRK [9, 10] on the basis of a combinatorial calculus, in the case of unimolecular reactions in molecules.

While interesting, they both suffer from a common defect: they are unable to justify the non-zero values observed for $k(E_0)$.

This is obviously due to the approximations involved in using the semiclassical partition function instead of the quantum one, and also to the counting of vibrational levels from their zero point energies. It will be shown now, how the latter can be compensated for leading to much better expressions of $k(E)$.

It has to be pointed out, however, that expressions 4.42 and 4.46 have been widely used up until very recently in order to discuss many aspects of ionization phenomena in gases, and that they are at least acceptable when excitation energies are rather high.

4.4 Critiques and Improvements of the Classical Approximation to $N_{vr}(E)$ and $G_{vr}(E)$

4.4.1 The Marcus-Rice and Whitten-Rabinovitch Approximations

The problem of counting harmonic vibrational and uncoupled rotational quantum states was approximated in two ways: firstly, the semiclassical partition function was used, which gives a rather bad approximation at low energy, improving for higher and higher energy values; secondly, the counting of states was considered from the zero-point energies of all oscillators. In the case of rather complex molecules the total zero point energy is the sum of an appreciable number of zero-point energies which are excluded from the total number of quantum states. Two successive modifications of the $N_v(E)$ and $G_v(E)$ expressions have been successively introduced in order to correct these expressions for this fact.

These are expressed as

$$N_v(E) = \frac{(E + aE_Z)^{v-1}}{\Gamma(v) \prod_{i=1}^{v} hv_i} \qquad \qquad 4.47$$

$$G_v(E) = \frac{(E + aE_Z)^v}{\Gamma(v+1) \prod_{i=1}^{v} hv_i}. \qquad \qquad 4.48$$

In these expressions, E_Z stays for the total zero point energy $\frac{1}{2} \sum_{i=1}^{v} hv_i$.

When a is equal to unity, we refer to this as the Marcus-Rice approximation [11]; when a is considered as a function of E ($a < 1$), we refer to this as the Whitten-Rabinovitch approximation [12].

The latter authors evaluate a as follows

$$a = 1 - \beta w; \qquad \qquad 4.49$$

For $E > E_Z$: $\log_{10} w = -1.0506 \left(\frac{E}{E_Z}\right)^{1/4}$,

for $E < E_Z$: $\frac{1}{w} = 5 \left(\frac{E}{E_Z}\right) + 2.73 \left(\frac{E}{E_Z}\right)^{1/2} + 3.51$, $\qquad \qquad 4.50$

$$\beta = v_d \frac{v-1}{v}; \quad v_d = \frac{\langle v^2 \rangle}{\langle v \rangle^2}; \quad \langle v^2 \rangle = \frac{1}{v} \sum_{i=1}^{v} v_i^2; \quad \langle v \rangle = \frac{1}{v} \sum_{i=1}^{v} v_i.$$

If ω_i values are used instead of v_i, it is easy to show that

$$\beta = (v-1) \frac{\sum \omega_i^2}{(\sum \omega_i)^2}. \qquad \qquad 4.51$$

Importantly, in this approximation, with $a = f(E)$, the relation between $G_v(E)$ and $N_v(E)$ is no longer that of a function and of its derivative.

In the case of rotational and vibrational states, the problem of the E correction persists. Equations 4.38 and 4.39 are also corrected as expressed by Eq. 4.52 below

$$\left. \begin{matrix} N_{vr}(E) \\ k=0 \\ \\ \\ \\ \\ G_{vr}(E) \\ k=1 \end{matrix} \right\} \frac{Q'_r(E + a_k E_Z)^{v-1+k+\frac{r}{2}}}{\Gamma\left(v + k + \frac{r}{2}\right) \sum_{i=1}^{v} hv_i}, \qquad 4.52$$

$$\text{with } a_k = 1 - \beta_k w; \quad \beta_k = \frac{v_d \, (v-1) \left(v - 1 + k + \dfrac{r}{2}\right)}{v^2};$$

v_d remains defined as above.

If the performances of these successive approximations are taken into account in the case of any molecule, one shall see that the classical approximation gives, to low $G_v(E)$ values an improvement when E increases. The improved $G_v(E)$ values, when $a = 1$, gives gross overestimation, as expected; the $G_v(E)$ values with $a = f(E)$ performs rather well if compared with the results of exact count, as is illustrated in Table 4.1 for cyclopropane (reproduced from [3]).

Table 4.1. $G_v(E)$ for cyclopropane by various methods.

E_0 (kcal mol^{-1})	Classical	Marcus-Rice	Whitten-Rabinovitch	Exact count [13]
10	0.00	$5.45 \cdot 10^2$	$7.17 \cdot 10^2$	$8.02 \cdot 10^2$
30	0.02	$20.9 \cdot 10^6$	$2.65 \cdot 10^6$	$2.69 \cdot 10^6$
50	$8.2 \cdot 10^2$	$21.5 \cdot 10^8$	$6.15 \cdot 10^8$	$6.12 \cdot 10^8$
100	$1.72 \cdot 10^9$	$9.94 \cdot 10^{12}$	$5.90 \cdot 10^{12}$	$5.84 \cdot 10^{12}$
150	$8.58 \cdot 10^{12}$	$4.03 \cdot 10^{15}$	$3.02 \cdot 10^{15}$	$3.00 \cdot 10^{15}$
200	$3.61 \cdot 10^{15}$	$4.27 \cdot 10^{17}$	$3.56 \cdot 10^{17}$	$3.54 \cdot 10^{17}$

It is easy to verify that, as far as the molecules are considered as collections of harmonic oscillators, this appears to be a drastic approximation, the method of Whitten and Rabinovitch performs very nicely for the evaluation of $G_v(E)$.

4.4.2 The Inversion of the Quantum Partition Function

The inversion of the quantum vibrational partition function could be a better way for the evaluation of $G_v(E)$ and $N_v(E)$ by the proper elimination of the approximation of the semiclassical partition function.

For one harmonic oscillator $(v = 1)$ of frequency v, the quantum partition function $Q_{v=1}$ is written if $s = (kT)^{-1}$

$$Q_{v=1} = \frac{1}{1 - e^{-hvs}} \qquad\qquad 4.53$$

or, if one writes u for hvs

$$Q_{v=1} = \frac{1}{1 - e^{-u}}.$$

Recalling that, by definition,

$$\sinh{(u/2)} = \frac{1}{2}(e^{u/2} - e^{-u/2}) = \frac{1}{2}(1 - e^{-u})\,e^{u/2},$$

from which, one obtains

$$1 - e^{-u} = e^{-u/2}(2\sinh{(u/2)}),$$

Q_v can be written

$$Q_{v=1} = \frac{e^{\frac{shv}{2}}}{2\sinh{(shv/2)}}. \tag{4.54}$$

For a collection of j oscillators:

$$Q_v = \frac{1}{2^v}\prod_{j=1}^{j-v} = \frac{e^{\frac{shv_j}{2}}}{2\sinh{(shv_j/2)}} = I(s). \tag{4.55}$$

In this case $I(s)$ is sufficiently complicated to require the use of another property of Laplace transform in order to be inverted.

The Fourier-Mellin integration theorem will be used, which quotes that for any function $I(s)$:

$$\lambda^{-1}\left\{\frac{I(s)}{s^k}\right\} = \frac{1}{2\pi i}\int_{c-i\infty}^{c+i\infty}\frac{I(s)\,e^{sE}\,ds}{s^k} = I$$

It would be too long to describe the methods leading to the evaluation of this integral, owing to their mathematical complexity. These methods, easily adapted for computers, (Cauchy's residue theorem, method of steepest descent) are described in specialized books and provide a method of counting in the case of anharmonic oscillators [4].

Notably, if one examines the comparative results obtained by all these methods, one can deduce the following conclusions, already stated by Forst and Prasil [14]; good results are obtained by the method of Whitten and Rabinovitch, by inversion of series and the method of steepest descent. A truncature of the inverse transformation of series known as the Haarhof approximation [15] is also rather useful, but performs better at rather high excitation energies.

4.5 Tests for the Validity of RRKM Calculations

Four principle tests for the validity of RRKM calculations are possible:
1. a comparison between measured and calculated breakdown diagrams and between calculated and measured mass spectra;
2. a comparison between calculated and measured $k(E_0)$ values;
3. a comparison between calculated and measured $k(E)$ values;
4. a comparison between calculated and measured kinetic energy distribution of dissociating pairs.

The results of these four tests will be successively exposed and discussed in the next four paragraphs.

4.5.1 Breakdown Diagrams — Mass Spectra

If one goes back to Sections 1.3.3, 1.3.4, and 1.3.5, one will see that the most probable scheme for the dissociation of $C_2H_6^+$ appears to be expressed by the simultaneous occurrence of many concurrent and consecutive reactions, such as:

$$C_2H_6^+ \begin{array}{l} \xrightarrow{k_1} C_2H_5^+ + H \xrightarrow{k_1'} C_2H_3^+ + H_2 \text{ (or } CH_3-CH^+ + H) \\ \xrightarrow{k_2} C_2H_4^+ + H_2 \xrightarrow{k_2'} C_2H_2^+ + H_2 \\ \xrightarrow{k_3} CH_3^+ + CH_3, \end{array}$$

each of the reactions being characterized by a rate constant k_i or k_i'. If N_0 molecular ions are created at time $t = 0$, the concentration of $C_2H_6^+$ at time t is given by

$$[C_2H_6^+] = N_0 \exp\left(- \sum_{i=1}^{3} k_i t\right). \qquad 4.56$$

If the consecutive reactions are now considered, such as:

$$C_2H_6^+ \xrightarrow{k_1} C_2H_5^+ + H \xrightarrow{k_1'} C_2H_3^+ + H_2,$$

the amount of $C_2H_6^+$, $C_2H_5^+$, and $C_2H_3^+$ present at time t, are given by

$$[C_2H_6^+] = N_0 \exp\left(- k_1 t\right) \qquad 4.57$$

$$[C_2H_5^+] = N_0 \frac{k_1}{k_1' - k_1} \left(\exp\left(- k_1 t\right) - \exp\left(- k_1' t\right)\right) \qquad 4.58$$

$$[C_2H_3^+] = N_0 - [C_2H_6^+] - [C_2H_5^+]$$

$$= N_0 \left\{ 1 - \frac{k_1'}{k_1' - k_1} \exp\left(-k_1 t\right) + \frac{k_1}{k_1' - k_1} \exp\left(-k_1' t\right) \right\}. \qquad 4.59$$

If all the k_1 and k_1' have been calculated for successive aliquot values of the excitation energy E, the breakdown diagram of the molecular ion can be calculated in the following way, for a mean time value t, consistent with the transit times characterizing the mass dispersive instrument (generally 10^{-5} s):

● the fraction of parent ions remaining at time t is calculated from Eq. 4.40 for successively given values of excitation energy E,

● the amounts of various products formed were then computed using Eqs. 4.58 and 4.59 and the rate constants already calculated,

● it may be recalled that the excitation energy of a fragment ion, determining its decomposition rate, could be calculated from the excitation energy of its precursor. The numerical values of the decomposition rate constants of fragment ions were hence tabulated along with values for the excitation energy of the parent ion.

Based on this scheme, the breakdown diagram of $C_2H_6^+$ has been calculated [18] by using the counting formula of Haarhof [15], including corrections for anharmonicities. It appears from Fig. 4.5 [19], where a comparison is made with the results of

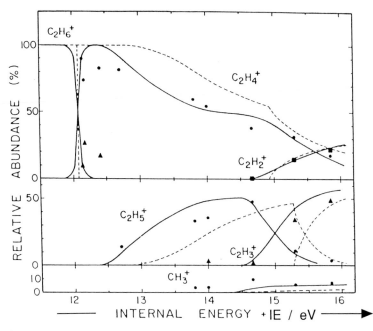

Fig. 4.5. Comparison between the calculated breakdown diagram [18], the results of ZKE-ion coincidences [19], and those of CE as given in [19], (Part 1, [57]).

ZKE-ion coincidences and those of charge exchange, that the general agreement between calculated and measured breakdown diagram is rather fair and gives to all the approximations included in the evaluation of $k(E)$, a comfortable support.

It is now the moment to point out that from the breakdown diagram, the mass spectrum for a given electron or photon energy can be calculated at the expense of an hypothesis made about the energy transfer function or energy deposition function. One would need to postulate a distribution function for the parent ions formed with various amounts of internal excitation energy as a result of electron or photon impact. So many ways have been proposed [4, 20] in order to solve this problem, without really solving it, that they will not be exposed here. In contrast with the early days of theoretical mass spectrometry, the mass spectrum itself no longer appears as a good check of the theory because, as it has been shown in Chapter 1, excellent experimental methods can immediately give the breakdown diagram. The problem of the energy deposition function appears consequently as able to be fruitfully reversed: the availability of excellent experimental breakdown diagrams and of mass spectra observed on the same instrument used in conjunction with parametric energy deposition functions, would be able to give information about these functions. Surprisingly, such determinations have not seemed to have aroused any interest up to now.

4.5.2 The Threshold Value of the Rate Constant

A more severe test of the validity of statistical calculations lies in the comparison between $k(E_0)$, the minimum rate constant value for a given dissociation process and the possible observation or non observation of "metastable ions". In this respect, it appears that minimum rate constant calculations did not support the occurrence, in the mass spectrum of C_2H_6, of detectable metastable $C_2H_6^+$ evoluting in $C_2H_5^+$ and $C_2H_4^+$, but provides for the occurrence of $C_2H_6^+$ evoluting metastably in $CH_3^+ + CH_3$, which does not seem to have been studied up to now.

The calculated values of k_{min} are given in Table 4.2 for $C_2H_6^+$ and $C_2D_6^+$ [18].

As is shown in the table, the calculated minimum values of the rate constants are too high in C_2H_6 and just sufficient in C_2D_6 to explain the occurrence of metastable

Table 4.2. Calculated minimum rate constant values in the decomposition of $C_2H_6^+$ and $C_2D_6^+$.

Process	k_{min} (s^{-1}) X = H	k_{min} (s^{-1}) X = D
$C_2X_6^+ \rightarrow C_2X_5^+ + X$	$2.08 \cdot 10^8$	$3.35 \cdot 10^6$
$C_2X_6^+ \rightarrow CX_3 - CX^+ + X_2$	$1.46 \cdot 10^8$	$1.86 \cdot 10^6$
$C_2X_6^+ \rightarrow C_2X_4^+ + X_2$	$7.45 \cdot 10^{10}$	$8.68 \cdot 10^8$
$C_2X_6^+ \rightarrow CX_3^+ + CX_3$	$3.28 \cdot 10^5$	$8.10 \cdot 10^3$

ions, at the exception of the last quoted dissociation process. A similar difficulty is encountered for the process.

$$CH_4^+ \rightarrow CH_3^+ + H,$$

regularly long discussed. A metastable transition observed for this process at the threshold energy for the appearance of CH_3^+ is difficult to explain by RRKM calculations. In this case, the occurrence of a metastable ion has been shown to be relevant to the tunneling of the H-atom through a rotational energy barrier, the problem of the tunneling rates can be treated theoretically in different satisfactory ways, [21, 22]. One of them will be described in Chapter 5.

The same phenomenon could probably also explain the processes involving either H or H_2 elimination from the parent ion.

4.5.3 Calculated and Measured $k(E)$ Values

A test of RRKM calculations that is at least as demanding as the $k(E_0)$ test, is the comparison between the experimentally determined shape of $k(E)$ curves, allowed for the first time by CE experiments [23], and more recently by the use of ion-ZKE coincidence measurements [24], and the shape of calculated $k(E)$ curves.

This test is obviously restricted to processes experimentally shown to depend on a single-exponential decay, and not to some of those for which two-component decay rates have been detected [25].

Two different groups of dissociation processes will be chosen in order to illustrate this point. The first one results from the measurements of rate constants available from ZKE-ion coincidences measurements [25] in the case of the processes:

$$C_6H_5X^+ \rightarrow C_6H_5^+ + X \quad (X = Cl, Br, I).$$

Figure 4.6 reproduces the experimentally measured and theoretically calculated $k(E)$ values for these processes, by using RRKM calculations in the Vestal approximation [26]. These calculations postulate with the exception of the C–X stretching vibration that becomes the reaction coordinate, that all vibrations in the molecular ion and in the transition state complex are considered as unchanged from those of the molecule.

The conclusion to be drawn from the agreement expressed by the figure between experimental and calculated values is that, in this case, the statistical energy redistribution is complete, allowing all oscillators to participate equally. In addition, refinement of the model considering anharmonic oscillators apparently is not necessary.

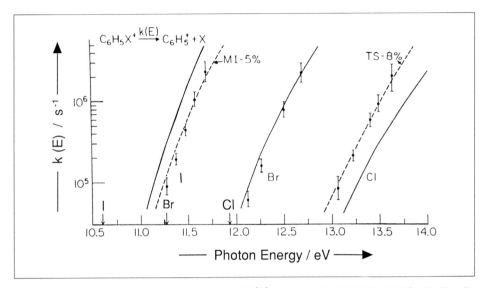

Fig. 4.6. Behavior of experimental and calculated $k(E)$ variations for $C_6H_5X^+ \rightarrow C_6H_5^+ + X$; $(X = Cl, Br, I)$ [24].

The second group goes back to the first rate-constant determination made by charge-exchange [23] for the following processes:

$$C_6H_6^+ \rightarrow C_6H_5^+ + H$$

$$C_6H_6^+ \rightarrow C_4H_4^+ + C_2H_2$$

$$C_6H_5CN^+ \rightarrow C_6H_4^+ + HCN.$$

If Fig. 1.26 is considered, it is clear that the two decay processes of $C_6H_6^+$ molecular ion are not in competition, owing to the very large differences observed between $k(E)$ values. This was confirmed by PIPECO measurements [26] and it appears therefore probable that these reactions take their origin in separate non-interacting, sets of electronic states or in a possible isomerization of $C_6H_6^+$, as suggested many years earlier [27].

In the case of the dissociation process of C_6H_5CN, it may be pointed out that an excellent agreement has been obtained between the experimental $k(E)$ curve and an RRKM calculation [29] assuming a "loose four-centered transition state", such as:

From the just quoted examples, it is quite easy to conclude that the validity of RRKM theory is quite acceptable concerning the behavior of the rate constant in function of the excitation energy.

Another problem remains: does the RRKM theory correctly describe the reaction products states, namely their kinetic energy distributions, which are the most accessible experimentally?

This problem will be treated in the next section.

4.5.4 Kinetic Energy Distributions of Fragmentary Ions

A broad description has been given of the methods used in the determination of translational energy distributions of ions either with or without a time window (see Section 2.2.5 and 2.1.2). In the first case (the case of metastable ions) the kinetic energy distributions are deduced from a suitable removal of instrumental effects by deconvolution of the measured metastable peak shapes. Three methods have been used in this respect [30–33] and the results are satisfactorily comparable. Whatever the situation is (with or without time window), the deduced or directly observed distributions pertain to one of the two types, illustrated in Fig. 4.7. In type A, the maximum of the distribution is realized for $E_T = 0$; in type B, on the contrary, the maximum of the distribution happens for an $E_T \neq 0$, the probability for the appearance of ions with $E_T = 0$ being less than this value or possibly zero.

As is shown in Fig 4.8, kinetic energy distributions [34] are very often dependent on the energy E absorbed by the molecular ion, except in some cases where they occur from a single repulsive potential energy surface, like that realized for the wide kinetic energy distribution observed in the dissociation of CH_3F^+ (Fig. 3.3, p. 145).

It is now appropriate to ask if the RRKM theory can reproduce qualitatively and quantitatively all these characteristics.

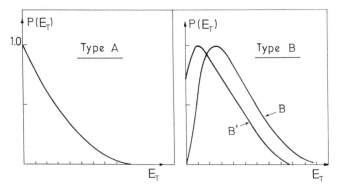

Fig. 4.7. Type A and B kinetic energy distributions (see text).

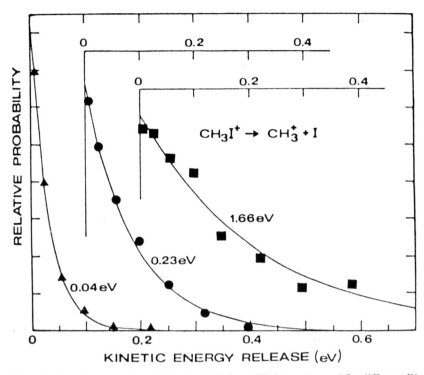

Fig. 4.8. Kinetic energy distribution for CH_3^+ from CH_3I^+ as observed for different E^* values [34].

In order to deduce the probability that a given ion of internal energy E dissociates into a pair (ion + neutral) of total kinetic energy E_T, one must consider that E_T arises from the conversion of vibrational energy in the reaction coordinate to translational energy, in crossing the energy barrier. In this respect, the probability per unit energy $P(E, E_T)$ that a molecular ion of internal energy E from the transition complex crossing, has a distribution of translational energy E_T is given by:

$$P(E, E_T) = \frac{N^*(E - E_0 - E_t)}{G^*(E - E_0)}.$$

4.60

In this expression the numerator expresses the number of possible cases and the denominator the total number of possibilities. If $(E - E_0)$ is replaced by E^* (the internal energy of the transition state), one writes equivalently:

$$P(E^*, E_T) = \frac{N^*(E^* - E_T)}{G^*(E^*)}.$$

4.61

If Eqs. 4.35 and 4.36 are used, recalling that the transition complex has $(v - 1)$ degrees of freedom, it is easy to express Eq. 4.61 as

$$P(E^*, E_T) = (v - 1) \frac{(E^* - E_T)^{v - 2}}{(E^*)^{v - 1}}.$$

4.62

It is easy to see that for $E_T = 0$, $P(E^*, E_T)$ is maximum and equal to

$$P_{max}(E^*, E_T) = \frac{v - 1}{E^*}.$$

4.63

If $P(E^*, E_T)$ is normalized with respect to P_{max}, one gets

$$I(E_T) = \left(1 - \frac{E_T}{E^*}\right)^{v - 2}.$$

4.64

One example of the behavior of $I(E_T)$ as a function of E_T at a given E^* is given in Fig. 4.9, for a molecular ion with $v = 12$ and $E^* = 1$ eV.

It is clear that for type A distributions, the theory represents qualitatively the observed distribution. Is this representation also quantitative? As far as the semiclassical approximation for the counting of states is used, the answer is gener-

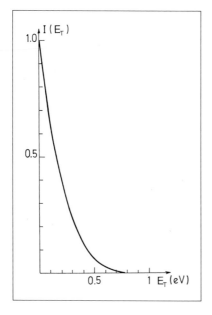

Fig. 4.9. RRKM calculated E_T distribution for E^* equal to 1 eV and $v = 12$.

ally negative. From Eq. 4.46 it is easy to show that the calculated values of the mean kinetic energy[4] $\langle E_T \rangle$ are equal to

$$\frac{E^*}{v}.$$
4.65

From the measurements presently published [35], it is easy to show that $\langle E_T \rangle$ values are more generally expressed by

$$\frac{E^*}{\alpha v} \text{ with } \alpha < 1.$$
4.66

This has been interpreted as a real number of active degrees of freedom, but may also be due to the bad approximation used for the counting of states.

A more severe criticism arises from the frequent occurrence of type-B distributions, impossible to reproduce on the pure basis of RRKM theory. As has been pointed out many times [36, 29], if RRKM theory is excellently appropriate for the rate-constant calculations in the region of the transition states on the potential energy surface, it will be inadequate for the quantitative description of the system in the products region of the energy surface. In this region, for the appearance of a dissociating pair, the number of vibrational degrees of freedom is $3n-12$, compared to the $3n-6$ for the molecular ion, and $3n-7$ for the transition state, a circumstance not included in the RRKM calculations. Similarly, the RRKM theory does not take into account the conservation of angular momentum nor the eventual long-range potential between the ion and the neutral fragment. As will be shown below, both these factors strongly affect the shape of the kinetic energy distributions.

4.5.5 A Statistical Description of Kinetic Energy Distributions after Dissociation

Above it was pointed out that the RRKM calculations are more able to give correct values of the dissociation rate constants than correct kinetic energy distributions of the products of molecular ion decompositions. Before engaging the reader in more sophisticated versions of the theory, it appears appropriate to compare experimentally measured kinetic energy distributions and a priori statistically calculated distributions of the reaction products. The way to the evaluation of such prior

---

[4] By definition, the mean kinetic energy $\langle E_T \rangle$ is given by

$$\frac{\int_0^{E^*} E_T \, P(E^*, E_T) \, dE_T}{\int_0^{E^*} P(E_T) \, dE_T}.$$

distributions is described in numerous papers and synthetized by their authors in a very interesting book [37]. A general treatment of the problem will not be given here, but instead only an example of the following type of dissociation:

A (polyatomic) → B (polyatomic) + C (atom).

If the non-fixed energy E^* of A, equal to $E - E_0$, is considered as partitioned over A and B in translational energy E_T of the A + B pair, in vibrational energy E_V and rotational energy E_R of the polyatomic radical B, one will write the a priori probability $P^0(E_T, E_V, E_R/E^*)$ to see this energy E^* statistically partitioned, as

$$P^0(E_T, E_V, E_R/E^*) = \frac{N_T(E_T) \, N_V(E_V) \, N_R(E_R)}{N(E^*)} \, \delta(E^* - E_T - E_V - E_R). \quad 4.67$$

In this expression the $N_m(E_m)$ are the densities of each particular kind of energy levels, $N(E^*)$ being the total density of states at energy E^*.

The δ function restricts the range of final states to those of a given energy E^*. The expressions of $N_m(E_m)$ are:

$$N_T(E_T) = C_T E_T^{1/2} \quad \text{from [38, 39]} \qquad 4.68$$

$$N_R(E_R) = C_R E_R^{r/2} \quad \text{from Eq. 4.2.14} \qquad 4.69$$

$$N_V(E_V) = C_V E_V^{s-1} \text{ from Eq. 4.2.19}. \qquad 4.70$$

In these expressions the C_m are a combination of constants; r is the number of rotational degrees of freedom of A; s is the number of vibrational degrees of freedom of A. By introducing the Eqs. 4.68, 4.69, and 4.70 in Eq. 4.67, one obtains

$$P^0(E_T, E_V, E_R/E^*) = K E_T^{1/2} E_R^{r/2} E_V^{s-1} \cdot \delta(E^* - E_T - E_R - E_V); \qquad 4.71$$

K being written for $C_T C_R C_V / N(E^*)$.
Integration of Eq. 4.73 over E_R yields

$$P^0(E_T, E_V/E^*) = K E_T^{1/2} E_V^{s-1} \int_0^{E^* - E_T - E_V} E_R^{(r/2)-1} \cdot \delta(E^* - E_T - E_R - E_V) \, dE_R . \quad 4.72$$

By application of the property of δ function, providing that

$$\int f(x) \, \delta(x - x_0) \, dx = f(x_0), \qquad 4.73$$

Eq. 4.72 gives

$$P^0(E_T, E_V/E^*) = K E_T^{1/2} E_V^{s-1} (E^* - E_T - E_V)^{(r/2)-1}. \qquad 4.73$$

Integration of Eq. 4.73 over E_V, leads to

$$P^0(E_T/E^*) = K E_T^{1/2} \int_0^{E^* - E_T} E_V^{s-1} (E^* - E_T - E_V)^{(r/2)-1} dE_V.$$ 4.74

As

$$\int_0^b (x - a)^m (b - x)^n dx = (b - a)^{m+n+1} \frac{\Gamma(m+1)\,\Gamma(n+1)}{\Gamma(m+n+2)} = K'(b-a)^{m+n+1};$$

when $m, n > -1$ and $b > 0$, Eq. 4.74 gives:

$$P^0(E_T/E^*) = KK' E_T^{1/2} (E^* - E_T)^{s+(r/2)-1} = C E_T^{1/2} (E^* - E_T)^{s+(r/2)-1}.$$ 4.75

If a better approximation is needed for the evaluation of $N_V(E_V)$, the Whitten-Rabinovitch approximation will be used. In this case,

$$P^0(E_T/E^*) = C E_T^{1/2} \int_0^{E^* - E_T} (E_V + a E_Z)^{s-1} (E^* - E_T - E_V)^{(r/2)-1} dE_V.$$ 4.76

As a is a rather complex function of E_V, it is easier to computerize Eq. 4.76. The function $(1/C)(P^0(E_T/E^*))$ is zero when $E_T = 0$ and for $E^* = E_T$; it goes through a maximum when:

$$E_T = \frac{E^*}{2s + r - 1}.$$ 4.77

This function has been calculated for a process such as:

$$XY_3Z \rightarrow XY_3 + Z,$$

admitting $E^* = 1$ eV; in this case $s = 6$, $r = 3$, $s + (r/2) - 1 = 6.5$ and $(E_T)_{max} = (1/14)$ eV. It is drawn, after normalization to its maximum, in Fig. 4.10. From this figure it is clear that if the statistics after dissociation are able to furnish $P(E_T)$ distribution curves of type B; it does not provide type-B' or type-A distributions of Fig. 4.7. It is, therefore, clear that the partitioning of E^* between all degrees of freedom of the products will be, quite generally, non-statistical. The comparison between the prior calculated statistical distributions and the observed distributions will be an interesting way for the classification of the energy partition modes between reaction products. Such comparisons, expressed by what is called the "surprisal" of a distribution (or how much it is surprising to observe a non-statistical distribution) are numerous in the case of molecules [37], and rather rare for the dissociation of ions. An example of the latter type has been given [40] in the case of process:

$$CH_4^+ \rightarrow H^+ + CH_3.$$

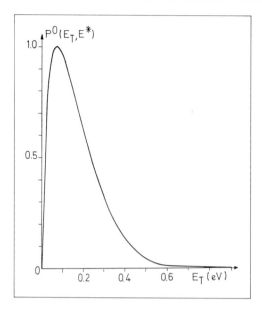

Fig. 4.10. Statistical E_T distribution after dissociation of XY_3Z in XY_3 and Z for an E^* value of 1 eV.

The occurrence of non-statistical distributions of E_T after dissociation imply that some constraints are effective in the dynamics of the dissociation process, reflected by the non-statistical sharing of E^* between products.

These last remarks strongly suggest that improvements needs to be introduced to the theoretical treatment of dissociation processes. These approachs will now be shortly described and discussed.

4.6 Some Improvements to the RRKM Theory

The discussion given in Section 4.5.4 and 4.5.5 suggests that improvements to RRKM theory are needed especially for the calculation of product state distribution, i. e., product kinetic energy distributions.

As pointed out above, such improvements would allow a better description of the situation in the products region of the energy hypersurfaces. It would be able to take into account the conservation of angular momentum and the long-range potential between the ion and its neutral counterpart. It would be too long, in this introductory text, to describe all these theoretical refinements in detail. Some of them will be briefly introduced in the following. The cited references will be of value for a more detailed treatment.

4.6.1 Improvements in the Localization of Transition-State Configuration

When a potential energy barrier for a unimolecular reaction is sufficiently high, with a well-defined saddle point (see, for instance, point M in Fig. 4.3), the position of the transition state is also quite clear: it is the usual saddle point. Quite frequently, however, the reaction takes place at a threshold energy value practically equal to the dissociation energy. This means that the activation energy, except for an eventual rotational barrier, is mainly zero. In such cases, two procedures have been offered for choosing the critical configuration.

The traditional approach is to assume the critical configuration as being located at the top of the rotational barrier, for a breaking bond extension of about four times its normal length [41]. At such extensions, the dissociating fragments are considered as free rotators, which give rise to a so-called loose transition state structure, often referred to as the Gorin model [42].

The second procedure results from trajectory calculations of model triatomics. In their paper [43], the authors proposed that the critical configuration is located where the density of states along the reaction coordinate is minimized. This is the criterion of "minimum local density of states". The proposal included in this criterion is that this "bottleneck" should represent the critical transition state configuration of the RRKM theory. It is generally thought that, for unimolecular reactions with a well-defined potential energy barrier, the criterion will locate the transition state configuration near the top of the barrier. This is not the case for unimolecular simple bond cleavage reactions. In model studies where the reaction coordinate is represented by a Morse function, the criterion gives a value of r^* at the transition state configuration, considerably less than the rotational barrier position [43–46]. The rotational barrier is located in the product region and the Gorin model would give an overestimate, since the energized reactant molecules and transition-state configuration would not be in equilibrium. A successful application of the criterion to the rate calculation of C_2H_6 into CH_3 radicals has been made by this method [47]. Surprisingly, this particularly suitable method has not yet been used for the theoretical calculations of rate constants for unimolecular dissociations in ions.

4.6.2 An RRKM Theory without Transition State

In a modification of the RRKM theory described in many papers from 1964 to 1976 [48], the only required properties are those of the separated fragments. This choice would normally lead to a better description of dissociation products states, being designed for a better description of the kinetics in the dissociation region. Its formulation rests on the quasi-equilibrium hypothesis and is based on microscopic reversibility.

It excludes the separability of reaction coordinates and, as it starts from the products to the energized molecular ion, it is able to make an explicit use of angular momentum conservation. The starting point of the theory is Eq. 4.78, giving $P_i(E)$ the probability of the products appearing in state i, when they realize a total energy E

$$P_i(E) \sim \left(\frac{\sigma}{\pi\lambda^2}\right)_i g_i .$$

4.78

In this equation:
λ is the de Broglie wavelength associated with the relative velocity of the fragments;
g_i is a degeneracy factor of the separated states;
σ is the cross-section for the rebuilding of initial species at a total energy E.

The evaluation of σ_i implies the use of a collision model and the conservation of angular momentum. The collision model very often used is the Langevin collision model, as in the theory of ion-molecule reactions [49]. The best success of the theory lies in the a priori calculation of kinetic energy distributions for fragmentary ions. It may be recalled that, as shown in Fig. 4.7, the measured kinetic energy distributions are either of type A, equivalent to a bidimensional Maxwell-Boltzmann distributions or of type B, equivalent to tridimensional MB distributions.

In the first model proposed for calculations [48] it is assumed that no restrictions are imposed by the Langevin cross-section model, but conservation of angular momentum is taken into account. One could argue that such a limiting case could hardly be realistic and would be realized only for strongly attractive potential surfaces with no barrier between reactants and products. The calculated translational energy distribution is expressed by

$$P(E_T) \sim \int_0^{E^* - E_T} N_V(E) \, (E^* - E_T - E)^{s-1} \, dE ,$$

4.79

with, for example, $s = \dfrac{r-1}{2}$.

In Eq. 4.79 $N_V(E)$ is the density of vibrational states of the products and r the number of final degrees of freedom.

Such $P(E_T)$ differs from zero for $E_T = 0$ and is an approximately decreasing function of E_T. For this reason, it is, loosely speaking, termed a two-dimensional Maxwell-Boltzmann distribution [48] since it is cast in the form [34]

$$P(E_T) \sim e^{-E_T/\beta} ,$$

4.80

where β is the average released energy.

It reproduces quite well the observed type-A distributions as, for instance, in the case of $CH_3^+ + I$ from CH_3I^+ [34].

In the second proposed model, the restrictions imposed by the Langevin collision model are taken into consideration, namely when the long-range potential is only weakly attractive and the centrifugal potential plays an important role. This model leads to a three-dimensional Maxwell-Boltzmann distribution:

$$P\left(E_T\right) \sim E_T^{1/2}\, e^{-E_T/\beta}.$$ 4.81

Such distributions have been measured and satisfactorily calculated by the second model [48, 50] for the following processes:

$$C_6H_5CN^+ \rightarrow C_6H_4^+ + HCN$$

$$C_4H_6^+ \rightarrow C_3H_3^+ + CH_3$$

$$C_6H_6^+ \rightarrow C_6H_5^+ + H.$$

These agreements constitute an encouraging proof for the validity of the quasi-equilibrium hypothesis within a given electronic potential energy surface. It is noteworthy to add that the RRKM theory without transition state (most often called Phase Space Theory [51]) can also reproduce experimental kinetic energy distributions arising from the decay of short lived collision complexes [48], as in the case:

$$C_2H_4^+ + C_2H_4 \rightarrow (C_4H_8^+)^* \rightarrow C_3H_5^+ + CH_3.$$

4.6.3 An Interesting Step on the Way to Microscopic Reversibility: the Ion-Dipole Transient Complexes

The application of the principle of microscopic reversibility to molecular-ion dissociation implies to rebuild the molecular ion from its ionized and neutral counterparts, as it is proposed by the RRKM theories without transition states. As shown above this needs the building of a model for the reverse reaction. Rather frequently, in such a model, the ionic fragment will interact with a polar neutral counterpart. This situation will, therefore, give rise to a "transient dipole complex", possibly a "hydrogen bonded complex". This situation was postualted in the late 1970s as an explanation of some molecular ion dissociations. One of the first examples was described in 1978 [52], for the decomposition of the oxonium ion A (Fig. 4.11) considered as resulting from the occurrence in the mass spectrometer of "weakly coordinated carbonium ions". The authors show that A and B ions, equilibrated prior to dissociation into $i - C_3H_7^+ + CH_2O^+$, presumably via the ion dipole complexes C and D and the proton dimer E.

During the last decade, many examples of ion-dipole complexes have been discovered [53]. As will be shown, their role is also exemplified when reaction paths are

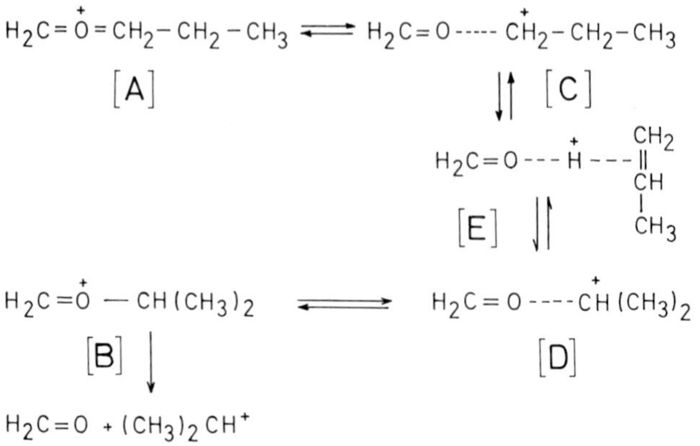

$$H_2\overset{+}{C}=O=CH_2-CH_2-CH_3 \rightleftharpoons H_2C=O\text{-----}\overset{+}{C}H_2-CH_2-CH_3$$

[A]   ↓↑ [C]

$$H_2C=O\text{---}\overset{+}{H}\text{---}\begin{array}{l}CH_2\\ \| \\ CH \\ | \\ CH_3\end{array}$$

[E] ↓↑

$$H_2\overset{+}{C}=O-CH(CH_3)_2 \rightleftharpoons H_2C=O\text{----}\overset{+}{C}H(CH_3)_2$$

[B] ↓   [D]

$$H_2C=O + (CH_3)_2CH^+$$

Fig. 4.11. Example of the decomposition of the oxonium ion [A] through the occurrence of weakly coordinated carbonium ions [C], [D], and [E].

deduced from ab initio calculations of potential energy surfaces, giving stronger support to subsequent RRKM calculations.

4.6.4 RRKM Calculations Supported by Ab Initio Calculated Potential Energy Surfaces

As pointed out in Chapter 3, the ab initio quantum calculations give good representations of ion structures. Their continuous refinements now allow the disentanglement of the successive evolution steps of a molecular ion along its dissociation path. A recently published study of two dissociation processes of the acetone molecular ion, will be given shortly, as an example [54].

The problem concerns the two parallel reactions observed for $C_3H_6O^+$ ions:

$$CH_3-CO-CH_3^+ \rightarrow CH_3-CO^+ + CH_3 \quad (1)$$

$$CH_3-CO-CH_3^+ \rightarrow CH_2=CO^+ + CH_4 \quad (2).$$

Experimental data collected on (1) and (2) are:
a. reaction (2) is the major channel for metastable ion fragmentation in the microsecond lifetime range;
b. reaction (1) becomes the major reaction channel at high internal energy of the reactant ion;
c. this appears surprising, owing to the almost equal appearance energies for both channels;
d. observation of an unusually large isotope effect, associated with methane loss from CD_3COCH_3: the loss of CD_3H is 70 times more probable than the loss of CH_3D [55].

Before any RRKM calculations, the reaction path on the CH_3–CO–CH_3^+ hypersurface was deduced from ab initio calculations.

The evolution of the system is described by the four principal structures [A], [B], [C], [D], schematically drawn below.

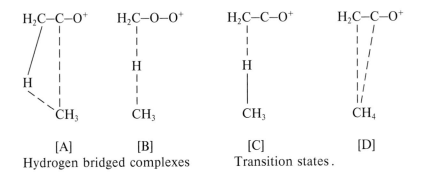

| [A] | [B] | [C] | [D] |
Hydrogen bridged complexes   Transition states.

Each of these structures is represented by a minimal or maximal energy in the evolution diagram of the system, drawn in Fig. 4.12.

From the ground-state structure of $C_3H_6O^+$, one observes a sharp increase in the potential energy, up to 16 kcal mol^{-1} where structures [A] and [B] are formed at the same energy. These species corresponding to the various conformations of the ion-dipole complex (CH_3CO^+/CH_3) may certainly interconvert freely. Some 5 kcal mol^{-1} higher, they will give rise to dissociation channel (1). With respect to the for-

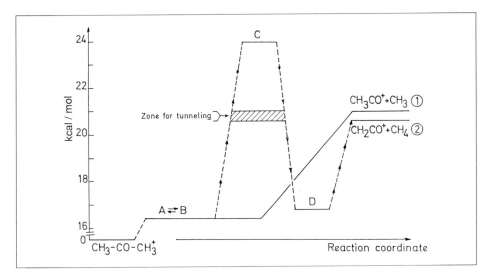

Fig. 4.12. Results of the ab initio calculations of the dissociation channels of acetone molecular ion.

mation of CH_2CO^+ and CH_4, structure [B] can be viewed as the leading conformation. Hydrogen transfer through transition state [C], some 3.4 kcal mol^{-1} above dissociation channel (2) leads to a weakly bound complex [D] some 3.8 kcal mol^{-1} below its components.

It is now appropriate to ask if this theoretical picture, used by RRKM calculations, can match the experimental situation. The calculations show that:

a. hydrogen transfer via the transition state [C] is effectively determining;

b. methyl loss occurs, in a continuously endothermic way, from complex [B];

c. reaction (2), slightly less endothermic than (1), is kinetically unfavorable due to the barrier associated with the former process;

d. a problem remains, since only CH_4 loss has been observed as a metastable process within the microsecond range.

The question that arose in d. was solved by accepting that a tunneling zone between 20.6 and 21 kcal mol^{-1} can lower the rate constant for reaction (2) in the range of $1.6 \cdot 10^4$ to $2.8 \cdot 10^4$ s^{-1}, in agreement with the observation of a metastable transition. Their metastable dissociation takes place over a limited energy range since, at the threshold for channel (1), the methyl loss is 75 times faster than the methane elimination which, from that energy, is completely suppressed.

The same tunneling phenomenon will also explain the surprising isotope effect for reaction (2), associated with the hydrogen transfer step.

Conclusions

A complete survey of improvements to RRKM statistical calculations of the unimolecular microcanonical reaction-rate constant does not fall within the scope of this introductory text. Some of these have been briefly reported on which the most recent developments have been based. Two excellent reviews about these developments were published recently, and include an extensive selection of important papers [56]. The developments of RRKM-based theoretical interpretations of the dissociation of internally excited molecular ions, constitute a central field in chemical kinetics and is a unique stimulation for the conception of more refined experimental methods. From improved experimental results, growth of the theory has also been observed. The continuous fitting of both these conditions makes study of the decay of excited molecular ions an outstanding research field.

References

1. Marcus RA (1952) Unimolecular Dissociations and Free Radical Recombination Reactions. J Chem Phys 20:359–364
2. Rosenstock HM, Wallenstein MB, Wahrhaftig AL, Eyring H (1952) Absolute Rate Theory for Isolated Systems and the Mass Spectra of Polyatomic Molecules. Proc Nat Acad Sci 38:667–678
3. Robinson PJ, Holbrook KA (1971) Unimolecular Reactions, Wiley Interscience, New York
4. Forst W (1973) Theory of Unimolecular Reactions. Academic, New York

5. Rice OK (1967) Statistical Mechanics, Thermodynamics and Kinetics. Freeman, San Fancisco
6. Setser DW (1966) Calculated Unimolecular Reaction Rates for Thermally and Chemically Activated Ethylene Oxide $- d_0$ and $- d_4$ and Acetaldehyde $- d_0$ and $- d_4$ Molecules. J Phys Chem 70:826-840
7. Doetsch G (1966) Guide to the Applications of Laplace Transforms. Saunders, Philadelphia
8. Wylic Jr CR (1951) Advanced Engineering Mathematics. McGraw-Hill Book Company Inc, New York
9. Rice OK, Rampsperger HC (1927) Theories of Unimolecular Gas Reactions at Low Pressures. J Am Chem Soc 49:1617-1629
10. Kassel LS (1932) Kinetics of Homogeneous Gas Reactions. Chemical Catalog Co, New York
11. Marcus RA, Rice OK (1951) Session of Free Radicals. The Kinetics of the Recombination of Methylradicals and Iodine Atoms. J Phys Colloid Chem 55:894-908
12. Whitten GZ, Rabinovitch BS (1963) Accurate and Facile Approximation for Vibrational Energy-Level Sums. J Chem Phys, 2466-2473
13. Schlag EW, Sandsmark RA (1962) Computation of Statistical Complexions as Applied to Unimolecular Reactions. J Chem Phys 37:168-171
14. Forst W, Prásil Z (1969) Comparative Test of Approximations for Calculation of Energy-Level Densities. J Chem Phys 51:3006-3012
15. Haarhoff PC (1963) The Density of Vibrational Energy Levels of Polyatomic Molecules. Mol Phys 7:101-117
16. Bouma WJ, Macleod JK, Radom L (1980) An Ab Initio Molecular Orbital Study of the CH_2O^+. Isomers: The Stability of the Hydroxymethylene Radical Cation. Int J Mass Spectrom Ion Proc 33:87-93
17. Osamura Y, Goddard JD, Schaefer III J (1981) Near Degenerate Rearrangement between the Radical Cations of Formaldehyde and Hydroxymethylene. J Chem Phys 74:617-621
18. Prásil Z, Forst W (1967) Application of the Statistical Theory of Mass Spectra to the Decomposition of $C_2H_6^+$ and $C_2D_6^+$. J Phys Chem 71:3166-3177
19. Stockbauer R (1973) Threshold Electron-Photoion Coincidence Mass Spectrometric Study of CH_4, CD_4, C_2H_6 and C_2D_6. J Chem Phys 58:3800-3815
20. Meisels GG, Chen CT, Giessner BG, Emmel RH (1972) Energy Deposition Functions in Mass Spectrometry. J Chem Phys 56:793-800
21. Flamme JP, Momigny J (1978) Theoretical Study of the Metastable Peaks Corresponding to the Rotational Predissociation of the CH_4^+ and CD_4^+ Ions. Chem Phys 34:303-309
22. Illies AJ, Jarrold MF, Bowers MT (1982) Fragmentation of Metastable CH_4^+ Ions and Isotopic Analogues. Kinetic Energy Release Distributions and Tunneling through a Rotational Barrier: Experiment and Theory. J Am Chem Soc 104:3587-3593
23. Andlauer B, Ottinger Ch (1971) Unimolecular Ion Decompositions: Dependence of Rate Constants on Energy from Charge Exchange Experiments. J Chem Phys 55:1471-1472
24. Baer T, Tsai BP, Smith D, Murray PT (1976) Absolute Unimolecular Decay Rates of Energy Selected Metastable Halobenzene Ions. J Chem Phys 64:2460-2465
25. Tsai BP, Werner AS, Baer T (1975) A Photoion-Photoelectron Coincidences (PIPECO) Study of Fragmentation Rates and Kinetic Energy Release in Energy Selected Metastable Ions. J Chem Phys 63:4384-4392
26. Vestal ML (1965) Theoretical Studies on the Unimolecular Reactions of Polyatomic Molecule Ions. I. Propane. J Chem Phys 43:1356-1369
27. Eland JHD, Schulte H (1975) Unimolecular Ion Decompositions: Rate Constants as a Function of Exitation Energy. J Chem Phys 62:3835-3836
28. Momigny J, Brakier L, d'Or L (1962) Comparaison entre les Effets de l'Impact Electronique sur le Benzène et sur les Isomères du Benzène en Chaîne Ouverte. Acad Roy Belgique-Bull Cl Science 48:1002-1015
29. Chesnavich WJ, Bowers MT (1977) Statistical Phase Space Theory of Polyatomic Systems. Application to the Unimolecular Reactions $C_6H_5CN^+ \rightarrow C_6H_4^+ + HCN$ and $C_4H_6^+ \rightarrow C_3H_3^+ + CH_3$. J Am Chem Soc 99:1705-1711

30. Terwilliger DT, Beynon JH, Cooks FRS, Cooks RG (1974) Kinetic Energy Distributions from the Shapes of Metastable Peaks. Proc R Soc London 341A:135-146
31. Smyth KC, Shannon TW (1969) Energetic Metastable Decompositions. J Chem Phys 51:4633-4642
32. Holmes JL (1977) Metastable Ion Studies (VIII): An Analytical Method for Deriving Kinetic Energy Release Distributions from Metastable Peaks. Int J Mass Spectrom Ion Proc 23:189-200
33. Mändli H, Robbiani R, Kuster Th, Seibl J (1979) Automatic Aquisition and Shape Analysis of Metastable Peaks. Int J Mass Spectrom Ion Proc 31:57-64
34. Mintz DM, Baer T (1967) Kinetic Energy Release Distributions for the Dissociation of Internal Energy Selected CH_3I^+ and CD_3I^+ Ions. J Chem Phys 65:2407-2415
35. Franklin JL (1976) Energy Partitioning in the Products of Ionic Decomposition. Science 193:725-732
36. Marcus RA (1973) General Discussion. Farad Discuss Chem Soc 55:381
37. Ben-Schaul A, Haas Y, Kompa KL, Levine RD (1981) Lasers and Chemical Change. Springer Series in Chemical Physics vol 10. Springer Verlag, Berlin
38. Messiah A (1961) Quantum Mechanics. North Holland, Amsterdam
39. Kinsey JL (1971) Microscopic Reversibility for Rates of Chemical Reactions Carried out with Partial Resolution of the Product and Reactant States. J Chem Phys 54:1206-1217
40. Momigny J, Locht R, Caprace G (1986) Mechanism for the Appearance of H^+ by Electroionization of CH_4. A Surprisal Analysis. Chem Phys 102:275-280
41. Forst WC (1973) Theory of Unimolecular Reactions. Academic, New York
42. Gorin E (1938) Photolysis of Acetaldehyde in the Presence of Iodine. Acta Physicochim USSR 9:681-696
43. Bunker DL, Pattengill M (1968) Monte Carlo Calculations. VI. A Re-Evaluation of the RRKM Theory of Unimolecular Reaction Rates. J Chem Phys 48:772-776
44. Marcus RA (1966) On the Theory of Chemical Reaction Cross Sections. I. A Statistical-Dynamical Model. J Chem Phys 45:2630-2638
45. Quack M, Troe J (1974) Specific Rate Constants of Unimolecular Processes. II. Adiabatic Channel Model. Ber Bunsenges Phys Chem 78:240-252
46. Hase WL (1972) Theoretical Critical Configuration for Ethane Decomposition and Methyl Radical Recombination. J Chem Phys 57:730-733
47. Hase WL (1976) The Criterion of Minimum State Density in Unimolecular Rate Theory. An Application to Ethane Dissociation. J Chem Phys 64:2442-2449
48. Klots CE (1964) Statistical Theory of Kinetic Energy of Fragmentation in Certain Unimolecular Dissociations. J Chem Phys 41:117-122
   Klots CE (1971) Reformulation of the Quasi-Equilibrium Theory of Ionic Fragmentation. J Phys Chem 75:1526-1532
   Klots CE (1972) Quasi-Equilibrium Theory of Ionic Fragmentation: Further Considerations. Z Naturforsch 27a:553-561
   Klots CE (1973) Thermochemical and Kinetic Information from Metastable Decompositions of Ions. J Chem Phys 58:5364-5367
   Klots CE (1973) Theory of Ionic Fragmentations: Recent Developments. Adv Mass Spectrom 6:969-974
   Klots CE (1976) Rate Constants for Unimolecular Decomposition at Threshold. Chem Phys Lett 38:61-64
   Klots CE (1976) Kinetic Energy Distributions from Unimolecular Decay: Predictions of the Langevin Model. J Chem Phys 64:4269-4275
49. Gioumousis G, Stevenson DP (1958) Reactions of Gaseous Molecule Ions with Gaseous Molecules. V. Theory. J Chem Phys 29:294-299
50. Chesnavich WJ, Bowers MT (1977) Statistical Phase Space Theory of Polyatomic Systems: Rigorous Energy and Angular Momentum Conservation in Reactions Involving Symmetric Polyatomic Species. J Chem Phys 66:2306-2315

51. Light JC (1964) Phase-Space Theory of Chemical Kinetics. J Chem Phys 40:3221–3229
    Pechukas P, Light JC (1965) On Detailed Balancing and Statistical Theories of Chemical Kinetics. J Chem Phys 42:3281–3291
    Light JC, Lin J (1965) Phase-Space Theory of Chemical Kinetics. II. Ion Molecule Reactions. J Chem Phys 43:3209–3219
    Pechukas P, Light JC, Rankin C (1966) Statistical Theory of Chemical Kinetics: Application to Neutral-Atom-Molecule Reactions. J Chem Phys 44:794–805
    Lin J, Light JC (1966) Phase-Space Theory of Chemical Kinetics. III. Reactions with Activation Energy. J Chem Phys 45:2545–2559
    Light JC (1967) Statistical Theory of Bimolecular Exchange Reactions. Disc Farad Soc 44:14–29
52. Bowen RD, Stapleton J, Williams DH (1978) Non-Concerted Unimolecular Reactions of Ions in the Gas Phase: Isomerisation of Weakly Co-Ordinated Carbonium Ions. J Chem Soc Chem Comm 24–26
53. Heinrich N, Schwarz H (1989) Ion/Molecule Complexes as Central Intermediates in Unimolecular Decompositions of Metastable Radical Cations of some Keto/Enol Tautomers. In: Maier JP (ed) Ion and Cluster Ion Spectroscopy and Structure. Elsevier, Amsterdam
54. Heinrich N, Louage F, Lifshitz C, Schwarz H (1988) Competing Reactions of the Acetone Cation Radical: RRKM-QET Calculations on an Ab Initio Potential Energy Surface. J Am Chem Soc 110:8183–8192
55. Lifshitz C, Tzidony E (1981) Kinetic Energy Release Distributions for $C_3H_6O^+$. Ion Dissociations: A Further Test of the Applicability of the Energy-Randomization Hypothesis to Unimolecular Fragmentations. Int J Mass Spectrom Ion Proc 39:181–195
56. Lifshitz C (1978) Unimolecular Decomposition of Polyatomic: Ions Decay Rates and Energy Disposal. Adv of Mass Spectrom 7A:3–18
    Lifshitz C (1989) Recent Developments in Applications of RRKM-QET. Adv of Mass Spectrom 11A:713–729

5 Miscellaneous Useful Topics in Molecular Ion Dissociation Phenomena

5.1 Appearance of Fragmentary Ions from Isolated Electronic States

From Chapters 1 and 2 it is easy to deduce that the complete redistribution of electronic energy in internal energy of the ground state ion, through radiationless transitions, is a widely realized behavior for complex molecular ions. In some cases, however, this step is limited by the occurrence of some radiative desexcitation. As the reader now realizes, this randomization is the prerequisite to the statistical distribution of vibrational energy in the ground molecular ion state, allowing an RRKM treatment of its subsequent decays.

Another circumstance could possibly limit the complete randomization of electronic energy. This will be realized if some electronic states could dissociate in a time-window shorter or concurrent with the one used for radiationless transitions. It does not appear that this behavior is frequent, but its occurrence has occasionally been detected and care has to be taken when one has to decide whether the behavior of a given ion is complete or is not, due to electronic energy randomization.

Before detailing two examples of incomplete energy randomization, it may be quoted that a preferential dissociation of $C_2F_6^+$ into $C_2F_5^+ + F$ was reported for its first excited electronic state, whereas the ground or higher excited states fragment predominantly into $CF_3^+ + CF_3$ [1, 2].

Similar specific ionic fragmentations were reported in methylhalides [3]. Two component decays observed for energy-selected $C_2H_5Cl^+$ and $C_2H_4Cl_2^+$ as well as for $HC \equiv C - CH_2Cl^+$ and $HC \equiv C - CH_2Br^+$ have been interpreted as non-statistical fragmentations [4]. It is remarkable that for all these cases the lowest $\tilde{A}$ electronically excited state was the only one involved.

As it has been shown from the PIPECO spectra of $C_2H_3F^+$ [5] and of *cis-* and *trans-*$C_2H_2F_2^+$ (*cis*-DFE and *trans*-DFE) [6], an isolated state decay sometimes occurs from higher electronic states.

5.1.1 The Vinyl Fluoride Case

As shown from PIPECO spectra [5] the breakdown diagram of C_2H_3F is easily interpreted as resulting from dissociation of the ground state ion only, except the appearance of $C_2H_3^+$. As shown in the branching ratio curve of $C_2H_3^+$ (Fig. 5.1) an important increase of the abundance of $C_2H_3^+$ ions arises when the fourth photoelectron band is excited, showing that this state preferentially decays into vinyl ions instead of forming an energy depopulated ion by radiationless transitions.

This situation, where the branching ratio shows a strong memory of the non-resonant photoelectron spectrum, is characteristic of an isolated state decay.

5.1.2 The *Cis*- and *Trans*-Difluoroethylenes

Cis- and *trans*-DFE were chosen in Chapter 2, as examples of molecular ions for which a competition between radiative decay and radiationless transitions are straightforward. As is easy to deduce from Fig. 5.2, where the PIPECO mass spectrum is compared to the photoelectron spectrum of *trans*-DFE [6], another competition is present from isolated state decay of C and D electronic states, i. e., the loss of an F-atom. This is shown by the shaded area in the $C_2H_2F^+$ branching ratio. Similarly, a second isolated state decay appears from the E and F states and concerns the loss of a hydrogen atom. This is visible in the shape of the branching ratio for $C_2H_2F_2^+$ ions, in agreement with the shape observed for bands 5 and 6 in the photoelectron spectrum. It is noteworthy that C and D states are described as resulting from the removal of πF and $\sigma(C-F)$ electrons, the F and G states being involved in

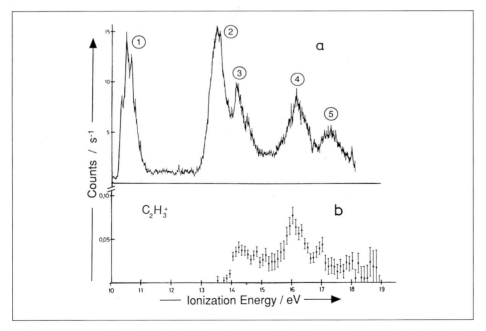

Fig. 5.1. PEPICO branching ratio observed for $C_2H_3^+$ ions from C_2H_3F compared with the photoelectron spectrum. The band (4) gives an isolated state decay contribution to the abundance of $C_2H_3^+$ ions [5].

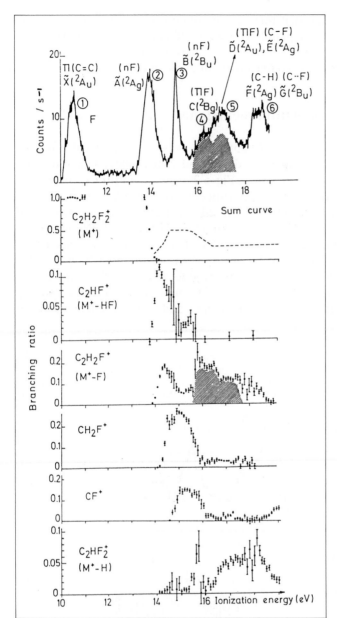

Fig. 5.2. PEPICO branching ratios observed in trans-DFE, compared with the photoelectron spectrum. Band (4) and (5) give isolated state decays to $C_2H_2F^+$; bands (4), (5), (6) give isolated state decays to $C_2HF_2^+$ [6].

the loss of a $\sigma(C–H)$ electron. It may be pointed out that such tentative correlations are rather more propensity rules than rigorous decay rules.

It may be added that the behavior of *cis*-DFE parallels the one observed for the *trans* isomer.

5.1.3 The Benzene Cation

The few specific examples of concurrently decaying electronic states, either radiationless or from isolated states decay, have to be completed by a short discussion of the problem encountered with the decay of benzene cations.

Does $C_6H_6^+$ decay in a purely statistical way or is it a typical case of non-communicating electronic states?

The origin of this debate goes back to 1962, when some authors [7] suggested by a comparative study of the mass spectra and ion energetics of the main fragmentary ions from benzene, hexadiynes, and butadienylacetylene (BDA), that excited benzene cations isomerize to BDA before any dissociation. The dissociation channels generally considered in $C_6H_6^+$ are:

$$C_6H_6^+ \rightarrow C_6H_5^+ + H \qquad (1)$$

$$C_6H_6^+ \rightarrow C_6H_4^+ + H_2 \qquad (2)$$

$$C_6H_6^+ \rightarrow C_4H_4^+ + C_2H_2 \qquad (3)$$

$$C_6H_6^+ \rightarrow C_3H_3^+ + C_3H_3 \qquad (4).$$

For these four channels, the threshold appearance energies with respect to the ground state of the molecular ion (9.25 eV), have been experimentally determined at 3.6 and 3.69 eV for channels (1) and (2); 4.16 and 4.19 eV for channels (3) and (4).

The fundamental question arising from the unimolecular decay of the benzene cation is whether hydrogen loss (process (1) and (2)) or C-loss (process (3) and (4)) originates from different electronic states. Charge-exchange measurements [8] revealed that the rate constants for process (1) and (3) are so different that both dissociation channels were non-competing (see Section 1.5.2.). This result was corroborated from photoionization yield curves [9] and from non-resonant PIPECO measurements [10]. More recently, however, it was concluded from ZKE photoelectron-photoion coincidence measurements [11] that the four channels are competing. This last observation appears as a challenge to the previous ones. Unfortunately, direct decay-rate constants measurements for both H- and H_2-loss channels have not been possible with PIPECO methods, due to the small mass changes, so that a direct test of whether C-loss channels and H-loss channels are competing or not, is still needed.

Such measurements have now been made by multiphoton ionization in a high-resolution time-of-flight mass spectrometer (reflectron TOF) [12].

By an appropriate two-photon absorption, benzene cations are prepared in their ground electronic and vibrational states, at 9.25 eV. The beam of $C_6H_6^+$ ions is subsequently enriched in energy by doubled frequency light pulses of a dye laser; they are continuously enriched in energy in a range between 5.1 and 5.5 eV.

By using the reflectron possibilities the rate constant for the four decay channels can be measured as a function of the excitation energy of $C_6H_6^+$. The results are shown in Fig. 5.3. It is evident from this figure that, at least in the quoted energy range, the four channels are competing. One is therefore induced to believe that the behavior of the $C_6H_6^+$ cation, is purely statistical. In other words, the electronically induced excited states of $C_6H_6^+$ decay to the ground state from which the four dissociation channels arise. In the same paper, the authors of this striking experiment perform RRKM calculations, in which transition-states vibrational frequencies have been chosen to fit the experimentally observed ones. This is an indirect agreement for the validity of the RRKM calculations in this particular case. It may be emphasized that these experiments are not inconsistent with the possible isomerization of benzene cations to BDA. One could suggest to study BDA through the same multiphoton ionization technique.

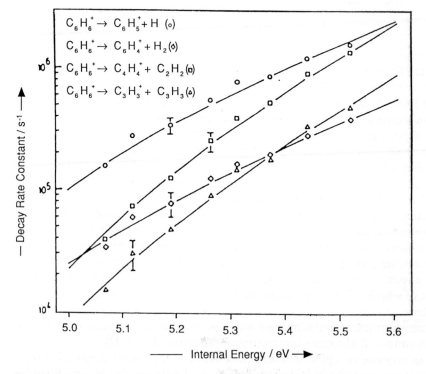

Fig. 5.3. Individual unimolecular rate constants $k(E)$ for the four competing channels of $C_6H_6^+$ cation as a function of the defined internal energy. The solid lines represent the results of an RRKM fit to the experimental data [12].

5.2 Indirect Population of Electronic States in Complex Molecular Ions

In the first part of this contribution, the occurrence of indirect population mechanisms of the electronic states of molecular ions, through the decay of superexcited states, has been emphasized. An example has been given of the population through this way of a dissociative channel of N_2O^+ (see Section 1.1.5.3). It was also shown that if charge exchange phenomena are unable to create superexcited states, Penning ionization does. It is not possible to review here all the numerous cases observed in triatomic molecules, but we will instead consider the possible occurrence of similar phenomena in more complex molecules. Many interesting examples of indirect population mechanisms are known in such molecules. Two of them will now be described here and discussed. It will be shown that if indirect population can lead to the realization of statistical decays, it is not excluded that it could also lead to the excitation of non-adiabatic channels.

5.2.1 The $CH_3F^+ \rightarrow CH_3^+ + F$ Channels

A recent paper on this problem reports about photoionization yield curves of CH_3^+ (Fig. 5.4) [13]. Three successive AEs are easily detected in the curves:

$AE_1 = 12.45 \pm 0.06 \text{ eV}$;

$AE_2 = 14.50 \pm 0.06 \text{ eV}$;

$AE_3 = 16.10 \pm 0.06 \text{ eV}$.

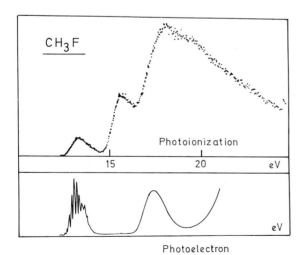

Fig. 5.4. A comparison between the photoionization yield curve for CH_3^+ from CH_3F with the photoelectron spectrum shows that the second threshold in CH_3^+ is due to an indirect population process [13].

With the well established thermochemical data, the threshold energies for CH_3^+ can be calculated as

$$CH_3F \rightarrow CH_3^+ + F^- \qquad \Delta H_0 = 10.99 \text{ eV} \qquad (1)$$

$$CH_3F \rightarrow CH_3^+ + F \qquad \Delta H_0 = 14.38 \text{ eV} \qquad (2).$$

The first step observed in the photoionization yield curve with an AE_1 value of 12.45 eV is obviously due to the ion pair process. As is easy to see in Fig. 2.4, the kinetic energy release in this process is quasithermal. This implies that the 1.45 eV observed in excess of ΔH_0 appears entirely as vibrational energy of CH_3^+.

The second step observed in the photoionization yield curve starts when the photoelectron band corresponding to the $\tilde{X}^2E$ state of CH_3F^+ has disappeared. The comparison between ΔH_0 calculated value for process (2) and the experimental AE_2 together with Fig. 3.4 (discussed in Chapter 3) shows that the first dissociation threshold of CH_3^+ in CH_3F is reached only by the decay of superexcited states lying between 14.5 eV and 16 eV. Their decay indirectly populates the first dissociation limit. This conclusion is supported by the lack of any CH_3^+ ions in the PIPECO non-resonant mass spectrum in this energy range, and by the lack of any cross-section in the photoelectron spectrum [15].

The third observed step starts at 16.10 eV when the $\tilde{A}^2A_1 + \tilde{B}^2E$ states of CH_3F^+ appear in the photoelectron spectrum. These CH_3^+ ions are ground state ions, as in the second step; their appearance is directly populated and their very wide kinetic energy distribution is practically similar from the threshold up to 18 eV. This observation suggests that they appear from a transition to a completely repulsive state.

From this example, it is clear that the decay of superexcited states lying between 14.5 and 16 eV is the only process to populate the $CH_3^+ + F$ dissociation channel, correlated with the ground state molecular ion. This circumstance appears as a way to fill the conditions requested for an RRKM behavior of $\tilde{X}$ CH_3F state, at least for the mentioned dissociation process.

5.2.2 The $C_2H_3^+$ + Cl and $C_2H_2^+$ + HCl Channels from C_2H_3Cl

The most abundant fragmentary ions observed under electron or photon impact in C_2H_3Cl correspond to the following two processes:

$$C_2H_3Cl \rightarrow C_2H_3^+ + Cl \qquad (1)$$

$$C_2H_3Cl \rightarrow C_2H_2^+ + HCl \qquad (2).$$

From photoionization measurements [16], it was shown that both processes have practically equal AEs: 12.48 $\pm$ 0.04 for (1) and 12.47 $\pm$ 0.1 eV for (2).

From measurements made in our laboratory, under electron impact, it was shown that, at their AEs, both processes form metastable ions. If one considers the photoelectron spectrum in Fig. 5.5 [17] of C_2H_3Cl as well as the photoionization yield curve for $C_2H_3^+$ and $C_2H_2^+$, it is clear that these ions arise in an energy region (between 12 and 13 eV) where the photoelectron spectrum is completely flat.

It is therefore evident that both the dissociation limits (1) and (2) are populated through the decay of superexcited states lying in this energy range (terms of Ryberg series converging to 13 eV would be good candidates).

The question arising now concerns the validity of a statistical treatment of the dissociation processes (1) and (2). One may regret that the PEPICO spectra of C_2H_3Cl have not been studied up to now. As ab initio calculations of potential energy hypersurfaces of $C_2H_3Cl^+$ are not available, we will take this occasion to treat the problem with the help of "correlation diagrams".

If channels (1) and (2) are written more explicitly as:

$$C_2H_3Cl \rightarrow C_2H_3^+ \, (^1A_1) + Cl \, (^2P_u) \qquad (1)$$

$$C_2H_3Cl^+ \rightarrow C_2H_2^+ \, (^2\Pi_u) + HCl \, (^1\Sigma^+) \qquad (2),$$

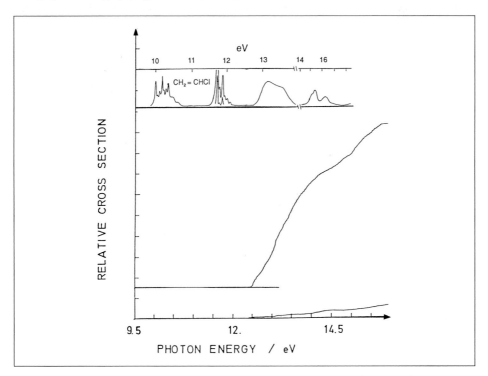

Fig. 5.5. A comparison between the photoionization yield curves for $C_2H_3^+$ and $C_2H_2^+$ from C_2H_3Cl and the photoelectron spectrum of C_2H_3Cl showing threshold population due to an indirect mechanism [16, 17].

and if it is pointed out that $C_2H_3Cl^+$ pertains to C_s symmetry, one can write (1) and (2):

$$C_2H_3Cl^+ \rightarrow C_2H_3^+ \, (^1A') + Cl\,(^2A'', \, ^2A', \, ^2A') \qquad (1)$$

$$C_2H_3Cl^+ \rightarrow C_2H_2^+ \, (^2A'', \, ^2A') + HCl\,(^1A') \qquad (2).$$

The products of channel (1) can arise form $^2A''$, $^2A'$, $^2A'$ states of $C_2H_3Cl^+$; those of channel (2) from $^2A''$ and $^2A'$ states of $C_2H_3Cl^+$.

The higher molecular orbitals of C_2H_3Cl are:

$$(7a')^2 \, (1a'')^2 \, (8a')^2 \, (2a'')^2 \, ,$$

the four first electronic states observed in the photoelectron spectrum are:

$$\tilde{X}\,(^2A''), \; A\,(^2A'), \; B\,(^2A''), \; C\,(^2A') \, .$$

Correlating all this information enables to propose in Fig. 5.6, a highly probable correlation diagram. In Fig. 5.6, the direct ionization processes observed in the photoelectron spectrum are indicated by rectangles. It is therefore clear that the occurrence of superexcited states around 12.5 eV is the only way to populate channels (1) and (2) at their threshold, correlated with $\tilde{X}^2A''$ and $\tilde{A}^2A'$ states. If the initial population of $\tilde{A}$ state can decay by radiationless transition to the $\tilde{X}$ state from which both channels proceed, the conditions are fulfilled for a statistical treatment.

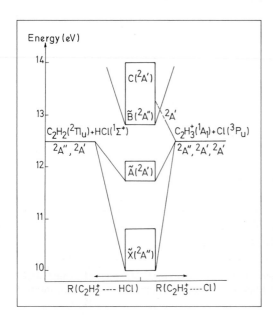

Fig. 5.6. Correlation diagram for $C_2H_3Cl^+$ showing how the two dissociation processes are populated by decay of superexcited states. One could predict that a PEPICO mass spectrum will at least need the excitation of the $\tilde{B}^2A''$ band to observe the $C_2H_3^+$ appearance through a predissociation process.

It may be noticed that light emission from $\tilde{A}$ to $\tilde{X}$ has never been reported. In this case the indirect population results in the possibility for the system to evolute in a statistical way. From the correlation diagram it is expected that the AEs of $C_2H_3^+$ in PEPICO spectra will lie somewhere between 13 and 14 eV, and will result from non-adiabatic behavior: the probable predissociation of $\tilde{B} + \tilde{C}$ states through the second $^2A'$ state correlated with the $\tilde{X}$ and $\tilde{A}$ states. This $^2A'$ state certainly corresponds to a doubly excited configuration of $C_2H_3Cl^+$.

Many other examples have been quoted in the literature including the threshold dissociations $CH_2 = CO \rightarrow CH_2^+ + CO$ [18], $CH_3OH \rightarrow CH_2OH^+ + H$ [19], and the threshold dissociation of allene and cyclopropene [20]: $C_3H_4^+ \rightarrow C_3H_3^+ + H$.

5.3 Tunneling through Potential Barriers – The Rotational Predissociation

In Chapter 4, when the general expression for $k(E)$ was derived from statistical considerations, attention was paid (Section 4.1.2.3) to the possibility of a tunneling of atoms or radicals through potential energy barriers. A first example of the influence of a tunneling possibility on the rate-constant value was described in 4.6.4 concerning the dissociation of CH_3–CO–CH_3^+ ions into $CH_2O^+ + CH_4$.

Such situations have only been scarcely mentioned in the literature, except H or D tunneling due to predissociation.

The phenomenon of rotational predissociation is best described if diatomic molecules are considered. It results from the occurrence of an effective potential energy $U(R, J)$ where the customarily used $U_0(R)$ potential energy is combined with the rotational energy of the molecule:

$$U(R, J) = U_0(R) + \frac{h(J + 1)J}{8\pi^2 c I}. \qquad 5.1$$

I is the moment of inertia of the diatomic rotator, and c the speed of light.

The use of $U(R, J)$ shows that vibrorotational levels of the molecule can be stable above the dissociation limit, because potential barriers arise from the combination of both terms in Eq. 5.1 as is shown in Fig. 5.7 for a hypothetical AH^+ ion.

For any energy level resulting from the combination of a vibrational with a rotational level, realized above the dissociation limit, one can expect H or D atom tunneling through the potential barrier within a time period defined by quantum mechanics as:

$$\tau = \frac{1}{2} \tau_0 \exp\left[\frac{4\pi}{h} \int_{R_1}^{R_2} \sqrt{2m(U - E)} \, dR\right]. \qquad 5.2$$

In this expression τ_0 is the time of vibration realized by the rovibrator on the way to dissociation, m is the mass of the tunneling particle, and the domain of integration ranges from R_1 to R_2, the distances corresponding to the barrier crossing.

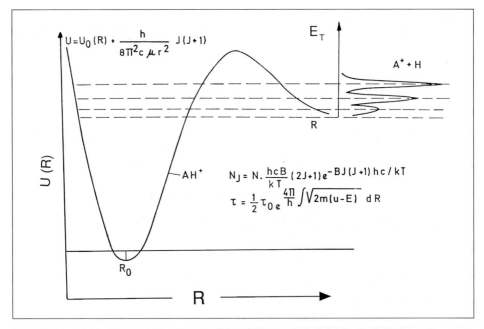

Fig. 5.7. A model for the vibrational potential energy curve of a hypothetical AH^+ ion. The position in energy of some quasibound states shows how the kinetic energy distribution of A^+ is expected to be modulated when produced by tunneling through the potential barrier.

If the diatomic molecules is a hydride ion, the vibrational quantum levels at a given temperature will modulate the fragmentary ion kinetic energy distribution as shown on the righthand side of Fig. 5.7.

If the lifetime of the AH^+ ion, due to the tunneling probability, corresponds to the time-window for the appearance of a metastable ion, one will see the metastable peak shape modulated through the rotational quantum levels compatible with the time-window of the ion-analyzing device.

As an example of this, Fig. 5.8 shows a momentum analysis for the rotational predissociation of HeH^+ [21], studied in the 10^{-5} to 10^{-6} s range. The cited paper ascribes the maxima to the following sets of v and J quantum numbers: (v, J); $(0,25)$ $(1,23)$ $(2,21)$, $(3,19)$ $(4,17)$.

In order to summarize the characteristics of a rotational predissociation, it can be assumed that:
● the kinetic energy distribution will be modulated through (v, J) values of the molecular ion able to justify a tunneling for a given time window;
● this distribution and its modulation will be temperature-dependent as is the rotational population.

Is such a model useful in the case of a polyatomic molecular ion? To our knowledge, a unique case has been investigated: the hydrogen or deuterium atom abstraction process from CH_4^+ and CD_4^+:

$$CH_4^+ (CD_4^+) \rightarrow CH_3^+ (CD_3^+) + H(D).$$

This process appears as a metastable process and, surprisingly, by using ICR experiments it was shown that its lifetime can extend to the millisecond range [22]. In a low and narrow temperature-variation range of 400 K to 500 K it was shown that the metastable intensity was temperature-dependent [23]. More recently, a weak but detectable kinetic energy modulation of the kinetic-energy distribution was observed between 145–423 K [24]. An experiment in which CH_4 and CD_4^+ were produced at about 1500 K in a monoplasmatron ion source [25] has shown strong kinetic energy modulation in the microsecond range, as shown in Fig. 5.8. The experimental kinetic energy modulations are given in Table 5.1.

Table 5.1. Kinetic energy observed for the microsecond decay of $CH_4^+ (CD_4) \rightarrow CH_3^+ (CD_3^+) + H(D)$ in meV.

$CH_4^+ \rightarrow CH_3^+ + H$	$CD_4^+ \rightarrow CD_3^+ + D$
0.0	0.0
1.7 ± 0.3	0.4 ± 0.2
5.8 ± 1.0	2.7 ± 0.8
23.0 ± 4.0	6.5 ± 1.2
56.0 ± 9.0	12.3 ± 2.0
	20.3 ± 2.7

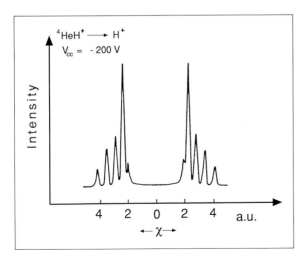

Fig. 5.8. Example of a forward-backward momentum distribution of H^+ fragment from $^4HeH^+$, in the microsecond lifetime range. χ stands for the center of mass momentum of H^+ in atomic units.

A theoretical model for CH_4^+ on the way to its dissociation has been proposed [26] which provides the means to calculate the position in energy of quasibound levels of CH_4^+ (CD_4^+) in the lifetime range of 10^{-7} to 10^{-5} s. This model depends on both J and K rotational quantum numbers of a symmetric top molecule. $U_0(R)$ is the long-range attractive potential energy between a point charge located on the C atom and the neutral H (D):

$$U_{J,K}(R) = -\frac{e^2 \alpha_{H(D)}}{R^4} + [J(J+1) - K^2]\frac{h^2}{8\pi^2 I_B}.$$ 5.3

In this formula, $\alpha_{H(D)}$ is the polarizability of H (D) atom, I_B is the moment of inertia for rotation occurring around an axis perpendicular to the symmetry axis; J and K are the rotational quantum numbers related to the absolute value of the rotational angular momentum and its projection on the symmetry axis. Tables 5.2 and 5.3 give the results of the calculations for CH_4^+ and CD_4^+. Calculated kinetic energy values compare favorably with the modulations observed in the experiment. The same model can also justify the occurrence of quasibound states with lifetimes in the milli-second range.

The validity of this rather crude model does not allow to conclude that the metas-table CH_4^+ (CD_4^+) ions are unexplainable, as was taught before by a statistical approach. In fact, a statistical approach where the tunneling process was homogene-ously introduced, reproduces quite well the observations reported between 145 K and 423 K [24]. Unfortunately, the statistical approach has not yet been tested at

Table 5.2. Calculated quasi-bound levels of CH_4^+ in the lifetime range 10^{-7}–10^{-5} s.

$v^{1)}$	J	K	E_T (meV)	τ (10^{-6} s)
12	12	4	1.4	1.8
12	15	10	1.4	1.8
12	21	18	1.1	9.2
11	15	5	3.6	1.0
11	22	17	4.0	0.3
10	19	0	12.2	3.2
10	19	1	11.9	6.0
10	20	6	13.3	0.8
10	21	9	12.4	2.2
10	22	11	13.5	0.6
10	23	13	13.0	1.0
9	25	4	39.9	0.2
9	25	5	37.1	0.6
9	25	6	33.7	3.0

[1]) v is the artificial vibrational quantum number of the level obtained from the diatomic-type potential; E_T is the kinetic energy released on the fragments.

Table 5.3. Calculated quasi-bound levels of CD_4^+ in the lifetime range 10^{-7}–10^{-5} s.

v	J	K	E_T (meV)	τ (10^{-6} s)
17	16	0	2.2	0.4
17	16	1	2.1	0.5
17	16	2	1.9	1.4
17	20	12	2.4	0.1
17	21	14	1.8	3.0
17	27	22	2.2	0.4
16	18	0	3.1	6.0
16	18	1	3.0	6.1
16	26	19	3.0	6.1
16	29	23	3.0	6.1
15	23	5	8.4	0.5
15	24	9	7.5	4.5
15	25	11	8.7	0.3
15	26	13	9.2	0.1
15	27	15	8.9	0.2
15	28	17	7.9	2.0
15	30	20	8.8	0.2
14	28	0	21.3	0.2
14	28	1	21.2	0.2
14	28	2	20.8	0.3
14	28	3	20.1	0.5
14	28	4	19.1	1.4
14	28	5	17.8	5.0
14	29	8	20.5	0.4
14	29	9	18.1	0.9
14	30	11	20.9	0.2
14	30	12	17.7	6.2

much higher temperatures in order to compare theoretical results with the results obtained with the monoplasmatron ion source.

5.4 Non-Adiabatic Spin-Forbidden Interactions

Referring back to Chapter 2, two main cases of non-adiabatic interactions between potential hypersurfaces were used to explain the procedure followed for a complete electronic energy redistribution into internal energy of the ground state molecular ion.

In some cases, such non-adiabatic interactions are at the origin of some dissociation processes involving spin-forbidden interactions that could, therefore, hardly been interpreted in the frame of RRKM theory.

It is not possible to report here all the dissociation processes of molecular ions where such effects have been shown to occur, either by the simple use of correlation diagrams, or as a result of ab initio quantum calculations of potential energy surfaces.

If one refers to Chapter 1 of this contribution, an example is shown in Fig. 1.11 about the behavior of N_2O^+ near the threshold for the appearance of NO^+ ions.

Two examples taken from more complex polyatomic ions will be considered here.

5.4.1 The Appearance of CD_2O^+ from CD_3OH^+

An intense metastable signal has been shown to occur at 16.2 eV, supporting the appearance of CD_2O^+ with a total kinetic energy ranging from 1.14 eV to 2.59 eV [27, 28]. The rather high appearance energy observed for the metastable transition indicates that the $\tilde{C}^2A''$ state of CD_3OH^+ is involved in the process. As it has been shown theoretically [29], above the $\tilde{A}^2B_1$ state of CH_2O^+, there exist $^2A_1$ (1) and $^2B_2$ (2) states at 1.76 and 2.3 eV, respectively. The first CH_2O^+ quartet state is quoted as lying near the $\tilde{A}(^2B_1)$ state.

With these values in mind one can draw the correlation diagram given in Fig. 5.9. From this diagram, it can be seen that the metastable process $CD_3OH^+ \rightarrow$

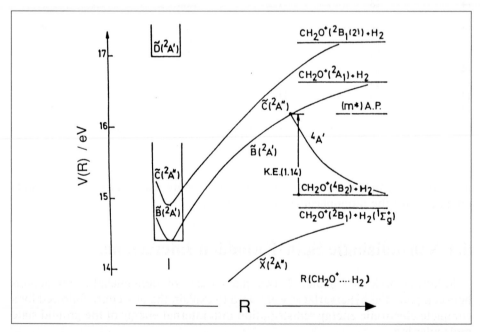

Fig. 5.9. Correlation diagram for the appearance of CD_2O^+ from CH_3OH^+ showing how the predissociation of the $\tilde{C}(^2A'')$ state though a $^4A'$ state arising from CD_2O^+ ($^4B_2$) + HD ($^1\Sigma_g^+$) explains the AE of CD_2O^+ (16.2 eV) and the kinetic energies observed from the corresponding metastable peak shape [28].

$CD_2O^+ + HD$ is due to a spin-forbidden, symmetry-allowed predissociation of the $\tilde{B}(^2A')$ state of CD_3OH^+ through a $^4A'$ repulsive state of the same ion, correlated with the CD_2O^+ $(^4B_2) + HD$ dissociation limit. Such a predissociation process can explain the large excess kinetic energy involved in this dissociation process.

5.4.2 The Dissociation of CHO_2^+ Ions

Another striking example concerns the decay of CHO_2^+ ions. It was shown that CHO_2^+ ions can exist either in the carboxylic structure $OCOH^+$ (I) or in the formate structure $O(CH)O^+$ (II) [30]. These ions loose an oxygen atom on the microsecond time scale by a composite mechanism undergoing a large isotope effect.

One of the components of the metastable peak corresponds to

$$OCOH^+ \, (^1A') \rightarrow O(CH)O^+ \, (^3A') \rightarrow HOC^+ + O \, ,$$

and appears with a small kinetic energy release.

The second component is due to a spin-orbit controlled, direct predissociation process:

$$OCOH^+ \, (^1A') \rightarrow OCOH^+ \, (^3A') \rightarrow HCO^+ + O \, .$$

In this second process, the kinetic energy release is large and is expected to be larger in the deuterated compound. This explains why the two components are resolved in the deuterated compound, but are hardly distinguishable in the hydrogenated species.

References

1. Simm IG, Danby CJ, Eland JHD (1973) Direct Observation of an Isolated State in the Ion $C_2F_6^+$: A Violation of the Quasi-Equilibrium Theory of Mass Spectra. J Chem Soc Chem Comm 832–833
   Simm IG, Danby CJ, Eland JHD (1974) The Fragmentation of $C_2F_6^+$ Ions Studied by Photoelectron-Photoion Coincidence Spectrometry. Int J Mass Spectrom Ion Proc 14:285–293
2. Inghram MG, Hanson GR, Stockbauer R (unpublished)
3. Eland JHD, Frey R, Kuestler A, Schulte H, Brehm B (1976) Unimolecular Dissociations and Internal Conversions of Methyl Halide Ions. Int J Mass Spectrom Ion Proc 22:155–170
4. Tsai BP, Werner AS, Baer T (1975) A Photoion-Photoelectron Coincidence (PIPECO) Study of Fragmentation Rates and Kinetic Energy Release in Energy Selected Metastable Ions. J Chem Phys 63:4384–4392
5. Dannacher J, Schmelzer A, Stadelmann J-P, Vogt J (1979) A Photoelectron-Photoion Coincidence Study of Vinylfuoride. Int J Mass Spectrom Ion Proc 31:175–186
6. Stadelmann J-P, Vogt J (1979) Isolated State Dissociations from Electronically Excited Radical Cations of Fluoroethanes Studied by Photoelectron-Photoion Coincidence Spectroscopy. Adv Mass Spectrom 8A:47–55

7. Momigny J, Brakier L, D'Or L (1962) Comparaison entre les Effets de l'Impact Electronique sur le Benzène et sur les Isomères du Benzène en Chaîne Ouverte. Acad Roy Belgique-Bull Cl Science 48:1002–1015
8. Andlauer B, Ottinger Ch (1972) Dissociation Lifetimes of Molecular Ions Produced by Charge Exchange. Z Naturforsch 27a:293–309
9. Rosenstock HM, Larkins JT, Walker JA (1973) Interpretation of Photoionization Thresholds: Quasi-Equilibrium Theory and the Fragmentation of Benzene. Int J Mass Spectrom Ion Proc 11:309–328
10. Eland JHD, Schulte H (1975) Unimolecular Ion Decomposition: Rate Constants as a Function of Excitation Energy. J Chem Phys 62:3835–3836
    Eland JHD, Frey R, Schulte H, Brehm B (1976) New Results on the Fragmentation of the Benzene Ion. Int J Mass Spectrom Ion Proc 21:209–211
11. Baer T, Willett GD, Smith D, Philips JS (1979) The Dissociation Dynamics of Internal Energy Selected $C_6H_6^+$. J Chem Phys 70:4076–4085
12. Kühlewind H, Kiermeier A, Neusser HJ (1986) Multiphoton Ionization in a Reflectron Time-of-Flight Mass Spectrometer: Individual Rates of Competing Dissociation Channels in Energy-Selected Benzene Cations. J Chem Phys 85:4427–4435
13. Locht R, Momigny J, Rühl E, Baumgärtel H (1987) A Mass Spectroscopic Photoionization Study of CH_3F. The CH_2^+, CH_3^+ and CH_2F^+ Ion Formation. Chem Phys 117:305–313
14. Rosenstock H, Drexl H, Steiner BW, Herron JT (eds) (1977) Energetic of Gaseous Ions. J Phys Chem Ref Data Vol 6 Suppl I
15. Karlsson L, Jadrny R, Mattsson L, Chau FT, Siegbahn K (1977) Vibrational and Vibronic Structure in the Valence Electron Spectra of CH_3 Molecules (X = F, Cl, Br, I, OH). Phys Scripta 16:225–234
16. Reinke D, Kräßig R, Baumgärtel H (1973) Photoreactions of Small Organic Molecules: I. Mass-Spectrometric Study of Vinylchloride, Vinylfluoride and 1,1-Difluorethylene in the Vacuum Ultraviolet. Z Naturforsch 28a:1021–1031
17. Lake RF, Thompson H (1970) Photoelectron Spectra of Halogenated Ethylenes. Proc Roy Soc London 315A:323–338
18. McCulloh KE, Dibeler VH (1976) Enthalpy of Formation of Methyl and Methylene Radicals from Photoionization Studies of Methane and Ketene. J Chem Phys 64:4445–4450
19. Brehm B, Fuchs V, Kebarle P (1971) Autoionization and Fragmentation Processes in Methanol and Ethanol. Int J Mass Spectrom Ion Proc 6:279–289
20. Parr AC, Elder FA (1968) Photoionization of 1,3-Butadiene, 1,2-Butadiene, Allene and Propyne. J Chem Phys 49:2659–2664
    Dannacher J, Vogt J (1978) Unimolecular Fragmentation of Energy Selected Allene Molecular Cations. Hel Chim Acta 61:361–372
21. Locht R, Maas JG, Van Asselt NPFB, Los J (1976) The Rotational Predissociation of a He H^+: Energy and Lifetime Measurements. Chem Phys 15:179–184
22. Smith RD, Futrell JH (1976) Slow Metastable Decomposition of Methane and Deuterated Methane Molecular Ions. Int J Mass Spectrom Ion Proc 20:425–427
23. Solka BH, Beynon JH, Cooks RG (1975) Metastable Methane Ions. Temperature Dependence of the Translational Energy Release. J Phys Chem 79:859–862
24. Illies AJ, Jarrold MF, Bowers MT (1982) Fragmentation of Metastable CH_4^+ Ions and Isotopic Analogues. Kinetic Energy Release Distributions and Tunneling through a Rotational Barrier: Experiment and Theory. J Am Chem Soc 104:3587–3593
25. Flamme JP, Wankenne H, Locht R, Momigny J, Nowak PJCM, Los J (1978) The Rotational Predissociation of CH_4^+ and CD_4^+ Ions. Chem Phys 27:45–49
26. Flamme JP, Momigny J (1978) Theoretical Study of the Metastable Peaks Corresponding to the Rotational Predissociation of CH_4^+ and CD_4^+ Ions. Chem Phys 34:303–309
27. Beynon JH, Fontaine AE, Lester GR (1968) Mass Spectrometry: The Mass Spectrum of Methanol. Part I: Thermochemical Information. Int J Mass Spectrom Ion Proc 1:1–24
28. Momigny J, Wankenne H, Krier C (1980) Correlation Diagram Approach to the Dissociative Ionization Mechanisms of Methanol. Int J Mass Spectrom Ion Proc 35:151–170

29. Pires MV, Galloy C, Lorquet JC (1978) Unimolecular Decay Paths of Electronically Excited Species: I. The H_2CO^+ Ion. J Chem Phys 69:3242–3249
30. Remacle F, Petitjean S, Dehareng D, Lorquet JC (1987) An Ab Initio Study of the Isomerisation and Fragmentation of CHO_2^+ Ions: An Example of Spin-Controlled Reactions? Int J Mass Spectrom Ion Proc 77:187–201

Appendix

The direct count of the number of vibrational states of CO_2, up to a total energy of 4002 cm^{-1}, is split into two tables; in the first one, energies of the lower vibrational states are calculated; in the second, the number of quantum states of CO_2 is calculated.

The formula giving the energies is:

$$E \text{ (cm}^{-1}) = 2349 \, v_1 + 1388 \, v_2 + 667 \, v_3.$$

The direct count of the number of states gives rise to combinatorial problem for the counting of degenerate states.

The number of ways to accomodate n quanta in q oscillators is given by:

$$W(E) = \frac{(n + q - 1)!}{n!\,(q - 1)!}.$$

As an example, the number of ways to accomodate 2 quanta in a triply degenerate level is

$$W(E) = \frac{4!}{2!\,2!} = \frac{4 \cdot 3 \cdot 2 \cdot 1}{2 \cdot 2} = 6.$$

In the case of CO_2, the occurrence of the doubly degenerate level v_3 will need the use of this combinatorial formula. One will calculate the following table:

$$
\begin{aligned}
n &= 1 \quad q = 2 \quad W(E) = 2 \\
n &= 2 \quad q = 2 \quad W(E) = 3 \\
n &= 3 \quad q = 2 \quad W(E) = 4 \\
n &= 4 \quad q = 2 \quad W(E) = 5 \\
n &= 5 \quad q = 2 \quad W(E) = 6 \\
n &= 6 \quad q = 2 \quad W(E) = 7
\end{aligned}
$$

v_1	v_2	v_3	E (cm^{-1})
0	0	0	0
		1	667
		2	1334
		3	2001
		4	2668
		5	3335
		6	4002
		7	4669 >

v_1	v_2	v_3	E (cm^{-1})
	1	0	1368
	1	1	2035
	1	2	2702
	1	3	3369
	1	4	4036 >
0	2	0	2736
0	2	1	3403
0	2	2	4070 >
0	3	0	4104 >
1	0	0	2349
1	0	1	3016
1	0	2	3683
1	0	3	4350 >
1	1	0	3707
1	1	1	4384 >
1	2	0	5085 >
2	0	0	4698 >
2	1	0	6086 >
3	0	0	7947 >

E (cm^{-1})	v_1	v_2	v_3	$W(E)$	$G(E)$
0	0	0	0	1	
667	0	0	1	2	3
1334	0	0	2	3	6
1368	0	1	0	1	7
2001	0	0	3	4	11
2035	0	1	1	3	14
2349	1	0	0	1	15
2668	0	0	4	5	20
2702	0	1	2	4	24
2736	0	2	0	1	25
3016	1	0	1	3	28
3335	0	0	5	6	34
3403	0	2	1	4	38
3683	1	0	2	4	42
3717	1	1	0	2	44
4002	0	0	6	7	51

An Example of an RRKM Calculation

It is proposed to calculate the near threshold behavior of the rate constants k_1 and k_2 for the isoenergetic processes:

$$C_2H_3Cl^+ \rightarrow C_2H_3^+ + Cl \qquad (1)$$

$$C_2H_3Cl^+ \rightarrow C_2H_2^+ + HCl \qquad (2).$$

The experimental measurements and results concerning both these processes have been discussed in 5.2.2. One will remember that (1) and (2) appear at 12.48 eV and that both processes are observed as metastables, including very low kinetic energy releases. Between their threshold and 13 eV, these dissociations results from the filling up of the molecular ion energy by an indirect population mechanism.

The successive steps for the calculation of k_1 and k_2 are described below.

1. The Vibrational Frequencies of the Energized Molecular Ion

The 12 fundamental vibrational frequencies of C_2H_3Cl are given in [1] and collected in Table 1.

Table 1. Normal vibrations of C_2H_3Cl.

Species	Description	Frequency (cm^{-1})
ν_1 (a')	symmetrical valence stretching in CH_2	3029
ν_2 (a')	valence stretching $C = C$	1608
ν_3 (a')	CH_2 deformation	1371
ν_4 (a'')	$C = C$ torsion	620
ν_5 (a')	$C - H$ stretching	3077
ν_6 (a')	CH_2 wagging	1038
ν_7 (a')	out of plane vibration	896
ν_8 (a'')	out of plane vibration	941
ν_9 (a')	antisymmetrical $C - H$ valence stretching	3120
ν_{10} (a')	wagging $C = C - Cl$	398
ν_{11} (a')	$C - Cl$ valence stretching	716
ν_{12} (a')	$C - H$ wagging in $CHCl$	1279

A regrouping of the frequencies is proposed for the molecule:

ν_1, ν_5, ν_9 (3075) (3); ν_2 (1608) (1); ν_3, ν_{12} (1325) (2); ν_6 (1040) (1);

ν_7, ν_8 (920) (2); ν_{11} (720) (1); ν_4 (620) (1); ν_{10} (400) (1).

From this and from the frequencies observed in the photoelectron spectrum [2] one proposes the following frequencies regrouping in $C_2H_3Cl^+$

ν_1, ν_5, ν_9 (3050) (3); ν_2 (1290) (1); ν_3, ν_{12} (1325) (2); ν_6 (1030) (1);

ν_7, ν_8 (920) (2); ν_{11} (720) (1); ν_4 (180) (1); ν_{10} (400) (1).

The ν_2 is taken from the photoelectron spectrum. The torsion frequency ν_4 is based on the following remark:

$$\nu(C_2H_4) = 800 \text{ cm}^{-1}; \quad \nu(C_2H_4^+) = 230 \text{ cm}^{-1}; \quad \varrho = \frac{\nu(C_2H_4^+)}{\nu(C_2H_4)} = 0.287$$

$$\nu(C_2H_3Cl^+) = 620 \cdot 0.287 = 178 \approx 180 \text{ cm}^{-1}.$$

2. The Choice of a Transition State for Process (1)

For this reaction, a loose transition complex will be used.

Its characteristics will be imposed by the ab initio calculated structure of $C_2H_3^+$ ion [3], as given in Fig. A 1.

The similarity of internuclear distances in CH_2 with those observed in C_2H_4 leads to the following attributions in the complex:

ν_1^* (a') (valence symmetric stretching):   3020 cm^{-1}

ν_9^* (a') (valence antisymmetric stretching):   3200 cm^{-1}

ν_3^* (a') (H$_2$C deformation):   1350 cm^{-1}.

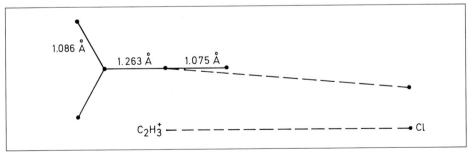

Fig. A 1. Transition state for the reaction $C_2H_3Cl^+ \rightarrow C_2 H_3^+ + Cl$.

If one accepts the validity of the rule $\omega_e\, R_e^2 = \text{Constant}$ for C_2H_3Cl and for $C_2H_3Cl^{+*}$, one can write:

$$\frac{R_e^{*2}\,\omega_e^*}{R_e^2\,\omega_e} = 1.$$

As $R\,(C=C)$ in C_2H_3Cl and in $C_2H_3^+$ are, respectively 1.332 Å and 1.263 Å, with $\omega_e\,(C=C)$ in C_2H_3Cl equal to 1608 cm^{-1}, one calculates for

$\quad v_2^*\,(a')$ (valence $C=C$ symmetric stretching): 1788 cm^{-1}.

A similar reasoning leads to

$\quad v_5^*\,(a')$ (valence $C-H$ stretching in CHCl):   3080 cm^{-1}

$\quad v_4^*\,(a'')$ ($C=C$ torsion):   250 cm^{-1}.

The in-plane deformation frequencies, corresponding to

$$\begin{array}{ccc}
\text{H} \quad \uparrow \; \uparrow & & \text{H} \quad\quad \uparrow \; \uparrow \\
\diagdown \text{C}-\text{C}-\text{H} & \quad\text{and}\quad & \diagdown \text{C}-\text{C}-\text{H}, \\
\text{H}^{\diagup} \qquad \downarrow & & \text{H}^{\diagup} \; \downarrow
\end{array}$$

have been chosen as equal to those observed for $C_2H_2^+$: $v_5\,(\pi_u) = 680$ cm^{-1} and $v_4\,(\pi_g) = 520$ cm^{-1}.

Therefore, $v_6^*\,(a') = 680$ cm^{-1} and $v_{12}^*\,(a') = 520$ cm^{-1}.

Two out-of-plane deformations are chosen as in C_2H_4 and C_2H_3Cl:

$\quad v_7^*\,(a'') = v_8\,(a'') = 900$ cm^{-1}.

A last vibrational mode has to be taken into account, the $C-C-Cl$ wagging in the complex (v_{10}^* a'). The rather big $C-Cl$ distance realized in the transition complex needs to postulate, for this vibrational mode, a very low frequency. This could be considered as an adjustable value in order to fit a reasonable excitation energy range for the occurrence of the metastable $C_2H_3Cl^+$ ion. As the minimum calculated value of k_1 is $4.1 \cdot 10^5$ s^{-1} one needs to see at least $G_v^*\,(E)$ equal to about 30 quantum states in order to increase the rate constant value outside the metastable range:

$\quad 4.1 \cdot 10^5 \cdot 30$ will give $k_1 = 1.2 \cdot 10^7$ s^{-1}.

These quantum states may represent a total energy of about 500 cm^{-1} compatible with the experimentally undiscernible AE's of $C_2H_3^+$ and of the metastable ion.

On these bases, a value for the $v_{10}^*\,(a')$ wagging $C-C-Cl$ frequency of 30 cm^{-1} appears as a good choice.

Table 2. Normal modes of the transition complex in reaction (1).

Normal mode	Frequency (cm^{-1})
v_1^* (a')	3020
v_2^* (a')	1790
v_3^* (a')	1350
v_4^* (a')	250
v_5^* (a')	3080
v_6^* (a')	680
v_7^* (a')	900
v_8^* (a')	900
v_9^* (a')	3200
v_{10}^* (a')	30
v_{11}^* (a')	React. Coord.
v_{12}^* (a')	520

From this table, one easily calculates:

$$E_Z^* = \frac{15720 \text{ cm}^{-1}}{2} = 7860 \text{ cm}^{-1} = 1.5612 \cdot 10^{-12} \text{ erg} .$$

$$\pi \omega_i^* = 1.545087 \cdot 10^{32} \text{ cm}^{-11}$$

$$\beta^* = 1.4685 .$$

3. The Choice of a Transition State for Process (2)

For the second reaction a four-center transition state needs to be selected (Fig. A 2). During the reaction, one goes from a C—C bond distance of 1.42 Å to a bond distance of 1.20 Å, as observed in $C_2H_2^+$. Simultaneously, the length of the C—H and C—Cl bonds increases and these bonds turn around the C atoms in order to realize a 90° intervalence angle, the H—Cl distance diminishing down to its value in HCl (1.27 Å). The HCl molecule to be lost increases its distance with respect to the C—C bond, up to a length for which its capture by the ion is no longer realized. The normal mode corresponding to this movement is the reaction coordinate.

The choice of the 11 other normal mode frequencies is made as follows:

a) Seven normal modes are considered analogous to the seven normal modes of $C_2H_2^+$:

$$v_1^* = 3370 \text{ cm}^{-1}; \; v_2^* = 1830 \text{ cm}^{-1}; \; v_3^* = 3280 \text{ cm}^{-1};$$

$$v_4^* \text{ (doubly degenerate)} = 7000 \text{ cm}^{-1}; \; v_5^* \text{ (doubly degenerate)} = 600 \text{ cm}^{-1}.$$

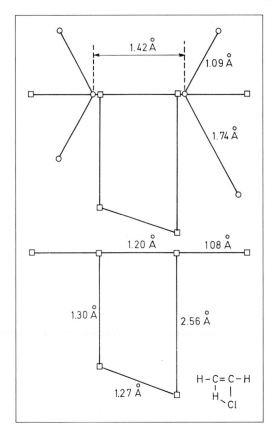

Fig. A2. Tansition state for the reaction $C_2H_3Cl^+ \rightarrow C_2H_2^+ + HCl$.

b) One normal mode is chosen for HCl with a frequency equal to the HCl vibrational frequency $v_6^* = 2290$ cm^{-1};

c) The frequency of two normal modes for the out-of-plane C–H vibrations is attributed by analogy with C_2H_4:

$v_7^* = v_8^* = 900$ cm^{-1} or $v_{7,8}^*$ considered as doubly degenerate ideal frequency, choosen in order to fit the experimental results. It will be considered that its frequency is imposed by the pseudo triple bond in $C_2H_2^+$: this means a value higher than the v_4 observed in $C_2H_3^+$ where the bond length is a bit higher than in $C_2H_2^+$. It is proposed $v_9^* = 300$ cm^{-1}.

From this table, one calculates:

$$E_Z^* = \frac{16170}{2} = 8085 \text{ cm}^{-1} = 1.606 \cdot 10^{-12} \text{ erg}.$$

$$\pi\omega_i^* = 2.5926 \cdot 10^{33} \text{ cm}^{-11}$$

$$\beta^* = 1.4462.$$

Table 3. Normal modes of the transition complex in reaction (2).

Normal mode	Frequency (cm^{-1})
ν_1^*	3370
ν_2^*	1830
ν_3^*	3280
ν_4^* doubly degenerate	700
ν_5^* doubly degenerate	600
ν_6^*	2990
ν_7^*	900
ν_8^*	900
ν_9^*	300
ν_{10}^*	Reaction Coordinate

4. Calculation of $k_1 (E)$ and $k_2 (E)$

The rate constant is written as

$$k(E) = \alpha \cdot \sigma \cdot \frac{G_v^* (E - E_0)}{h N_v (E)}.$$

It will be considered that the transmission coefficient α is unity.

Similarly, the statistical factor σ is also, in both cases, unity; there is only one way to produce the dissociation fragments.

The $N_v (E)$ values and $G_v (E - E_0)$ values will be calculated on the basis of the Whitten-Rabinovitch approximation (see 4.4.1).

For the energized molecule:

$$N_v (E) = \frac{(E + a E_Z)^{v-1}}{\Gamma(v) \prod_{i=1}^{12} h \nu_i} = \frac{(E + a E_Z)^{11}}{\Gamma(12) (hc)^{12} \prod_{i=1}^{12} \omega_i}.$$

The threshold energy for the decay of the energized molecule is AE ($C_2H_3^+$ or $C_2H_2^+$) − IE (C_2H_3Cl) = 2.48 eV or $2.00 \cdot 10^4$ cm$^{-1} \cong 3.97 \cdot 10^{-12}$ erg.

In the expression of $N_v (E)$:

$$a = 1 - \beta \omega$$

$$\beta = (v - 1) \frac{\sum \omega_i^2}{(\sum \omega_i)^2},$$

and as E will always be bigger than E_Z,

$$\log_{10} \omega = - 1.0506 \left(\frac{E}{E_Z} \right)^{1/4}.$$

Evaluation of $N_v(E)$ is given in Table 4 for some successive values of E. Similarly, $G_v^*(E - E_0)$ have been evaluated through the formula:

$$G_v^*(E - E_0) = \frac{(E - E_0 + a^* E_Z^*)^{v^*}}{\Gamma(v^* + 1) \prod_{i=1}^{11} h v_i^*} = \frac{(E - E_0 + a^* E_Z^*)^{11}}{\Gamma(12)(hc)^{11} \prod_{i=1}^{11} \omega_i^*}.$$

In this case $v^* = 11$, because one of the vibrational modes of the molecular ion becomes the reaction coordinate.

Table 4. Calculated values of $N_v(E)$, G_v^*, k_1 and k_2 for reactions (1) and (2).

E cm^{-1}	$N_v(E)$ erg^{-1}	$E - E_0$ cm^{-1}	$G_v^*(E - E_0)$ Reaction (1)	$G_v^*(E - E_0)$ Reaction (2)	k_1 s^{-1}	k_2 s^{-1}	k_1/k_2
20000	$0.3664 \cdot 10^2$	0	1	1	$4.11 \cdot 10^5$	$4.11 \cdot 10^5$	1
20500	$0.4464 \cdot 10^2$	500	41	3	$1.14 \cdot 10^7$	$1.18 \cdot 10^6$	9.6
21000	$0.5413 \cdot 10^2$	1000	169	14	$4.71 \cdot 10^7$	$3.90 \cdot 10^6$	12.0
22000	$0.7911 \cdot 10^2$	2000	1366	107	$2.60 \cdot 10^8$	$2.04 \cdot 10^7$	12.7
23000	$1.1411 \cdot 10^2$	3000	6754	516	$8.93 \cdot 10^8$	$6.82 \cdot 10^7$	13.0
24000	$1.6260 \cdot 10^2$	4000	25219	1888	$2.34 \cdot 10^9$	$1.75 \cdot 10^8$	13.3
25000	$2.2909 \cdot 10^2$	5000	79808	5819	$5.25 \cdot 10^9$	$3.83 \cdot 10^8$	13.7

The last column of Table 4 gives the variation of the relative abundances of $C_2H_3^+$ and $C_2H_2^+$. As is easy to see from Fig. 5.5, these show the expected variation and order of magnitude observed in the photoionization yield curve.

In Fig. A 3, the values of the calculated rate constants are given as function of E.

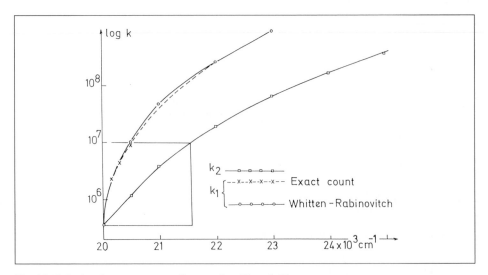

Fig. A3. Calculated rate constants for reaction (1) and (2)

A direct count for the transition state (2) shows a complete agreement with the values given in Table 4.

A direct count made for transition state (1) leads to the values given by $(\times)$ in the curve. These are only slightly different from the one given in Table 4. If one accepts that the metastable energy region extends up to 10^7 s^{-1}, the rectangular area shows the energy region for the appearance of metastable ions.

References

1. De Hemptinne M, Germain-Lefevre G, Van Riet R, Lenaerts D (1960) Spectre Raman et Spectre Infrarouge de Chlorures de Vinyle Deutérés. Acad Roy Belgique-Bull Cl Science, 5e Serie 46:310–335
2. Lake RF, Thompson H (1970) Photoelectron Spectra of Halogenated Ethylenes. Proc Roy Soc London 315A:323–338
3. Weber J, Yoshimine M, McLean AD (1976) A Cl Study of the Classical and Nonclassical Structures of the Vinyl Cation and their Optimum Path for Rearrangement. J Chem Phys 64:4159–4164

Part III Electron Attachment Processes in Molecules and Molecular Aggregates

Eugen Illenberger

1 Negative Ions in the Gas Phase

1.1 Stability of Negative Ions, Electron Affinity

It has been known for most of this century that many atoms and molecules can also exist as negatively charged ions in the gas phase [1, 2]. They can, for instance, be generated by electron transfer from neutrals or anions [3, 4] according to

$$A + M \rightarrow A^+ + M^- \qquad\qquad 1.1\,a$$

$$A^- + M \rightarrow A + M^-, \qquad\qquad 1.1\,b$$

or simply by attachment of free electrons

$$e^- + M \rightarrow M^-. \qquad\qquad 1.2$$

Since the binding energy of the extra electron to M is usually less than the ionization energy of A, Reaction 1.1a is generally endothermic and can only occur if A and M contain sufficient (kinetic and/or internal) energy.

Free electron attachment (Reaction 1.2) and the eventually subsequent decomposition of the molecular anion are the subject of this contribution. Before treating this problem in detail, it is necessary to recall some general facts concerning the stability of negative ions.

The possibility of an atom or a molecule to form a *thermodynamically* stable anion is expressed by the electron affinity. The *electron affinity* of a molecule is formally defined as the energy difference between the neutral (M) and the anion (M$^-$) in their respective ground states. By convention, the electron affinity of M is considered *posi-*

tive if the ground state of M⁻ lies *below* that of M, and *negative* if the ground state energy of M⁻ is *higher* than M. A positive value for the electron affinity indicates the existence of a stable anion in which the extra electron exists in a bound state. The electron affinity of a neutral particle thus corresponds to the ionization energy (or detachment energy) of the anion. (Ionization is commonly defined as a transition associated with the removal of one or more electrons.)

In molecules which are characterized by considerable geometry change between the neutral and the anion, one has to distinguish between the (adiabatic) electron affinity (EA) of M and the vertical detachment energy (VDE) of M⁻ (Fig. 1.1). The latter is the number which is experimentally obtained in a Franck-Condon transition in photodetachment from molecular anions. The problem is analogue to adiabatic and vertical ionization energies in photoionization.

A great many molecules, among them such ubiquitous compounds as N_2, CO_2, CH_4, ethylene, benzene, etc., are not able to form *thermodynamically stable anions*. By capture of free electrons, however, they can form temporary negative ions (TNIs).

Note that free electron attachment is an electronic transition which, in any case, initially creates an unstable (or metastable) negative ion with respect to ejection of the extra electron. This is true, regardless of the sign of the electron affinity. If it is positive, electron attachment generates an (electronically) excited state of the anion. A subsequent transition to the thermodynamically stable ground state is possible, provided that stabilization mechanisms are operative within the lifetime of the TNI. This is, however, generally not the case under single collision conditions (see below).

Figure 1.2 illustrates the situation for a neutral molecule and its cation and anion in terms of the total energy and the binding energy of the electrons.

While the ionization energy of most organic molecules lies around 10 eV, the electron affinity scale ranges to about 3.5 eV [5–9], and only particular molecules like PtF_6 and related compounds have higher values [10]. Figure 1.2 shows the situation

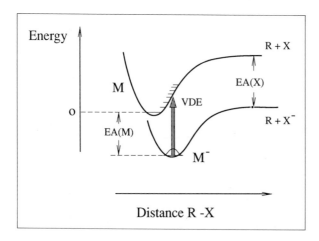

Fig. 1.1. Born-Oppenheimer potential energy curves illustrating electron affinity (EA) and vertical detachment energy (VDE).

for a positive and negative electron affinity. In the latter case the ground state negative ion is unstable with respect to autodetachment:

$$M^- \to M + e^-.$$ 1.3

Accordingly, the energy of the cation is, in any case, *above* that of the neutral, and the cation may recombine to the neutral molecule

$$e^- + M^+ \to M + E^*,$$ 1.4

provided that electrons are present in the vicinity of the ion. As in the case of electron capture by a neutral molecule with positive electron affinity, recombination is also a question of operative mechanisms to carry off the excess energy E^* (composed of the ionization energy and the energy of the colliding electron).

The energy difference between the highest occupied MO (HOMO) and the lowest unoccupied MO (LUMO) in the neutral corresponds to the lowest electronic excitation. The neutral molecule possesses many more unoccupied (virtual) MOs in which an excited electron can exist in a *bound* state. These are either valence type MOs or Rydberg type MOs converging to the respective ionization limit. Rydberg MOs

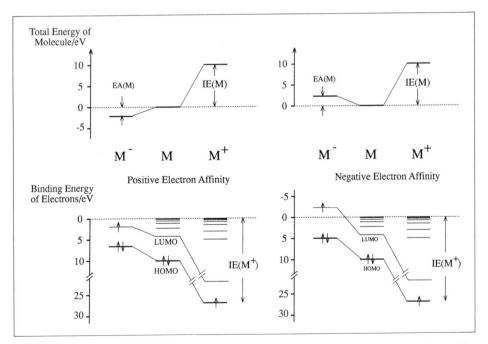

Fig. 1.2. Total energy and molecular orbital representation for a neutral molecule, its cation and its anion (for positive and negative electron affinity).

reflect the long range Coulomb interaction of the excited electron with the positive core:

$$V(r) \sim -\frac{e^2}{r},$$

1.5

leading to an infinite number of electronically excited states of the neutral molecule with the excited electron in a bound state.

On going to the positive ion, the energy to remove an electron from the HOMO is now much larger than in the neutral, since the electron interacts with a doubly charged positive core. According to the "rule of thumb" introduced by Tsai and Eland [11], the energy for double ionization is approximately 2.8 times that for single ionization. Like in the case of the neutral, the Coulomb interaction between the electron and the positive core results in an infinite number of electronically excited states.

The situation changes significantly on going to the anion. Here the extra electron interacts with a neutral molecule. At larger distances, the interaction can be approximated by the charge-induced dipole potential:

$$V(r) = -\frac{\alpha e^2}{2r^4};$$

1.6

(α is the polarizability of the neutral) which falls off much faster than the Coulomb potential. This bears important consequences for the nature of a negative ion in that the extra electron possesses a considerably lower binding energy if a bound state exists at all. Comparatively low binding energy and a very restricted number of electronically excited states (if existent at all) are the salient features of negative ions. This can easily be rationalized from elementary quantum mechanics. Consider the simple three-dimensional radial symmetric potential well:

$$V(r) = \begin{cases} \infty, & r = 0 \\ -V_0, & 0 < r \le r_0 \\ 0, & r \ge r_0 \end{cases}$$

1.7

which we will use for a crude approximation of the real problem (Fig. 1.3). The attractive polarization interaction is approximated by the step at $r = r_0$ and the strong repulsion due to the other electrons by setting $V = \infty$ at $r = 0$. The question is under which conditions do bound states ($E < 0$) exist?

The radial part of the Schrödinger equation is

$$\frac{d^2\psi}{dr^2} + \frac{2}{r}\frac{d\psi}{dr} + \frac{2m}{\hbar^2}(E - V_{eff})\psi = 0,$$

1.8

with $V_{eff} = V + \hbar^2 l(l + 1)/2mr^2$ being the effective potential, E the total energy of the electron, and $\hbar = h/2\pi$ the Planck constant.

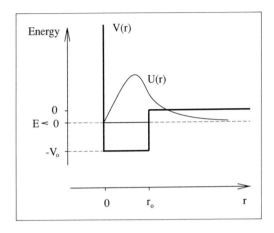

Fig. 1.3. Spherical potential well as a crude approximation of the interaction of an electron with a neutral molecule (see text).

If we restrict to $l=0$ and substitute $\psi(r) = U(r)/r$, an elementary calculation leads to

$$U'' + \frac{2m}{\hbar^2} (E - V) U = 0, \qquad\qquad 1.9$$

with $U'' = \mathrm{d}^2 U/\mathrm{d}r^2$.

Equation 1.9 corresponds formally to the one-dimensional treatment of the problem.

For $0 < r \leq r_0$, the Schrödinger equation is

$$U_1'' + \frac{2m}{\hbar^2} (E - V_0) U_1 = 0, \qquad\qquad 1.10\,a$$

and for $r \geq r_0$

$$U_2'' + \frac{2m}{\hbar^2} E\, U_2 = 0. \qquad\qquad 1.10\,b$$

With the substitutions $\dfrac{2m}{\hbar^2} (E + V_0) = k^2$ and $-\dfrac{2m}{\hbar^2} E = x^2$, we can write

$$U_1'' + k^2 U_1 = 0 \qquad\qquad 1.11\,a$$

$$U_2'' - x^2 U_2 = 0. \qquad\qquad 1.11\,b$$

Note that both k^2 and x^2 are positive numbers (V_0 is a positive, E a negative number, and $|E| < |V_0|$).

The differential equation 1.11 a has oscillating solutions of the form

$$U_1 = A_1 \sin kr + B_1 \cos kr,$$

and Eq. 1.11 b has solutions of the form

$$U_2 = A_2 \exp(-\varkappa r) + B_2 \exp(\varkappa r).$$

The normalization condition $\int_0^\infty \psi(r)\, dr = 1$ immediately requires $B_2 = 0$. The boundary conditions are

$$U_1(0) = 0$$
$$U_1(r_o) = U_2(r_o)$$
$$U_1'(r_o) = U_2'(r_o).$$

The first conditions requires $B_1 = 0$ and the remaining two lead to two homogeneous linear equations for the amplitudes A_1 and A_2.

$$A_1 \sin \varkappa r_o = A_2 \exp(-\varkappa r_o), \qquad\qquad\qquad\qquad 1.12\ a$$

$$A_1 k \cos kr_o = -A_2\varkappa \exp(-\varkappa r_o). \qquad\qquad\qquad 1.12\ b$$

These equations have a solution only if the determinant vanishes. This yields

$$\tan kr_o = -\frac{k}{\varkappa}, \qquad\qquad\qquad\qquad\qquad\qquad 1.13$$

which is the *eigenvalue condition*.

Equations 1.12 a and 1.12 b relate the amplitudes A_1 and A_2, and the wavefunctions U_1 and U_2 have the form

$$U_1 = A_1 \sin kr, \qquad\qquad\qquad 0 < r \le r_o \qquad\qquad 1.14\ a$$

$$U_2 = \frac{A_1 \sin kr_o}{\exp(-\varkappa r_o)} \exp(-\varkappa r), \qquad r \ge r_o. \qquad\qquad 1.14\ b$$

The remaining amplitude A_1 can be calculated from the normalization condition

$$\int_0^\infty \psi(r)\, dr = \int_0^{r_o} \frac{U_1}{r}\, dr + \int_{r_o}^\infty \frac{U_2}{r}\, dr = 1.$$

We will not carry out this calculation, but rather, return to the eigenvalue problem and the condition for stationary (i. e. bound) states.

If we define the "volume" of the potential well [12] as

$$C^2 = \frac{2m}{\hbar^2} V_0 r_0^2,$$

which is equivalent to

$$C^2 = (x^2 + k^2) r_0^2,$$

the eigenvalue condition can be written in the form

$$\tan kr_0 = - \frac{kr_0}{(C^2 - (kr_0)^2)^{1/2}} \cdot \qquad\qquad\qquad 1.13\,a$$

This equation can be solved graphically.

In Fig. 1.4 the graph of $\tan kr_0$ and $- kr_0/(C^2 - (kr_0)^2)^{1/2}$ versus kr_0 is plotted for different values of C^2. For $C^2 < (\pi/2)^2$ there is obviously no eigenvalue, for $(\pi/2)^2 < C^2 < (3\pi/2)^2$ there is one, and for $(3\pi/2)^2 < C^2 < (5\pi/2)^2$ we have two eigenvalues, etc.

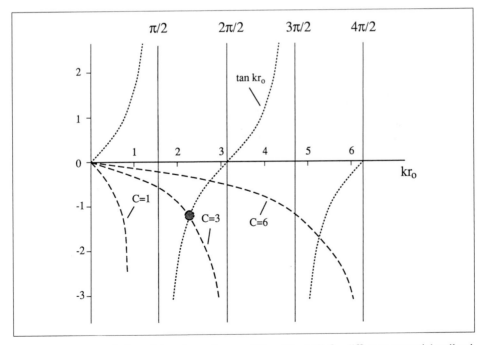

Fig. 1.4. Graphical solution of the eigenvalue condition (Eq. 1.13) for different potential well volumes.

This simple model potential as a crude approximation of the interaction between an electron and a neutral molecule clearly demonstrates that there must be some minimum volume of the potential well in order to allow a bound state for the excess electron.

For $C^2=9$, for example, Fig. 1.4 shows that Eq. 1.13 a is fulfilled for $kr_o=2.28$. From the definition of k^2, it follows that

$$(kr_o)^2 = \frac{2m}{\hbar^2}(E + V_o) \cdot r_o^2,$$

or

$$E + V_o = \frac{(kr_o)^2 \hbar^2}{2mr_o^2} = \frac{(kr_o)^2}{C^2} \cdot V_o.$$

The eigenvalue is given by

$$E = -V_o\left(1 - \frac{(kr_o)^2}{C^2}\right).$$

For the present example ($C^2 = 9$, $kr_o = 2.28$), we obtain

$$E = -0.42\,V_o.$$

The potential volume defined above can numerically be expressed as

$$C^2 = V_o(eV) \cdot r_o^2(\text{Å}) \cdot 0.264.$$

A potential volume of $C^2=9$ is thus obtained, e. g., for $V_o=5$ eV and $r_o=2.61$ Å. The electron has then a binding energy (or a negative eigenvalue) of 2.1 eV. We can compare these numbers with a real situation. The polarizability of the CCl_4 molecule is $\alpha = 11.2$ Å^3 [13]. In absolute numbers the polarization interaction (Eq. 1.6) can be expressed as

$$V(eV) = -7.2 \cdot \frac{\alpha(\text{Å}^3)}{r^4(\text{Å})}.$$

The potential between an excess electron and the neutral CCl_4 molecule is then $V(2.5$ Å$) = -2$ eV, $V(2$ Å$) = -5$ eV and $V(1.5$ Å$) = -16$ eV. Since the C–Cl bond length is 1.8 eV, considerable repulsion will counter balance the attractive interaction. Admitting $V(r) \approx -5$ eV for $r \le 2$ Å we have a situation roughly comparable to the model potential above. The experimental value for the electron affinity of CCl_4 is 2.0 ± 0.2 eV [3].

Although this value nearly coincides with that derived by our model potential, it is clear that an accurate calculation of electron affinities requires a much more sophisticated theoretical treatment and high level computational techniques than does a consideration of the simple charge-induced dipole interaction. The approximation, however, is sufficient to illustrate the essential features of a negative ion, namely, its restricted number of electronic states and the comparatively low binding energy of the extra electron.

1.2 Formation of Temporary Negative Ions by Free Electron Attachment

Consider the interaction of an electron with a molecule M in the gas phase under single collision conditions. This interaction may be divided into two classes, namely, *direct scattering* and *resonant scattering*.

In *direct scattering* the incident electron collides with the target molecule and will eventually be deviated from its original direction (Fig. 1.5). If the energy of the electron is unaffected by the scattering process, we have *elastic scattering*. Strictly speaking, due to momentum and energy conservation, the electron always loses an amount of the order m/M of its initial energy. Since this fraction is generally less than 10^{-5}, it is negligible in most cases. If the electron loses some energy due to excitation of internal degrees of freedom in the target molecule, we speak of *inelastic scattering*. According to the argument above $(m \ll M)$, excitation of rotational and vibrational

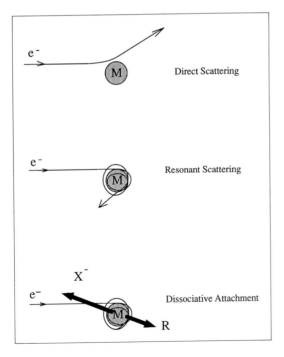

Fig. 1.5. Schematic representation of direct and resonant electron scattering.

energy is unlikely, so direct inelastic scattering predominantly causes electronic excitation of the target molecule. The electronic excitation, of course, may subsequently result in vibrational excitation of the ground electronic state through internal conversion etc.

Resonant scattering is considered to occur if the incoming electron is trapped for a certain time (considerably longer then the direct transit time) in the vicinity of the molecule to form a temporary negative ion (TNI). Electron capture can only happen if the energy of the incident electron fits that of the TNI. The process thus represents an electronic transition from a continuum state $(M + e^-)$ to a discrete (quasi bound) state of the anion (M^-). For this reason, the terms *temporary negative ion* and *resonance* are used synonymously. The lifetime of a TNI varies on a large scale depending on the energy of the resonance and the size of the molecule. It ranges from less than a few vibrational periods $(10^{-14}$ s), as in N_2^- [14, 15], to the μs range for larger polyatomic molecules such as SF_6^- and many other (often perfluorinated) compounds [8, 16–18]. For comparison, an electron of 1 eV takes $5 \cdot 10^{-16}$ s to travel a distance of 3 Å, as is the case in direct scattering. According to Heisenberg's uncertainty principle, the lifetime of a TNI is connected with an energy width given by

$$\Gamma \approx \frac{\hbar}{\tau},$$

        1.14

with $\hbar = h/2\pi = 6.6 \cdot 10^{-16}$ eV·s, the Planck constant. A resonance with a mean lifetime of 10^{-14} s has thus a natural linewidth of ≈ 66 meV. We will see below that, in molecules, the width of a resonance is generally controlled by the Franck-Condon transition rather than by its lifetime.

A TNI may either decay by emission of the extra electron or by fragmentation into thermodynamically stable, negatively charged and neutral products (dissociative electron attachment).

If the energy of the detached electron equals that of the incident electron, we speak of *resonant elastic scattering*; otherwise, we refer to *resonant inelastic scattering*. Due to the longer residence time of the extra electron, the neutral molecule may be left in a vibrationally excited state. This is the case when the TNI is characterized by a substantial geometry change, electron capture thereby providing a strong coupling between electronic and nuclear motion.

So far, we have considered direct and resonant scattering in a rather mechanistic way. The reader should recall that the de Broglie wavelength of a 5 eV electron is 5.5 Å and, thus, is considerably larger than a bond length in a molecule. A more rigorous treatment of electron scattering describes the incoming electron beam as a plane wave containing all components of angular momentum (partial waves). In resonant scattering there are always one or more partial waves which undergo *constructive interferences* within the target, while in direct scattering this is not the case [19].

We will now briefly consider the *electronic configuration* of temporary negative ions. If the electron is attached without changing the electronic configuration of the target molecule, i. e., if the state of the TNI is formally obtained by adding the extra electron into one of the empty (virtual) MOs, we speak of a *single particle* (1p) *resonance*. On the other hand, if electron attachment is accompanied by an electronic excitation of the molecule, we have a *two particle — one hole* (2p − 1h) *resonance*, with two electrons in normally unoccupied MOs. Such electronic states are also called *core excited resonances*. Note that formation of a 2p − 1h resonance represents a two electron transition, similar to the 2h − 1p electronically excited ("non Koopman's") states in positive ions (see Chapter 1 of the first contribution to this volume).

If the energy of the core-excited resonance lies *above* that of the associated electronically excited neutral molecule (the parent), we speak of an *open channel core excited resonance*; if it lies below, we refer to a *closed channel* or *Feshbach resonance*. A Feshbach resonance cannot decay into the associated excited neutral molecule via a one electron process. Its decay into the ground state of the neutral requires a change of the electronic configuration which often results in a longer lifetime.

Finally, if the incoming electron induces a strong coupling between electronic and nuclear motion, the transfer of electronic to vibrational energy (EVT) can prevent the extra electron from autodetaching. Such states are called *nuclear-excited Feshbach resonances*.

What is the mechanism responsible for trapping an incoming electron for times considerably longer than the direct transit time through the molecule's dimension? In the case when the incoming electron induces an electronic excitation in the target molecule that results in a Feshbach resonance, the extra electron is then simply trapped in the field of the excited molecule; its emission via a one electron process is energetically not possible.

The other mechanism which applies for single particle and open channel core-excited resonances describes the trapping by the *shape* of the effective interaction potential between the incoming electron and the molecule. Therefore, such resonances are called *shape resonances*.

The trapping mechanism can be pictured using the (spherically symmetric) polarization potential introduced above for the long-range interaction. The effective potential is then

$$V(r) = -\frac{\alpha e^2}{2r^4} + \frac{\hbar^2 l(l+1)}{2mr^2}, \qquad\qquad 1.15$$

with l being the angular momentum quantum number of the incoming electron. Combination of the attractive polarization and the repulsive centrifugal potential results in a potential barrier. At shorter distances the electron-electron repulsion dominates, leading to a potential curve which may have a form as illustrated in Fig. 1.6. If the molecule possesses an energetically accessible unfilled MO, characterized through its symmetry by a particular value of l (see Section 2.6 of the first contribution to this volume), then an incoming electron of that particular angular momentum quantum number can temporarily be captured within the barrier. The resulting TNI represents a discrete electronic state lying in the (ionization or detachment) continuum. Such states are called *quasi-bound states*; they only exist for $l \neq 0$.

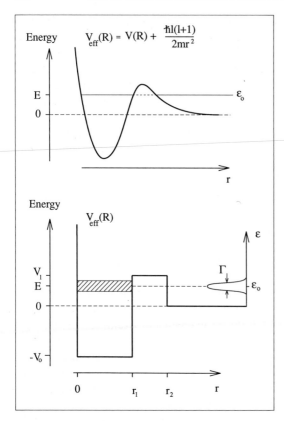

Fig. 1.6. Hypothetical potential diagram for the interaction of an electron of angular momentum quantum number l and a neutral molecule. Approximation of the effective potential by a step function (below).

At this point, it should be noted that the long-range interaction between an electron and a positive ion has the form

$$V(r) = -\frac{e^2}{r} + \frac{\hbar^2 l(l+1)}{2mr^2}, \qquad\qquad 1.16$$

which does not result in a centrifugal barrier necessary for the formation of shape resonances. However, it is well known that recombination of positive ions (Eq. 1.4) proceeds through resonances (in this case, autoionizing states of the neutral molecule) and subsequent dissociation into neutral fragments. The process is called *dissociative recombination* and is completely analogue to dissociative electron attachment, the latter having one more electron in the system. Dissociative recombination is likely to proceed via doubly excited states of the neutral [20] in analogy to the core-excited resonances in electron attachment.

On the other hand, shape resonances are well known features in the photoionization cross section [21]. This can only be explained by invoking higher order terms for the attractive interaction in the effective potential (Eq. 1.16).

The existence of quasi-bound states can also be rationalized by an explicit quantum mechanical calculation using the model potential shown in Fig. 1.5 b, the centrifugal barrier for a certain l is approximated by the step $(+ V_l)$ between r_1 and r_2. Although the treatment is straightforward and analogue to that for the model potential of Fig. 1.3, the resulting equations become significantly more complex and we will only consider the results.

A bound state $(E < 0)$ can only exist (as above) if the potential well possesses a certain minimum volume. For $E > 0$ there is no longer an eigenvalue condition, i. e., any energy $E > 0$ is allowed. However, the energy, characterized by the condition

$$\tan kr_1 = -\frac{k}{\varkappa}$$

$$k^2 = \frac{2m}{\hbar^2} (E + V_o), \, \varkappa^2 = \frac{2m}{\hbar^2} (V_l - E),$$

which is similar to the eigenvalue condition for the simple potential well (Eq. 1.13) has a special meaning in the way that the ratio of the amplitude of the wavefunction within the well $(0 < r \leq r_1)$ to that outside $(r \geq r_2)$ has a maximum. This is the characteristic feature of a *resonance phenomenon*. The square of the wavefunction represents the probability density yielding the width Γ of the resonance (Fig. 1.6). In the dynamic picture the width of the resonance is connected with the lifetime of the quasi-bound state through Heisenberg's uncertainty principle introduced above.

If the electronic ground state of the anion is characterized by such a shape resonance, the molecule possesses a *negative electron affinity*, if a bound state $(E < 0, l=0)$ exists, the molecule has a *positive electron affinity*. Figure 1.7 summarizes the different types of resonances.

Nitrogen and ethylene are examples of molecular systems with negative electron affinities. Both $N_2^- \, (^2\Pi_g)$ ions [14, 15, 22] and $C_2H_4^- \, (^2\Pi)$ ions [23, 24] are formed in

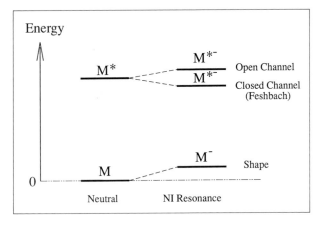

Fig. 1.7. Classification of resonances.

their electronic ground states by electron impact with energies of about 2 eV having autodetachment lifetimes of the order of a few vibrational periods. Core excited resonances have also been identified in nitrogen. One at 9.23 eV associated with $N_2 (B^3 \Pi_g)$ at 7.4 eV and another at 11.48 eV associated with the Rydberg state $N_2 (E^3 \Sigma_g^+)$ at 11.87 eV [25]. The first resonance is thus an open channel and the second one a closed channel (or Feshbach) resonance.

1.3 Decomposition Reactions of Temporary Negative Ions

We will now consider the fate of a polyatomic temporary anion formed by resonant electron capture. The molecular anion can react via the following channels

$$M^{-(*)} \xrightarrow{\tau_a} M^{(*)} + e^- \tag{1.17}$$

$$M^{-(*)} \xrightarrow{\tau_r} M^- + \hbar\omega \tag{1.18}$$

$$M^{-(*)} \xrightarrow{\tau_d} R + X^- \text{ or } R^- + X \text{ etc} . \tag{1.19}$$

Reaction 1.17 represents autodetachment, reaction 1.18 is radiative stabilization to the stable ground state which is only possible for compounds having a positive electron affinity, and 1.19 finally is the unimolecular decomposition into stable neutral and negatively charged fragments (dissociative electron attachment).

As mentioned above, the autodetachment lifetime (τ_a) may extend up to the μs scale for larger molecules. Such metastable anion states are typically generated within a narrow resonance near 0 eV where dissociation channels into stable fragments are not yet accessible. Owing to their long lifetime, they can easily be observed by mass spectrometric techniques.

Although radiative stabilization (Eq. 1.18) has been studied for a number of atoms [26, 27], data for molecules are virtually non existent. Since radiative lifetimes are of the order of 10^{-9}–10^{-8} s, it is indeed likely that radiative stabilization is slow compared with the competing channels 1.17 and 1.19. In addition, metastable anions may undergo internal conversion rather than radiative stabilization.

Finally, dissociative attachment typically proceeds in the time domain between 10^{-14} and 10^{-12} s, depending on the mechanism of the reaction. The decomposition can occur if: a) thermodynamically stable, negatively charged fragments exist for the respective compound, and b) these channels are energetically available at the energy of the TNI. If the dissociative attachment channel is accessible, it will generally strongly compete with autodetachment.

Let us consider the simple reaction leading to two fragments

$$e^- (\varepsilon) + M \rightarrow M^- (\varepsilon) \rightarrow R + X^- ; \tag{1.20}$$

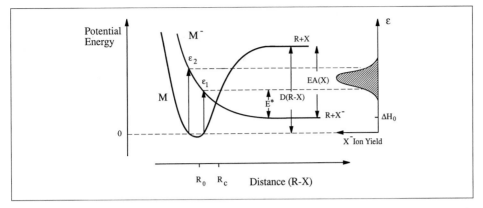

Fig. 1.8. Born-Oppenheimer potential energy curves associated with electron attachment and subsequent electronic dissociation.

ε is the electron energy. The process can be pictured in a Born-Oppenheimer potential energy diagram, as depicted in Fig. 1.8. Such curves are rigorously applicable only to diatomic molecules. For polyatomics they represent two-dimensional cuts through hyperdimensional surfaces along a reaction coordinate Q. However, they retain their simple meaning (even for polyatomic molecules) for a *direct electronic dissociation*. In this case, the extra electron has a strong and specific influence on the R$-$X bond or, in terms of MOs, the electron is captured into an MO with a significant R$-$X antibonding character. In accordance with the Franck-Condon principle, transitions from the continuum state $(M + e^- \, (r \to \infty))$ to M^- are only possible within the energy region between ε_1 and ε_2.

The molecular anion once formed may either dissociate into $R + X^-$ or lose the extra electron to regenerate the neutral molecule. Autodetachment can occur for $R \le R_c$, where R_c is the crossing point of the two potential energy curves. For $R \ge R_c$ the additional electron is localized and bound to fragment X to an extent that autodetachment of M^- is no longer possible.

The cross section for dissociative electron attachment σ_{da} can be written as

$$\sigma_{da} = \sigma_o \exp\left(- \tau_d/\tau_a\right), \qquad\qquad 1.21$$

with σ_o the attachment cross section; τ_d is the time it takes the products to dissociate until autodetachment becomes impossible and τ_a is the autodetachment lifetime introduced above. The exponential term in Eq. 1.21 is called the *survival probability*. It gives the dissociation probability of the TNI with respect to autodetachment.

The dissociation time is given by

$$\tau_d = \int_{R(\varepsilon)}^{R_c} \frac{dR}{v\,(R)}, \qquad\qquad 1.22$$

with $v(R)$ the radial velocity between R and X^-:

$$v(R) = \left(\frac{2(E^* - V_f)}{\mu}\right)^{1/2},$$

1.23

and μ their reduced mass. $V_f(R)$ is the potential energy curve for the anion and E^* is the excess energy of the process. In the case of diatomics, E^* is purely translational energy imparted to the fragments.

The line shape of the ion yield curve of fragment X^- is explained by the reflection principle [28, 29] already introduced in Section 2.4 of the first contribution to this volume. Here, the continuum wave frunction is approximated by a delta function at the classical turning point, so that the vibrational wave function of the neutral ground state is *reflected* at the repulsive potential $V_f(R)$, see Fig. 1.9.

A relation between the width of the ion yield curve (Γ_d) and the associated potential curves $V_i(R)$ and $V_f(R)$ can easily be obtained for a constant survival probability. Near the equilibrium distance R_o, the potential energy can be approximated by the harmonic form

$$V_i(R) = \frac{1}{2}\,\alpha\,(R - R_o),$$

1.24

with α the force constant. For the vibrational ground state ($E = \hbar\omega/2$), we can write

$$2|R_2 - R_o| = |R_1 - R_2| = 2|R_1 - R_o| = 2\left\{\frac{\hbar\omega}{\alpha}\right\}^{1/2},$$

1.25

where ω is the frequency of the oscillator:

$$\omega = (\alpha/\mu)^{1/2}$$

1.26

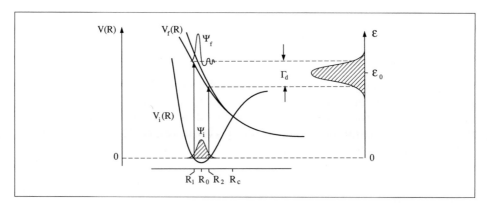

Fig. 1.9. Reflection principle to describe the peak shape of an ion yield curve.

With $\alpha = \partial^2 V_i(R)/\partial R^2 = V_i^{\parallel}$, the stiffness of the oscillator, we can write

$$|R_2 - R_1| = 2(\hbar^2/\mu V_i^{\parallel})^{1/4}. \qquad 1.27$$

If we set $V_f^{\mid} = \text{const}$ in the Franck-Condon region, the width of the resonance becomes

$$\Gamma_d = \frac{2V_f^{\mid}\hbar^{1/2}}{(\mu V_i^{\parallel})^{1/4}}. \qquad 1.28$$

The width of the resonance is thus directly proportional to the slope of the repulsive potential curve of the anion. Γ_d is the width obtained by reflecting the Franck-Condon region $(R_2 - R_1)$ at $V_f(R)$.

The wave function of the vibrational ground state of the harmonic oscillator is

$$\Psi_i = C \exp\left(-\mu\omega(R - R_o)^2/2\hbar\right),$$

with the probability density

$$\Psi_i^2 = C^2 \exp\left(-\mu\omega(R - R_o)^2/\hbar\right). \qquad 1.29$$

With Eqs. 1.25 and 1.26, the probability density at the classical turning points (R_1, R_2) with respect to the maximum (at R_o) becomes

$$\Psi_i^2(R_1) = \Psi_i^2(R_2) = \Psi_i^2(R_o)/e.$$

Γ_d derived in Eq. 1.28 is thus the full width of the ion yield curve at the position where its value is $1/e = 0.37$ of the maximum (Fig. 1.9).

The energy-dependent cross-section for dissociative attachment can then be written as

$$\sigma_{da}(\varepsilon) \sim \exp\left(-\frac{4(\varepsilon - \varepsilon_o)^2}{\Gamma_d^2}\right), \qquad 1.30$$

with ε_o being the electron energy at the resonance maximum.

Equation 1.30 holds if σ_o and P, the survival probability, are independent of the electron energy.

In a semiclassical treatment, O'Malley [29] derived a more general expression of the form

$$\sigma_{da}(\varepsilon) \sim \frac{1}{\varepsilon} G(\varepsilon) P(\varepsilon), \qquad 1.31$$

with $P(\varepsilon)$ the survival probability, and $G(\varepsilon)$ a Gaussian function similar to 1.30 with an additional term considering autodetachment. Equation 1.31 indicates that the cross-section for dissociative attachment has an overall reciprocal dependence with electron energy.

Of course, in a polyatomic molecule dissociative attachment may generally not proceed directly via a purely repulsive energy surface as in Fig. 1.8 and 1.9, but rather through more indirect processes such as electronic or vibrational predissociation or rearrangement in the precursor ion prior to dissociation.

Figure 1.10 illustrates vibrational predissociation. The coordinates Q_1 and Q_2 represent different molecular motions in a polyatomic molecule. The anion M^- formed on electron attachment has sufficient energy to dissociate but the energy is initially in vibrational modes that do not correspond to the reaction coordinate Q_2.

Figure 1.11 shows rearrangement in the parent anion prior to dissociation. As will be shown below, these different reaction mechanisms can often be identified and distinguished through the different energy partitioning among the fragments formed. The energetic position and width of an ion yield curve, however, in every case is an image of the primary step of electron capture by the target molecule. This is true regardless of the decomposition mechanism, provided that the dissociation limit and the potential energy along the reaction coordinate lie below the energy of the anion in the Franck-Condon region. For complex fragmentation reactions, in particular when competitive dissociation channels like

$$M^- \rightarrow R_1 + X^-$$
$$\rightarrow R_2 + Y^-$$
$$\rightarrow \ldots. \qquad\qquad 1.32$$

are available within the resonance, the cross-section for the formation of a specific ion is

$$\sigma(X^-) = \sigma_0 \cdot P(X^-),$$

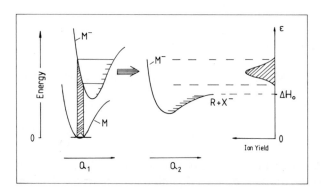

Fig. 1.10. Schematic potential energy curves illustrating dissociative electron attachment via vibrational predissociation. Q_1 and Q_2 represent different motions in a polyatomic ion.

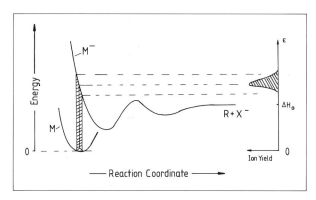

Fig. 1.11. Potential energy diagram for dissociative electron attachment associated with rearrangement in the TNI.

with σ_o the attachment cross-section introduced above. $P(X^-)$ is the probability for the formation of X^- with respect to other competitive channels (1.32), including auto-detachment. $P(X^-)$ will generally depend on the internal energy of the precursor ion (i. e., the incident electron energy) so the ion yield curve can no longer rigorously be determined by the reflection principle. The Gaussian profile will then be weighted by the energy dependence of the individual channel.

Figure 1.8 directly gives the energy balance for dissociative electron attachment

$$\varepsilon = D\,(R-X) - EA\,(X) + E^*;\qquad\qquad 1.33$$

D is the bond dissociation energy in the neutral molecule, and EA is the electron affinity of X and $\varepsilon_1 \le \varepsilon \le \varepsilon_2$. The minimum heat of reaction is

$$\Delta H_o = D\,(R-X) - EA\,(X).\qquad\qquad 1.34$$

Combining Eqs. 1.33 and 1.34 we can write for the total excess energy:

$$E^* = \varepsilon - \Delta H_o.\qquad\qquad 1.35$$

For reactions where ΔH_o is known, Eq. 1.35 gives the total excess energy for a given incident electron energy. In polyatomic molecules, E^* is generally shared among the different degrees of freedom (translational and internal energies of the fragments). If the kinetic energy of the ionic fragment is measured, the principle of conservation of linear momentum will give the total translational energy (E_T) imparted to both fragments. Furthermore, in the case where X^- is an atomic ion, energy amounting to $(E^* - E_T)$ must be deposited as internal energy in the remaining neutral fragment. Measurement of the ion kinetic energy as a function of the incident electron energy yields detailed information on the distribution of excess energy in the unimolecular decomposition of the TNI.

Conversely, in cases where ΔH_o is not known, experimental determination of the appearance energy and the kinetic energy release of the ionic fragment will help to provide more detailed information on thermodynamic quantities such as bond dissociation energies and electron affinities. By that, the method represents a simple and powerful way of studying energy distributions in unimolecular decompositions. Analogous information (e. g., from photoionization) is only accessible through coincidence techniques as described in the first contribution to this volume.

1.4 Experimental Methods to Study Electron Attachment

In Chapter 2 of the first contribution to this volume, we have already introduced two important techniques for the study of electron attachment processes in the gas phase, namely, electron attachment spectroscopy (EAS) and electron transmission spectroscopy (ETS). In the first method, an electron beam of defined energy is crossed with a molecular beam and the resulting negative ions are recorded as a function of the incident electron energy by means of a mass spectrometer. This method is applicable to molecules containing electronegative atoms or groups like halogenated molecules, alcohols or cyano compounds. Such molecules are generally characterized by low lying channels for dissociative attachment giving stable negative fragments.

However, as mentioned above, in a great many molecules the ground state of the anions is unstable with respect to autodetachment in the gas phase. The N_2^- ion, for example, cannot form a thermodynamically stable fragment and in ethylene or benzene none of the possible dissociation channels (like C_2^- or H^-) are accessible at the energy of the TNI. These resonances can be observed in electron transmission. In ETS an electron beam passes a cell containing the gas of interest and the *transmitted* electron current versus incident energy is recorded. Since the transmitted current represents the unscattered part of the primary intensity, the ratio of incident to transmitted current is a measure of the *total electron scattering cross-section*. The formation of TNIs can be identified in the transmission spectrum through resonances on the slowly varying background of direct scattering.

EAS and ETS are mostly used to obtain *relative* cross sections as a function of the incident electron enery and only few data on *absolute* numbers are available. The reason is that determination of absolute numbers is by no means a straightforward procedure in beam experiments. The absolute cross-section can principally be obtained from the Lambert-Beer law (see also Section 2.3 of the first contribution to this volume):

$$I = I_o \exp\left(-\sigma \cdot n \cdot l\right), \qquad\qquad 1.36$$

which relates the primary (I_o) and unscattered part of the electron intensity (I); σ is the cross-section, n the gas density, and l the distance of interaction between the electron beam and the target gas ("scattering length"). For a dissociative attachment reaction yielding fragment X^-, e. g., Eq. 1.36 becomes

$$Z(X^-) = I_0 \left(1 - \exp\left(-\sigma_{da}(X^-)\, n \cdot l\right)\right), \qquad\qquad 1.37$$

with $Z(X^-)$ the number of X^- ions formed per unit time and $\sigma_{da}(X^-)$ the cross-section for dissociative attachment yielding X^-. In Eq. 1.37 I_0 is measured in number of primary electrons per unit time. Under single collision conditions, $\sigma \cdot n \cdot l \ll 1$, thus the cross-section is given to a good approximation by

$$\sigma_{da}(X^-) = \frac{Z(X^-)}{I_0\, n\, l}. \qquad\qquad 1.38$$

Of these quantities a reliable determination of n and $Z(X^-)$ is difficult. Measurement of the absolute gas density in the interaction volume in a beam experiment is a delicate problem (which we will not consider here), and the ion count rate is strongly controlled by various parameters like draw out field, transmission of the mass filter, detector efficiency, etc.

Absolute values for electron attachment cross-section can be derived from *electron swarm techniques*. In such experiments electrons are produced in a drift tube by photoionization, thermoionic emission, etc. They travel under the influence of a uniform electric field and make collisions with the gas consisting of the target gas diluted in an inert carrier gas [8, 30, 31]. The product $p \cdot d$ (p: gas pressure, d: overall drift distance) must be large enough for the electrons to attain an equilibrium energy distribution within a distance $x \ll d$, regardless of their initial energy of liberation. The distribution of the electron energy is then characterized by a distribution function $f(v)$ which depends on the carrier gas and E/p or E/n, the "pressure reduced" electric field $(E/n \sim TE/p)$. In a swarm experiment the absolute rate k of electron attachment is a directly obtainable quantity. It is related to the electron attachment cross-section by

$$k = \int_0^\infty \sigma_0 f(v)\, v\, dv; \qquad\qquad 1.39$$

$f(v)$ is the normalized velocity distribution, so that

$$k = \bar{\sigma}_0 \bar{v} \qquad\qquad 1.40$$

with $\bar{\sigma}_0$ and $\bar{v}$ being the averaged quantities.

Electron velocity distributions are known for a number of non-attaching gases such as N_2, Ar, C_2H_4 and allow the determination of the attachment cross-section averaged over the velocity distribution $f(v)$. This can be done over a wide range of E/n (10^{-20}–10^{-16} Vcm^2) corresponding to mean electron energies from subthermal energies to several eV.

Another method to obtain absolute numbers for the electron attachment rate uses the electron cyclotron resonance (ECR) technique in a flow tube [32, 33]. Both methods record the *decrease* of electrons in the presence of the scavenger gas and have provided reliable and fairly consistent data. They do, however, not provide information on the products ultimately formed as a result of the capture process.

References

1. Thomson JJ (1911) Rays of Positive Electricity. Philosoph Mag 21:225–249
2. Thomson JJ (1921) On the Structure of the Molecule and Chemical Combination. Philosoph Mag 41:510–544
3. Lacmann K (1980) Collisional Ionization. In: Lawley KP (ed), Potential Energy Surfaces. Wiley, New York
4. Kebarle P, Chowdhury S (1987) Electron Affinities and Electron-Transfer Reactions. Chem Rev 87:513–534
5. Jordan KD, Burrow PD (1987) Temporary Anion States of Polyatomic Hydrocarbons. Chem Rev 87:557–588
6. Jordan KD, Burrow PD (1977) Studies of the Temporary Anion States of Unsaturated Hydrocarbons by Electron Transmission Spectroscopy. Acc Chem Res 11:341–348
7. Janousek K, Brauman JI (1979) Electron Affinities. In: Bowers MT (ed) Gas Phase Ion Chemistry, vol 2. Academic Press, New York
8. Christophorou LG (ed) (1984) Electron Molecule Interactions and Their Applications, vols 1 and 2. Academic Press, New York
9. NIST Standard Reference Database 19 B (1990) US Department of Commerce
10. Korobov MV, Kuznetsov SV, Sidorov LN, Shipachev VA, Mit'kin VN (1989) Gas-Phase Negative Ions of Platinum Metal Fluorides. II. Electron Affinity of Platinum Metal Hexafluorides. Int J Mass Spectrom Ion Proc 87:13–27
11. Tsai BP, Eland JHD (1980) Mass Spectra and Doubly Charged Ions in Photoionization at 30.4 and 58.4 nm. Int J Mass Spectrom Ion Proc 36:143–165
12. Flügge S (1976) Rechenmethoden der Quantenmechanik. Springer, Heidelberg
13. CRC Handbook of Chemistry and Physics, 66th edn. (1986) CRC Press, Boca Raton, Florida
14. Birtwistle DT, Herzenberg A (1971) Vibrational Excitation of N_2 by Resonance Scattering of Electrons. J Phys B: Atom Molec Phys 4:53–70
15. Berman M, Estrada H, Cederbaum LS, Domcke W (1983) Nuclear Dynamics in Resonant Electron-Molecule Scattering Beyond the Local Approximation: the 2.3-eV Shape Resonance in N_2. Phys Rev A 28:1363–1381
16. Fenzlaff M, Gerhard R, Illenberger E (1988) Associative and Dissociative Electron Attachment to SF_6 and SF_5Cl. J Chem Phys 88:149–155
17. Fenzlaff M, Illenberger E (1989) Energy Partitioning in the Unimolecular Decomposition of Cyclic Perfluororadical Anions. Chem Phys 136:443–452
18. Lifshitz C, MacKenzie Peers A, Grajower R, Weiss M (1970) Breakdown Curves for Polyatomic Negative Ions. J Chem Phys 53:4605–4619
19. Massey HSW, Burhop EHS, Gilbody HB (1969) Electronic and Ionic Impact Phenomena, vols I–IV. Oxford University Press
20. McGowan JWm, Mitchell JBA (1984) Electron-Molecular Positive-Ion Recombination. In: Christophorou LG (ed) Electron Molecule Interactions and Their Applications, vols 1 and 2. Academic Press, New York
21. Eland JHD (1984) Photoelectron Spectroscopy. Butterworths, London
22. Krauss H, Mies FH (1970) Molecular Orbital Calculation of the Shape Resonance in N_2^-. Phys Rev A 1:1592–1598

23. Walker IC, Stamatovic A, Wong SF (1978) Vibrational Excitation of Ethylene by Electron Impact: 1-11 eV. J Chem Phys 69:5532–5537
24. Bowman CR, Miller WD (1965) Excitation of Methane, Ethane, Ethylene, Propylene, Acetylene, Propyne and 1-Butyne by Low-Energy Electron Beams. J Chem Phys 42:681–686
25. Schulz GJ (1973) Resonances in Electron Impact on Diatomic Molecules. Rev Mod Phys 45:423–486
26. Mück G, Popp HP (1968) Quantitative Ausmessung des Chlor-Affinitätskontinuums. Z Naturforsch A 23:1213–1220
27. Boldt G (1958) Rekombinations- und "Minus"-Kontinuum der Sauerstoffatome. Z Phys 154:319–329
28. Taylor HS (1970) Models, Interpretations, and Calculations Concerning Resonant Electron Scattering Processes in Atoms and Molecules. In: Prigogine I, Rice SA (eds) Advances in Chemical Physics, vol XVIII. Interscience Publishers, New York
29. O'Malley TF (1966) Theory of Dissociative Attachment. Phys Rev 150:14–29
30. Christophorou LG (1971) Atomic and Molecular Radiation Physics. Wiley, London
31. Hunter SR, Carter JG, Christophorou LG (1989) Low Energy Electron Attachment to SF_6 in N_2, Ar and Xe Buffer Gas. J Chem Phys 90:4879–4891
32. Mothes KG, Milhelcic D, Schindler RN (1971) Zur Messung von Elektroneneinfangquerschnitten mit Hilfe der Elektronenzyklotronresonanz. Ber Bunsenges Phys Chem 75:9–14
33. Mothes KG, Schindler RN (1971) Die Bestimmung absoluter Geschwindigkeitskonstanten für den Einfang thermischer Elektronen durch CCl_4, SF_6, C_4F_8, C_7F_{14}, N_2F_4 und NF_3. Ber Bunsenges Phys Chem 75:938–945

2 Electron Attachment Spectroscopy

2.1 Experimental Considerations

In EAS a beam of electrons produced by an electron monochromator in crossed with the target gas beam and the anions are recorded with a mass spectrometer. Figure 2.1 shows a configuration which uses a trochoidal electron monochromator (TEM) and a quadrupole mass filter [1] (see also Sections 3.2.6 and 3.3.4 in the first contribution). The monochromator is made of molybdenum in order to reduce surface problems. Non-magnetic stainless steel is used for the other components and for entire ultra-high vacuum system.

The operating principle of the TEM is, in short, as follows: electrons emitted by a tungsten filament are aligned by a homogeneous magnetic field generated by a pair of Helmholtz coils located outside the vacuum system and collimated by a series of electrodes. The three electrodes following the filament (B_1, B_2, and B_3) have holes drilled 3 mm off-center so that the electrons enter the deflection region (C_1, C_2) off axis. The deflection region is defined by the crossing of the magnetic field $\vec{B}$ (z direction) with a small electric field $\vec{E}$ (y direction). The electric field is generated by applying a potential difference between the two parallel plates C_1 and C_2 which are spaced 3 mm apart. In the deflection region, the electrons describe a trochoidal or cycloidal path, depending on whether or not they possess velocity components perpendicular to the z axis when entering the region. The trochoidal or cycloidal motion of the electrons implies motion of their guiding center with a constant velocity

$$\vec{v} = \frac{[\vec{E} \times \vec{B}]}{|\vec{B}^2|}$$

in the x direction (i. e., along an equipotential plane). This implies a dispersion of the electrons according to their z velocity and only those electrons which reach the axis are transmitted through the axial hole of the electrode following the deflection region (S_1) (Fig. 2.1). The energy-selected electrons are then accelerated or decelerated by the electrodes S_2 and S_3 into the reaction chamber which is kept at a fixed electrical potential.

Calculations of the electron trajectories and the electron energy distribution have been carried out by several authors [2, 3]. The TEM was first realized by Stamatovic and Schulz [4, 5] in 1968 and has since been employed by many groups, mostly in ETS [6–9].

Ions formed in the reaction volume (as defined by the crossing of the electron beam with the target gas beam effusing from a capillary) are extracted by a small electric field (less than 1 V cm^{-1}) and accelerated by a series of parallel electrodes onto the entrance hole of a commercial quadrupole mass filter. The mass-selected ions

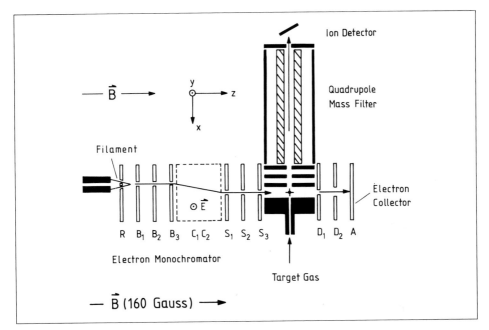

Fig. 2.1. Schematic representation of the electron attachment spectrometer.

are accelerated to the first dynode of a 17-stage CuBe multiplier. The multiplier pulses are capacitively decoupled from the high voltage output of the multiplier, then amplified and formed by standard counting electronics and stored in a laboratory computer. The computer drives the TEM and thus the electron energy via digital/analog outputs in combination with an operational amplifier.

After passing through the reaction chamber the electrons are accelerated towards the electron collector (A) which serves as a beam intensity monitor. The pressure in the reaction chamber is kept low (less than 10^{-4} mbar) in order to ensure single-collision conditions. This establishes that the electron intensity measured at the collector (A) is essentially equal to the primary intensity of the beam. We note that in ETS one uses higher gas pressures and larger interaction paths so that the interaction of the electrons with the gas molecules can be made visible in the *attenuation* of the transmitted electron current (see Chapter 3).

The electron energy distribution can be obtained either by applying a stopping voltage to electrode S_3 and recording the electron retarding curve or by measuring the associative electron attachment to SF_6, which is known to be sharply peaked. The SF_6 resonance is used to calibrate the energy scale. The line shape and energetic position of the SF_6 resonance is well known from "threshold photoelectron spectroscopy by electron attachment" [10–12] which yields a value of 30 meV for the full width at half-maximum (fwhm) and a peak maximum which is effectively at 0 eV.

Figure 2.2 shows the ion yield curve for SF_6^- ion formation and the corresponding electron beam intensity measured at collector A. In the example shown the fwhm is less than 0.1 eV. Values down to 0.04 eV have been obtained with this configuration. The energy resolution, however, is often determined by instabilities caused by the interaction of the compound under investigation with the electrode surfaces. As a result of such effects, most of the ion yield curves presented here were measured with an energy resolution between 0.1 and 0.2 eV (and at times even worse!). Nonetheless, despite the limited energy resolution, the present configuration - the combination of a TEM with a quadrupole mass filter — is particularly suited to the study of electron attachment reactions in the very low energy region since it combines several important necessary properties:

1) The axial magnetic field prevents spreading of the electron beam at low energies so that reasonable intensities can be achieved even near 0 eV (see Fig. 2.2). Beam spreading is a major problem when electrostatic devices are used.

2) The TEM can be operated in a continuous mode. Because of the presence of the magnetic field, the energy of the electrons entering the reaction chamber is not influenced by the presence of the (continuous) ion draw-out field. The ion draw-out field only causes a slight deflection of the electrons along a plane of constant electric potential in the x direction. The deflection, however, is negligible even at very low electron energies.

3) The alignment of the electron beam by the magnetic field allows a separation of electrons and negative ions at the secondary electron mulitplier.

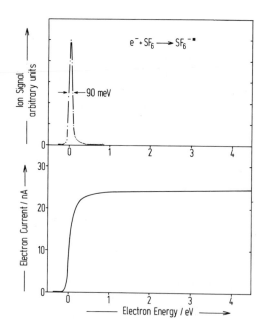

Fig. 2.2. Ion yield curve for SF_6^{-*} formation and electron beam intensity monitored at collector A.

2.2 Ion Yield Curves

2.2.1 SF_6 and SF_5NCO

SF_6 is the compound which has probably been the most intensively studied with regard to low-energy electron attachment. It possesses one of the highest attachment cross-sections known and has, for this reason, found extensive application as a gaseous dielectric in high-voltage power devices and as an effective electron scavenger in gases and liquids [13–16]. The rate constant for electron attachment has been measured with the electron swarm [17] and ESR technique [18]. Values between $2 \cdot 10^{-7}$ and $3 \cdot 10^{-7}$ cm^3 s^{-1} have been derived at room temperature. Assuming a Maxwell-Boltzmann distribution for the electrons ($v = (8kT/\pi m)^{1/2} = 1.2 \cdot 10^7$ cm s^{-1} at 300 K), Eq. 1.40 yields an electron attachment cross section near 10^{-14} cm^2. For comparison, the photoionization cross-section is of the order of 10^{-17} cm^2 at 30 eV [19].

Although SF_6 has extensively been studied for some decades, a reliable value for the (adiabatic) electron affinity was established only recently ($EA(SF_6) = 1.05 \pm 0.10$ eV [20, 21]).

It is, however, well known that thermal electron attachment generates the anion in a metastable state. The autodetachment lifetime is long enough ($\approx 10^{-5}$ s) to allow detection of SF_6^- by standard mass spectrometric techniques. It is likely that capture of thermal electrons creates the anion in its electronic ground state. The MOs involved and the actual structure of the SF_6^- ion have yet to be established. Recent theoretical calculations [21, 27–29] consistently show that the HOMO in SF_6^- (point group O_h) is the totally symmetric $6a_{1g}$ orbital with S–F antibonding character associated with a substantial increase in the S–F bound length on proceeding from the neutral to the anion. Another possibility is capture of the additional electron into degenerate MOs with symmetry lower than a_{1g} leading to a Jahn-Teller distortion in the anion [30]. Lack of any photodetachment signal from SF_6^- also suggests a considerable geometry change between anion and neutral [31]. In any case, formation of metastable SF_6^- by capture of thermal electrons is thought to proceed via strong coupling between electronic and nuclear motion which prevents the extra electron from autodetaching for a certain time. SF_6^- thus represents a *nuclear excited Feshbach resonance*. For that reason it is often assigned as SF_6^{-*} (see Fig. 2.4).

Figure 2.3 shows some ion yield curves obtained in the energy range between 0–20 eV. The intensity ratio $SF_6^- : SF_5^- : F^- : F_2^-$ is $1000 : 80 : 1.5 : 0.05$ (count rate at the respective peak maximum) which shows that electron capture by SF_6 is essentially controlled by the formation of the parent anion. Apart from the ions shown in Fig. 1.3, the fragments SF_2^-, SF_3^- and SF_4^- can also be observed, each appearing with low intensity [26, 32].

SF_5^- exhibits a strong temperature dependence in that its intensity is increased by two orders of magnitude on going from room temperature to 800 K [33]. A closer inspection of the SF_5^- ion yield curve revealed that at low temperature it consists of two separate resonances peaking at 0.0 and 0.38 eV, with the first resonance showing

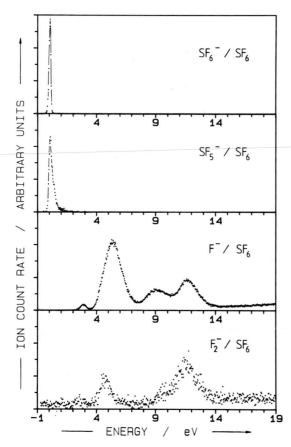

Fig. 2.3. Some selected ion yield curves observed in electron attachment to SF_6 (from [26]).

a strong temperature dependence. The explicit evaluation of this temperature dependence leads to $\Delta H_o = 0.2$ eV for the activiation energy of the reaction

$$SF_6^- \rightarrow SF_5^- + F. \tag{2.1}$$

Returning to Fig. 2.3, we see that SF_6 possesses a series of additional resonances at higher energies. Of these, F^- ion appears to arise from all of them, with the exception of the resonance at 0 eV. Since $D(F_5S-F) = 4.0 \pm 0.14$ eV [34, 35] and $EA(F) = 3.399$ eV [36] the thermodynamic threshold for reaction

$$e^- + SF_6 \rightarrow SF_5 + F^- \tag{2.2}$$

becomes $\Delta H_o = 0.65 \pm 0.14$ eV so that F^- formation is not expected near zero eV.

Electron attachment to SF_6 may then be pictured by a potential energy diagram plotted in Fig. 2.4: capture of 0 eV electrons generates SF_6^{-*} which is vibrationally excited. Coupling between electronic and nuclear motion prevents autodetachment for a certain time. With increasing temperature of the target molecule, decomposition into $SF_5^- + F$ becomes more likely. At higher electron energies, additional TNIs can be formed. These electronic states decompose into various negatively charged and neutral fragments. The asymptotic energies of some decomposition channels are indicated in Fig. 2.4. It is likely that the states peaking at 5.5 eV, 9 eV and 11.5 eV (Fig. 2.3) are core-excited resonances (SF_6^{*-} as opposed to vibrationally excited SF_6^{-*} formed at 0 eV) with two electrons in normally unoccupied MOs.

While SF_5^- arises only at low energy, its complement F^- ion is observed from resonances extending to higher energies (Fig. 2.3). This mode of behavior is a conspicuous feature of the decomposition of polyatomic negative ions to yield complementary fragment ions (such as SF_5^- and F^-) and can easily be rationalized by the different possibilities for excess energy distribution. TOF measurements of ionic fragments have shown [26] that all decomposition reactions in the present compound only yield low kinetic energy release so that the excess energy is virtually internal energy of the fragments. In the reaction yielding F^- ions the available excess energy can be deposited in the neutral fragment (SF_5) which may then undergo further decomposition reactions.

Conversely, if we detect SF_5^- ions the final state is uniquely determined and consists only of the two fragments $SF_5^- + F$. The available excess energy must then be deposited as internal energy of the ion, which above 3–4 eV becomes unstable with respect to dissociation *and* autodetachment (EA (SF_5) = 3.8 ± 0.14 eV [26]). It is thus likely that resonances at higher energies are coupled with dissociation channels con-

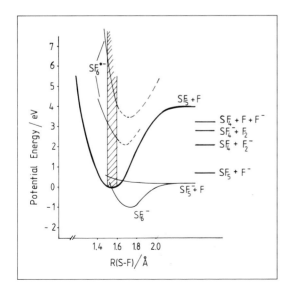

Fig. 2.4. Hypothetical potential energy diagram illustrating electron attachment to SF_6.

sisting of more than two fragments. This behavior is also shown by the compound SF_5NCO in Fig. 2.5. NCO, like the isoelectronic N_3 radical, is a pseudohalogen with an electron affinity near 3.8 eV [37]. In contrast to SF_6 and a further derivative, SF_5Cl [26], the present compound does not appear to form a metastable parent ion. Numerous fragmentation channels are observed in SF_5NCO. Figure 2.5 presents the ions yield curves for the complementary ion pairs F^-/SF_4NCO^- and NCO^-/SF_5^-. In both cases, the smaller ion is generated from all the negative ion resonances, while its larger counterpart arises only at low energies. Although the NCO radical can carry away internal energy there are distinctly more degrees of freedom in SF_5, thus making the observation of NCO^- ions at higher energies more probable than that of SF_5^- ions.

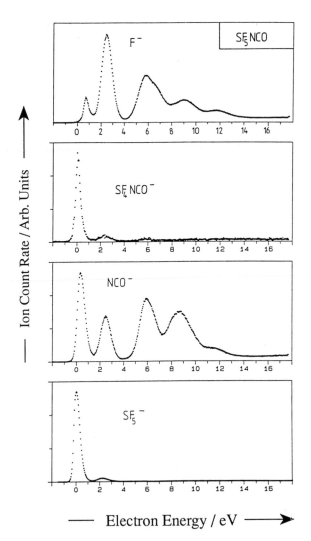

Fig. 2.5. Ion yield curves for the complementary ions F^-/SF_4-NCO^- and NCO^-/SF_5^- generated in SF_5NCO (from [37]).

Electron Energy / eV ⟶

2.2.2 Chlorofluoromethanes

Among the chlorofluorocarbons the most commonly used compounds are $CFCl_3$ and CF_2Cl_2, which still continue to serve as aerosol propellants, blowing agents for foams, and as refrigerants. They are known to possess remarkably high photochemical stability since they are transparent to wavelengths longer than 2300 Å [38, 39]. It is believed that their "infinite" tropospheric lifetime is responsible for the depletion of the stratospheric ozone layer [40-42]. While these compounds undergo photodissociation (chlorine abstraction) only for photon energies greater than 5.5 eV, they are very efficiently decomposed in the presence of low-energy electrons. Figure 2.6 presents ion yield curves observed in $CFCl_3$. As in all other halogenated methanes no parent anion is observed. $CFCl_3^-$ immediately decomposes following electron capture.

An estimate of the absolute attachment cross-section can be derived from measured attachment rates. With k ranging from $1 \cdot 10^{-7}$ cm^3 s^{-1} [43] to $2.4 \cdot 10^{-7}$ cm^3 s^{-1} [44] at 300 K one obtains a cross-section of the order of 10^{-14} cm^2.

Since attachment of thermal electrons exclusively yields Cl^- ions, the above number represents the cross section for *dissociative attachment*. It should be noted that

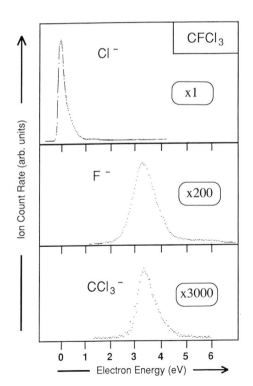

Fig. 2.6. Ion yield curves obtained in CF_3Cl (from [1]).

the cross section for *photodissociation* into $Cl + CFCl_2$ is by four orders of magnitude lower (at 7.6 eV photon energy which corresponds to the maximum in the absorption band [38]). The overall intensity ratio is $Cl^- : F^- : CCl_3^- = 1000 : 5 : 0.5$.

The other chlorofluoromethanes (including CCl_4) exhibit a similar behavior in that Cl^- is exclusively formed within a narrow resonance at low energy and additional negatively charged fragments arise from a broader resonance located at higher energy [1, 45]. The general trend is that the attachment cross section decreases as the number of F atoms in the molecule increases.

With regard to the ozone problem mentioned above, dissociative electron attachment has been considered as a possible tropospheric sink for chlorinated hydrocarbons [46]. Since free electrons are not ubiquitous in the troposphere, however, dissociative attachment, in spite of its high cross section, may not be operative. Heterogeneous decomposition reactions have been observed when these classes of compounds are adsorbed on solid surfaces [47–50] and it is likely that electron transfer is an elementary step here.

We have carried out semiempirical MO calculations [51] on the chlorofluoromethanes indicating that the lowest virtual MO is clearly $\sigma^*(C-Cl)$ in character. These calculations predict a substantial decrease in the C—Cl bond order on going from the neutral to the anion, e.g., from 0.88 to 0.25 in the case of CF_3Cl and from 0.92 to 0.63 for each of the three C—Cl bonds in $CFCl_3$.

The electronic ground state of the anion is then characterized by a significant localization of the additional charge onto the chlorine atom(s) while the charge density at the fluorine atoms remains virtually unaffected. The ability of a halogen in a neutral molecule to accommodate extra negative charges has been called *electron capacity* [52]. In contrast to the *electron affinity* and the *electronegativity* the electron capacity of the halogens increases on descending group 17 of the periodic table. Based on these MO calculations, low-energy electron attachment to chlorofluoromethanes may be interpreted as accommodation of an extra electron into a $\sigma^*(C-Cl)$ MO and subsequent direct electronic dissociation along a repulsive potential energy surface into $R + X^-$.

Interestingly, the anions CCl_4^-, $CFCl_3^-$, and $CF_2Cl_2^-$ are known to exist as stable compounds (with respect to autodetachment *and* dissociation). They have been observed in charge transfer experiments and electron affinities of 2.0 ± 0.2 eV, 1.1 ± 0.3 eV and 0.4 ± 0.3 eV have been derived for CCl_4, $CFCl_3$ and CF_2Cl_2, respectively [53]. These anions are not observable in electron attachment experiments since the transition to the anionic potential energy surface occurs above the dissociation limit. The situation is illustrated in Fig. 2.7. Lack of an operative mechanism to carry off the excess energy prevents the molecular anion to be observed in its relaxed configuration.

At this point, we refer to Chapter 4 where electron attachment experiments to van der Waals clusters of such compounds are discussed. Since an aggregate possesses a "built-in" three body stabilization mechanism, generation of a TNI within an aggregate represents a method to prepare anions in their relaxed form.

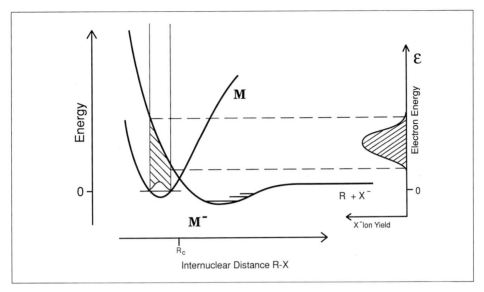

Fig. 2.7. Hypothetical potential energy curve illustrating low energy electron attachment to fluoro-chloromethanes (see text).

MO calculations also predict higher lying virtual MOs with $\sigma^*(C-F)$ character. Figure 2.6 indeed shows that the resonance at 3.3 eV yields the complementary ions F^- and CCl_3^- arising from the decompositions

$$CFCl_3^{*-} \rightarrow F^- + CCl_3$$

$$\rightarrow F + CCl_3^-$$

which are associated with the cleavage of a C–F bond. It requires a more sophisticated theoretical treatment, however, to judge whether electron impact at a few electronvolts energy may still be described within the one electron formalism.

In contrast to chlorine containing compounds, tetrafluoromethane captures electrons within a very broad resonance (Fig. 2.8) yielding the ions F^- and CF_3^-. From well established thermodynamic data the energetic threshold for reaction

$$e^- + CF_4 \rightarrow F^- + CF_3 \tag{2.3}$$

is calculated as $\Delta H_o = 2.26 \pm 0.2$ eV [54, 55]. At the experimental threshold (≈ 4.5 eV) we thus have already more than 2 eV of excess energy. Cl^- formation from $CFCl_3$, on the other hand, occurs near the thermodynamic threshold within a comparatively narrow resonance.

If the reaction is described in terms of a purely repulsive potential energy curve $V_f(R)$, the width Γ_d of the resonance is given by Eq. 1.28. Γ_d increases with the slope

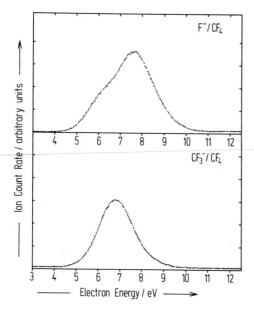

Fig. 2.8. Ion yield curves obtained in CF_4 (from [1]).

of the anionic potential energy surface in the Franck-Condon region and decreases with the stiffness and the reduced mass of the oscillator associated with the potential energy of the neutral. Since the difference in the stiffness of the C—F and C—Cl oscillators is partly compensated by their different reduced masses and both have to be taken with the exponent 0.25, the experimentally observed width of an ion yield curve in the fluorochloromethanes essentially reflects the slope of $V_f(R)$.

In the following section we will apply a TOF technique to obtain information on how the excess energy is distributed among the products and their degrees of freedom. As mentioned, E^* amounts to 2 eV at the experimental threshold of reaction 2.3.

2.3 Excess Energy

An explicit theoretical treatment of the problem of how the available excess energy is distributed among the fragments remains a rather complex problem for a polyatomic molecule, regardless of how accurately the potential energy surfaces are known. The most widely used approaches are based on statistical methods which are extensively discussed in the second contribution to this volume.

Franklin et al. [56, 57] derived a simple relationship for the distribution of kinetic excess energy $f(E_T)$ at a given total excess energy E^*

$$f(E_T) = \frac{(N-1)\,(E^* - E_T)^{(N-2)}}{E^{*(N-1)}}, \qquad\qquad 2.4$$

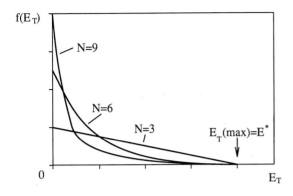

Fig. 2.9. Calculated distributions of translational energy imparted to the fragments according to the statistical treatment (Eq. 2.4).

with the mean translational energy

$$\overline{E_T} = \int_0^{E^*} dE_T \, f(E_T) \, E_T = \frac{E^*}{N} .$$

2.5

N denotes the number of *activated* oscillators in the transient negative ion. If N' denotes the number of vibrational normal modes, we can write $N = \alpha \cdot N'$ with $0 \le \alpha \le 1$. For a non-linear molecule with n atoms we have $N' = 3n - 6$.

In addition to the usual premises of a statistical treatment, Eq. 2.4 is based on the further assumption that the translational energy of the products arises entirely from the kinetic energy located in the reaction coordinate of the transition state and that rotational energy can be ignored.

Figure 2.9 shows kinetic energy distributions calculated according to Eq. 2.4 for different numbers of activated oscillators.

These are monotonically decreasing functions peaking at zero energy. More sophisticated treatments, like the "statistical adiabatic channel" model [58, 59], may also yield distribution functions peaking at some energy above zero. Kinetic energy distribution of fragment ions is comprehensively treated in Section 4.5.4 of the second contribution to this volume.

We will now apply a time-of-flight (TOF) technique to obtain information on the translational energy of fragment ions released in dissociative electron attachment.

2.3.1 Flight Times of Ions

In Section 3.3.3 of the first contribution to this volume, we considered the flight times of ions in a two field TOF mass spectrometer (Wiley-McLaren design). It was shown that the mass resolution of such an instrument is inherently limited by the translational energy of ions which is, at best, a Maxwell-Boltzmann distribution in the case of parent ions. We will now proceed the other way and obtain information on the translational energy of a fragment ion by measuring its flight time from the source to the detector.

Figure 2.10 shows the experimental arrangement. The electron beam is pulsed at one of the electrodes before the reaction chamber and the flight time of an ion is obtained by measuring the time difference between the electron pulse and the arrival time of the ion at the detector. The configuration consists of a draw-out region (ion draw-out field ε_1, distance travelled s_1), an acceleration region ε_2, distance travelled s_2) and a drift space ($\varepsilon_x=0$, s_3) represented by a quadrupole mass filter. (We use here the symbol ε for the electric field. Note that the same symbol was used for the electron energy in Chapter 1.) With respect to flight times, the instrument acts as a two field mass spectrometer.

The flight times within the different regions can be expressed as

$$t_{s_1} = \frac{(2M_i)^{1/2}}{ze\varepsilon_1} \left\{ (E_T^i + ze\varepsilon_1 s_1)^{1/2} \pm (E_T^i)^{1/2} \right\} \qquad 2.6$$

$$t_{s_2} = \frac{(2M_i)^{1/2}}{ze\varepsilon_2} \left\{ (E_T^i + ze\varepsilon_1 s_1 + ze\varepsilon_2 s_2)^{1/2} - (E_T^i + ze\varepsilon_1 s_1)^{1/2} \right\} \qquad 2.7$$

$$t_{s_3} = (2M_i)^{1/2} \frac{s_3}{2(E_T^i + ze\varepsilon_1 s_1 + ze\varepsilon_2 s_2)^{1/2}} \qquad 2.8$$

with M_i the mass of the fragment ion and E_T^i its initial translational energy. The + and −sign in Eq. 2.6 refers to initial velocity vectors directed away from (turn-around

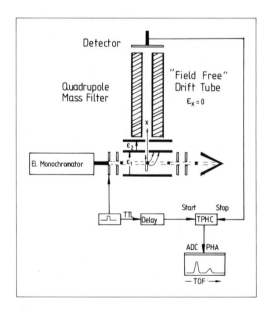

Fig. 2.10. Experimental setup for TOF analysis (from [1]).

ions) and toward the detector (direct ions), respectively. Note that only t_{s_1} has different values for direct and turn around ions; the latter will be decelerated, reversed, and accelerated, so that it will leave the draw-out region with the same velocity as the direct ion, but with a certain time delay ΔT given by

$$\Delta T = \frac{(8M_i E_T^i)^{1/2}}{ze\varepsilon_1}. \qquad\qquad 2.9$$

The initial kinetic energy of the fragment ion can, in principle, be obtained by measuring ΔT. In a conventional TOF mass spectrometer the aperture of the ion draw-out electrode and the applied draw-out field are such that all ions are transmitted to the detector, regardless of their initial orientation. For ions which are ejected *isotropically* with a *discrete initial energy* the TOF distribution then has a rectangular shape [60] and ΔT is then the width of the rectangular TOF distribution.

   In the present configuration, a comparatively low ion draw-out field is used ($\varepsilon_1 = 4$ V cm^{-1}) so that *discrimination* against velocity vectors perpendicular to the flight tube axis occurs. The amount of discrimination depends on ε_1 and E_T^i. This is illustrated in Fig. 2.11 for an ion having an initial energy of $E_T^i = 0.4$ eV. With $s_1 = 0.5$ cm and $r_1 = 0.15$ cm (radius of the electrode aperture) all direct ions having initial velocity vectors within a cone angle of 29° are detected. For turn around ions the corresponding angle is only 12°. Figure 2.12 shows the calculated TOF spectrum ($M_i = 1$ amu) for the present configuration ($\varepsilon_2 = 29$ Vcm^{-1}, $s_2 = 1.5$ cm, $s_3 = 22.5$ cm). $\Delta\tau$ is the broadening of the "direct" peak due to perpendicular velocity components. The ratio $\Delta\tau/\Delta T$ increases with decreasing E_T^i until for $E_T^i < 0.05$ eV discrimination no longer occurs, i.e. ions emitted in the entire solid angle are detected. T_o in Fig. 2.12

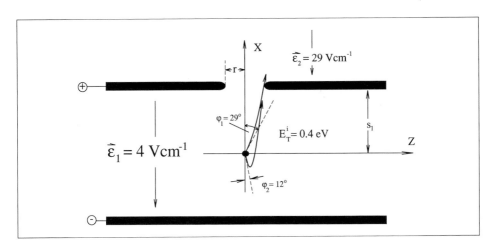

Fig. 2.11. Ion trajectories in the draw out region for two initial velocity vectors. The kinetic energy is 0.4 eV. The cone angle indicates the initial velocity vectors accepted by the aperture of the draw-out electrode.

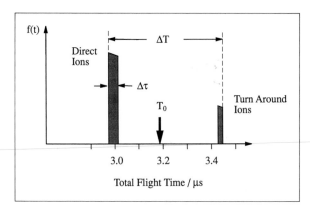

Fig. 2.12. Calculated TOF spectrum for ions ($M_i = 1$) with discrete initial energy of 0.4 eV. T_o is the flight time of an ion ($M_i = 1$) with zero kinetic energy.

assigns the flight time of an ion with zero initial energy. Note that the "direct" and "turn around" peak are not exactly symmetric around T_0. The time a direct ion gains (over an ion with $E_T^i = 0$) is less than a turn-around ion loses [61].

2.3.1.1 Thermal Ions

We will now derive an expression which relates the width of the TOF distribution with the mean energy of ions having a Maxwell-Boltzmann distribution

$$f(E_T^i)\, dE_T^i \sim (E_T^i)^{1/2} \exp\left(- E_T^i/kT\right) dE_K^i, \qquad 2.10$$

with $\overline{E_T^i} = 3/2\, kT$ the average energy. Near room temperature ($\overline{E_T^i} = 0.04$ eV at $T = 300$ K) we have no discrimination against perpendicular components. The corresponding velocity distribution is

$$f(v)\, dv \sim v^2 \exp\left(- M_i v^2/2kT\right) dv, \qquad 2.11$$

and its projection onto the flight tube axis is

$$f(v_x)\, dv_x \sim \exp\left(- M_i v_x^2/2kT\right) dv_x. \qquad 2.12$$

The distribution function in the variable t can easily be obtained in the scale around $T_o = 0$. For $E_T^i \ll ze\varepsilon_1 s_1 = 2$ eV (which is well fulfilled for thermal ions), we have $(E_T^i + ze\varepsilon_1 s_1)^{1/2} \approx (ze\varepsilon_1 s_1)^{1/2}$, thus it follows from Eq 2.6 that

$$t = \pm \frac{(2M_i E_T^i)^{1/2}}{ze\varepsilon_1} = \pm\, v_x\, \frac{M_i}{ze\varepsilon_1}$$

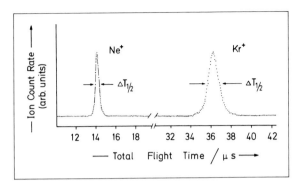

Fig. 2.13. Time-of-flight spectra for Ne$^+$ and Kr$^+$ ionized by electron impact.

in the scale $T_o = 0$. This is a linear relation between t and v_x, i.e., $dt/dv_x = $ const, and the TOF distribution becomes the Gaussian form

$$f(t) \, dt \sim \exp\left(- t^2 \frac{(ze\varepsilon_1)^2}{2M_i \, kT}\right) \, dt \, .$$ 2.13

The full width at half maximum is

$$\Delta T_{1/2} = \frac{2 \, (2M_i \, kT \, \ln 2)^{1/2}}{ze\varepsilon_1} \, .$$

Setting $\overline{E_T^i} = 3/2 \, kT$, we obtain

$$\Delta T_{1/2} = \frac{(3.7M_i \, \overline{E_T^i})^{1/2}}{ze\varepsilon_1} \, .$$ 2.14

Equation 2.14 relates the mean energy of a Maxwell-Boltzmann distribution with the width of the corresponding time-of-flight peak. Fig. 2.13 shows experimentally observed TOF peaks for Ne$^+$ and Kr$^+$. The peak widths and shapes coincide with those predicted by Eqs. 2.13 and 2.14 for $T = 400$ K, the approximate temperature of the reaction chamber. The result proves that the HF quadrupole field does not affect the flight times of the ions, at least not to an extent relevant for the present experiments. This is, however, only achieved by increasing the transmission energy from the standard value of about 10 eV to > 40 eV.

It should finally be noted that ε_1 and ε_2 was adjusted in order to obtain space focusing.

2.3.1.2 Ions with Excess Energy

When the ions are formed with considerable translational energy $E_T^i \gg kT$, the instrument essentially detects ions emitted along the flight tube axis, resulting in a

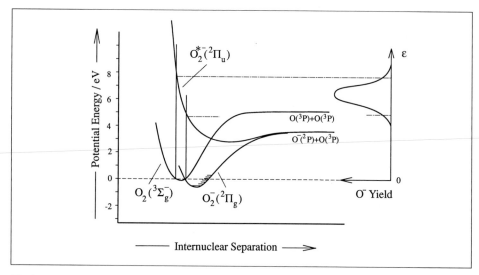

Fig. 2.14. Potential energy curves involved in electron attachment to isolated O_2 (see also Fig. 4.13).

TOF doublet due to direct and turn around ions. Within the approximation of complete discrimination against perpendicular velocity components, the kinetic energy can be calculated from the experimentally observed time difference according to Eq. 2.9 as

$$E_T^i = \frac{(\Delta Tze\varepsilon_1)^2}{8M_i}. \qquad\qquad 2.15$$

We will now illustrate this technique for two examples, dissociative electron attachment to O_2 and CF_4.

The first process has been studied extensively in the past [62–64]. Figure 2.14 shows the ground state potential energy curves for the neutral and the anion, as well as the repulsive potential energy curve O_2^{*-} involved in dissociative attachment. Low- energy electron attachment to O_2 generates $O_2^- (^2\Pi_g)$ ions in vibrationally excited states. The autodetachment lifetime, however, is too short for the ion to be detectable with a mass spectrometer. For the generation of thermodynamically stable O_2^- from oxygen clusters, see Section 4.5.

Figure 2.15 presents the O^- ion yield curve and the corresponding O^- TOF spectra. As in all of the following TOF spectra, the flight-time scale refers to an ion with zero kinetic energy, i.e. $T_0 = 0$. In that scale, direct ions have negative flight times and turn around ions have positive ones.

In Figure 2.16 the kinetic energy of O^- ions calculated according to Eq. 2.15 from the experimentally observed flight time difference ΔT is plotted. This gives, as

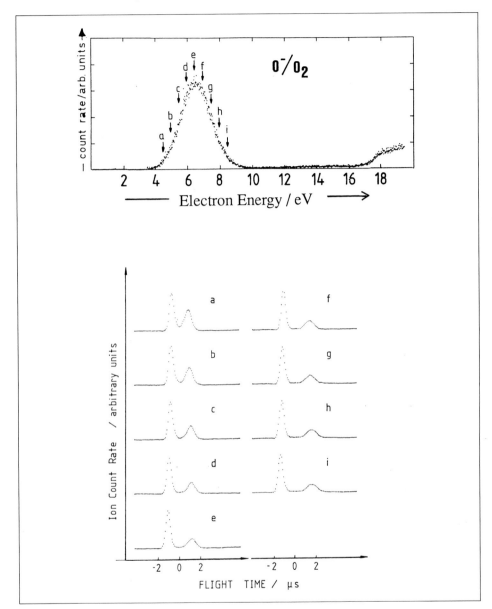

Fig. 2.15. O⁻ ion yield curve and TOF spectra taken at different electron energies.

expected, a straight line with a slope of 0.5 since the translational energy is released in equal amounts to O^- and O. Extrapolation of the line to $E_T = 0$ yields the thermodynamic threshold for the reaction in good agreement with the value calculated from known thermodynamical data ($\Delta H_o = 3.60$ eV from $D\,(O-O) = 5.16$ eV and

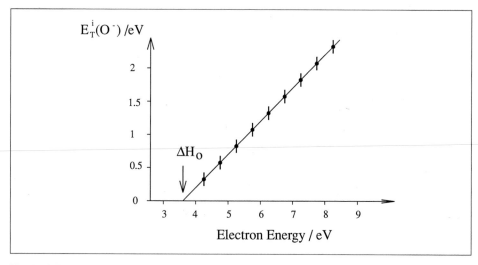

Fig. 2.16. Translational energy of O^- as a function of electron energy (from [1]).

$EA(O) = 1.46$ eV [36]). The continuous increase in the O^- ion yield above 17 eV is likely to be a result of the non resonant ion pair formation process

$$e^- + O_2 \rightarrow O_2^{**} + e^-$$
$$\qquad\qquad \longrightarrow O^+ + O^-, \qquad\qquad\qquad\qquad\qquad\qquad 2.16$$

which has an energetic threshold of 17.2 eV.

2.3.2 Tetrafluoromethane. Energy Distribution of Products

Figures 2.17 and 2.18 present the ion yield curves and TOF spectra for CF_3^- ions and F^- ions. In the case of CF_3^- we observe an incompletely separated TOF doublet at low energies which becomes more separated with increasing electron energy and, hence total energy. F^- formation, on the other hand, exhibits qualitatively different behavior: at low energies a doublet is visible signalizing F^- ions released with significant kinetic energy. Above 6 eV an additional peak at $T = 0$ appears which becomes dominant in the TOF spectrum on further increasing the electron energy. This feature indicates the concomitant formation of thermal or near-thermal F^- ions. We suggest the following mechanism for this effect: Electron attachment to CF_4 occurs via two overlapping resonances associated with the parent anion in its electronic ground state and an electronically excited state. The ground state ion decomposes into the complementary channels:

$$CF_4^- \rightarrow CF_3 + F^- \qquad\qquad 2.17$$

$$\rightarrow CF_3^- + F, \qquad\qquad 2.18$$

with high kinetic excess energy release in both cases. The CF_4^{*-} ion, in contrast, is only correlated with F^- ion formation. This reaction may proceed according to

$$CF_4^{*-} \rightarrow F^- + CF_3^*$$
$$\phantom{CF_4^{*-} \rightarrow F^- +}\;\longrightarrow CF_2 + F, \qquad\qquad 2.19$$

or directly by

$$CF_4^{*-} \rightarrow F^- + CF_2 + F. \qquad\qquad 2.20$$

The thermodynamic threshold for this multiple fragmentation process is 6.0 eV (as calculated using standard thermodynamic data). Since little is known about the energy of the first electronically excited state of CF_3 [65, 66], it is not certain whether the reaction proceeds synchronously or via the excited trifluoromethyl radical.

Returning to the channels 2.17 and 2.18 correlated with the ground state of the parent ion, we may now ask how much of the available excess energy E^* is released as kinetic energy of the two products. Since the precursor is a 5 atomic ion, F^- and CF_3^- formation is expected to be associated with a *distribution* of kinetic energy. This distribution can in principle be obtained by an explicit computer-simulation procedure [67]. The evaluation "by hand" (simply taking ΔT from the peaks in the doublet) then yields the most probable kinetic energy for the ion. The TOF spectra unambiguously reveal that the distribution of kinetic energy for both F^- and CF_3^- reaches a maximum at some finite value of E_T^i, and not at zero energy as in distributions by the simple statistical model (Eq. 2.4 and Fig. 2.9, page 275).

If the final dissociation channel consists of two fragments, as in Reaction 2.17 and 2.18, the principle of conservation of linear momentum allows to calculate E_T, the translational energy imparted to both fragments

$$E_T = E_T^i \cdot M/M_n, \qquad\qquad 2.21$$

with M the mass of the target molecule and M_n that of the neutral fragment.

An evaluation of the TOF spectra (Fig. 2.17 and Fig. 2.18 (high energy component)) reveals that the mean kinetic energy release to $CF_3^- + F$ and $CF_3 + F^-$, respectively, increases linearly with the electron energy and hence, with E^* [1, 68]. The slope of the line is 0.33 for F^- ion formation and 0.62 for CF_3^- formation. In other words, in reaction 2.17, 33 % of the available excess energy is released as kinetic energy while it is 62 % for channel 2.18.

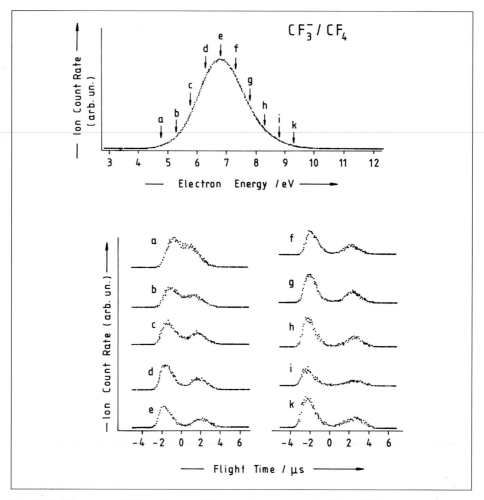

Fig. 2.17. TOF spectra of CF_3^- taken at different electron energies (from [80]).

How do these results correlate with the statistical picture of the unimolecular decomposition? In both channels, the same *fraction* of E^* arises as translational energy. In terms of activated classical oscillators, this leads to $\alpha=0.18$ for CF_3^-+F and $\alpha=0.33$ for F^-+CF_3. The kinetic energy *distribution*, however, differs substantially in both cases from that predicted by the statistical treatment. These energy distributions (Fig. 2.9) would give one single peak in the TOF spectrum. This indeed appears to be the case for the decomposition of the excited parent ion (channel 2.19 or 2.20).

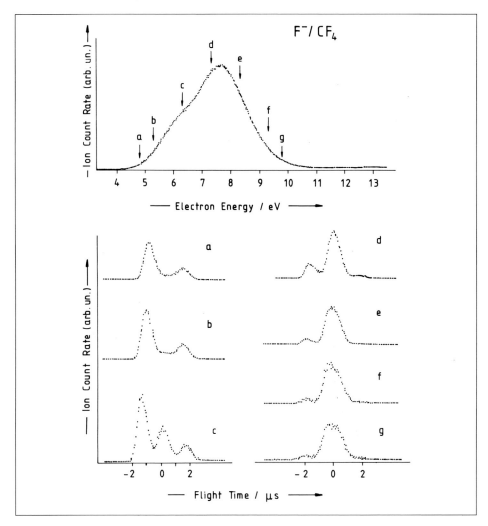

Fig. 2.18. TOF spectra of F⁻ taken at different electron energies (from [80]).

2.3.3 CF₃Cl and CF₃CN. An Example of Two Completely Different Decomposition Mechanisms

Kinetic energy release for the various channels in dissociative attachment to CF_3Cl and other halogenated methanes has recently been studied in detail [1, 68, 69, 80]; we will focus only on the reaction yielding Cl^- ions. Figure 2.19 shows

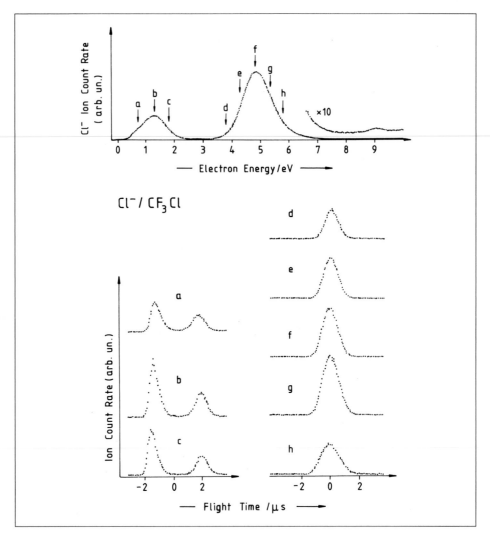

Fig. 2.19. TOF spectra of Cl^-/CF_3Cl taken at different electron energies (from [80]).

the Cl^- ion yield curve and, below that, the corresponding TOF spectra. For energetic reasons the TNI formed at low energies must decompose according to

$$CF_3Cl^- \rightarrow Cl^- + CF_3 . \qquad\qquad 2.21$$

The resonance near 5 eV may lead to electronic excitation of CF_3^* and/or further fragmentation. Cl^- ion formation from CF_3Cl is thus similar to F^- ion formation from CF_4, with the distinction that both resonances are now completely separated. The complementary reaction (CF_3^- + Cl) is energetically inaccessible from the low-energy resonance. The CF_3^- ion is only observed with extremely low intensity within the second resonance.

Figure 2.19 clearly indicates that the Cl^- ion is generated with considerable excess energy from the first resonance and with low excess energy from the second resonance. An analysis of the translational energy imparted to both fragments vs. electron energy is shown in Fig. 2.20. We see that at the experimental threshold (0.12 eV) virtually 100 % of the available excess energy is converted into kinetic energy of the fragments. Any additional excess energy results only in a small further increase in the translational energy. A similar completely unbalanced excess energy distribution has recently been observed in dissociative attachment to CF_3I [70].

Let us now replace chlorine by a CN group. The cyano radical is a well known pseudohalogen with an electron affinity (3.82 eV) even greater than that of the halogen atoms [71]. The cyanide ion (like the Cl^- ion from CF_3Cl) is formed within a resonance between 0.6 eV and 3 eV [72] (not shown here). The F_3C-CN bond dissociation energy has not yet been established and ΔH_o for CN^- ion formation is not known exactly. (The present result yields $\Delta H_o \leq 0.6 \pm 0.2$ eV.) We can say, though, that if the CN^- ion is generated at the high energy side of the resonance, the reaction products must contain a total excess energy of nearly 2 eV (as in Cl^-/CF_3Cl). A TOF analysis of CN^- ions, surprisingly, results in single, narrow peaks, indicating CN^- ion formation with low kinetic energy (Fig. 2.20). A TOF analysis of F^- ions which also appears from the low-energy resonance in CF_3CN also indicates low translational energy release.

Although the systems CF_3Cl and CF_3CN are similar in that the energy of the transient negative ions and the energy of the dissociation limits (Cl^- + CF_3 and CN^- + CF_3) nearly coincide, they behave completely different in the manner in which the excess energy is distributed. This phenomenon can easily be rationalized by the different electronic structures of the two anions.

According to MNDO calculations (and a treatment on a higher level of theory may not change the basic argument), the first virtual MO in CF_3Cl is a strongly local-

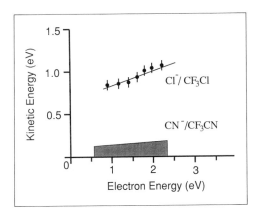

Fig. 2.20. Kinetic energy imparted to Cl^- + CF_3 and CN^- + CF_3, determined from TOF measurements (from [1]).

ized $\sigma^*(C-Cl)$ orbital. Dissociative electron attachment to CF_3Cl at low energies can be considered to proceed via accommodation of the additional charge into this MO and subsequent direct electronic dissociation along a repulsive "Σ" surface.

In contrast, in CF_3CN the lowest unoccupied molecular orbital (LUMO) has essentially $\pi^*(C-N)$ character. Electron attachment then occurs via capture of the excess electron in the π^* MO. The subsequent decomposition into the energetically accessible channels ($CN^- + CF_3$ and $F^- + CF_2CN$) is only possible through transfer to the reaction coordinate of the energy initially deposited in the $C-N$ coordinate. The reaction generating F^- ions is additionally associated with transfer of the excess charge initially localized on the CN group to the fluorine atom. Decomposition of the transient CF_3CN^- ion is a reaction which falls within the class of vibrational predissociation introduced above. In contrast to direct electronic dissociation (where only one specific bond is initially influenced by the excess charge), in the present case the entire molecule is involved in the reaction resulting in highly effective energy randomization and hence low kinetic energy release. It has been shown that effective energy randomization is a general feature of electron capture by cyano compounds [72,73].

Direct electronic dissociation may be discussed in terms of the impulsive model with its two limiting cases, the "purely impulsive" (or "soft radical") and the "rigid radical" limit [74,75]. The former predicts the maximal amount of excess energy which can be transferred to internal energy.

In the soft radical limit it is assumed that, for example, in the case of CF_3Cl in the first step, the carbon atom recoils from the Cl^- ion with the full available energy E^*. (This is true if the $C-Cl$ repulsion is much stronger than the $C-F$ interaction in the CF_3 radical.) In the second step, the carbon atom runs into the rest of the radical, exciting the internal degrees of freedom which for reasons of symmetry are only vibrational modes in CF_3. The final translational energy of Cl^- ions and CF_3 is determined solely by conservation of energy and linear momentum leading to

$$E_T = E^* \frac{\mu_{C-Cl}}{\mu_{Cl-CF_3}},\qquad\qquad 2.22$$

with μ_{C-Cl} the reduced mass of chlorine and carbon at either end of the breaking bond, and μ_{Cl-CF_3} the reduced mass of chlorine and CF_3.

For CF_3Cl the soft radical limit predicts that 62% of E^* emerges as vibrational energy of the CF_3 radical and 38% as translational energy E_T. The Cl^- ion carries away 26% and CF_3 12% of E^*.

In the "rigid radical" limit the Cl^- ion is pictured as recoiling from a completely rigid CF_3 radical. In this limit no vibrations are excited in CF_3 and the available energy E^* is partitioned solely between product recoil energy and, possibly, rotational energy, subject to the constraint of angular momentum conservation. In the dissociation of CF_3Cl^- (C_{3v}) no rotational excitation is expected for symmetry reasons and so the Cl^- ion carries away 68% and CF_3 32% of E^*.

The present system behaves at the experimental threshold according to the rigid limit, while most of the *additional* energy is transferred to vibrational energy of CF_3. At the high-energy end of the resonance (see Fig. 2.20) approximately 50 % of E^* arises as internal energy. (It must be remembered that the maximum amount predicted for the soft radical limit is 62 %.) This may be qualitatively rationalized by the fact that, with increasing energy the, $C-Cl$ force constant becomes larger so that the Cl^- ion is no longer recoiling against a completely rigid CF_3 radical.

2.3.4 Methanol, Deuterated Methanol and Allylalcohol

In contrast to the compounds discussed above, methanol shows unusual behavior in that negative ions only appear at energies far above their energetic thresholds with a rather complex feature of kinetic energy release. The ions observed are O^-, OH^- and CH_3O^- [76]. From well established thermodynamic data [77] the minimum heat of reaction ΔH_o for the lowest channels can be calculated as

$$e^- + CH_3OH \rightarrow OH^- + CH_3, \qquad \Delta H_o = 2.1 \text{ eV} \qquad\qquad 2.23$$

$$\rightarrow O^- + CH_4, \qquad \Delta H_o = 2.4 \text{ eV} \qquad\qquad 2.24$$

$$\rightarrow CH_3O^- + H, \qquad \Delta H_o = 2.9 \text{ eV}. \qquad\qquad 2.25$$

On proceeding to the deuterated species CH_3OD, CD_3OH, CD_3OD and CH_2DOH, the deuterated fragments listed in Table 1 could be observed.

As an example Fig. 2.21 shows the results for CH_3OD.

While the CH_3O^- ion is formed in three resonances above 5.5 eV, OH^-, OD^- and O^- ions only appear within a prominent resonance peaking at 10.5 eV. From Fig. 2.21 it is apparent that the high-energy resonance simultaneously yields OH^- and OD^- ions which points to "hydrogen scrambling" in the TNI prior to decomposition. In contrast, the observation that the CH_3O^- ion, rather than the CH_2DO^- ion is formed indicates a direct cleavage of the OD bond (without hydrogen scrambling). By measuring the deuterated methanols it becomes apparent that OH^- (or OD^-) ion formation is always accompanied by hydrogen scrambling while formation of CH_3O^- ions (and the deuterated analogues) proceeds directly (no hydrogen scrambling). These results are summarized in Table 2.1.

A TOF analysis of the different fragment ion shows that in methanol and the deuterated methanols only O^- is formed with considerable translational energy. In the case of CH_3O^- (and its deuterated analogues, channel 2.25)) the light hydrogen (or deuterium) atom carries away 97 % (94 %) of the total translational energy so that a TOF study of CH_3O^- ions cannot provide any further information about the energy distribution of this channel. The O^- ion is formed at 9 eV with low kinetic energy and at 10.5 eV with a "quasi-discrete" translational energy (separated TOF doublet) corresponding to a mean energy of 1.0 ± 0.2 eV for the O^- ion.

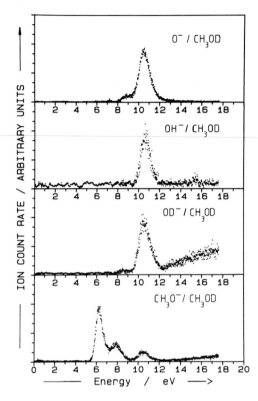

Fig. 2.21. Negative ion formation in CH$_3$OD (from [76]).

We thus have a rather complicated situation in methanol and the deuterated methanols: O$^-$ and OH$^-$ (OD$^-$) ions are formed from a resonance peaking at 10.5 eV which is roughly 8 eV above their energetic thresholds. It is very likely that the associated fragmentation channels consist of more than two fragments. In contrast, formation of CH$_3$O$^-$ ions (and the corresponding deuterated analogues) proceeds directly. The unfavorable mass ratio of hydrogen to CH$_3$O$^-$ ions, unfortunately, precludes any detailed information on the excess energy distribution using TOF methods. Owing to the presence of the magnetic field in our experimental configuration, H$^-$ (and D$^-$) ions cannot be detected, thus ruling out the study of the complementary reaction to 2.25.

The sort of behavior shown here is not necessarily a particular feature of compounds containing OH groups [78]. Fig. 2.22 shows that in allyl alcohol the ions OH$^-$ and C$_3$H$_5$O$^-$ are formed within a resonance peaking at 1.7 eV. Electron attachment to allyl alcohol is expected to proceed via the lowest virtual MO, which is π^* (C−C). This picture is strongly supported by our MNDO calculations which indicate that the LUMO in allyl alcohol (as in ethylene) has distinctly π^* (C−C) character with an eigenvalue near that of ethylene [76,79]. Ethylene in fact captures electrons near 1.8 eV to form the anion in the ground state (see next chapter).

Table 2.1. Relative intensities for negative formation in methanol and the different deuterated methanols taken at the resonance maximum (10.5 eV) and normalized in each compound to the intensity of O^-.

	O^-	OH^-	OD^-	CH_3O^-	CD_3O^-	CH_2DO^-	CHD_2O^-
CH_3OH	100	14	–	10	–	–	–
CD_3OD	100	–	17	–	10	–	–
CH_3OD	100	7	14	10	–	0	–
CD_3OH	100	7	4	–	9	–	0
CH_2DOH	100	8	3	0	–	15	–

The interpretation of negative ion resonances in methanol, on the other hand, is by no means straightforward. Our MNDO calculations indicate high lying (compared to allyl alcohol) virtual MOs with mixed $\sigma^*(C-H)$, $\sigma^*(O-H)$ and $\sigma^*(C-O)$ antibonding components. However, the generally low translational energy release and the observation of hydrogen scrambling supports the picture of a core excited resonance at 10.5 eV which decomposes through rather indirect processes like internal conversion followed by vibrational predissociation. Since the ionization energy of CH_3OH is 10.85 eV, the core-excited resonance may contain two electrons in Rydberg type MOs.

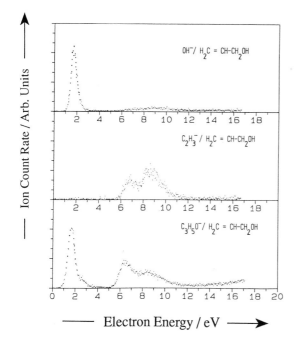

Fig. 2.22. Negative ion formation in allylalcohol (from [76]).

In summary, methanol like other oxygen containing compounds [78] is character-ized by high-energy resonances associated with low cross-sections and hydrogen scrambling in the TNI prior to dissociation. Allylalcohol is a particular exception because of its π_{C-C} system which allows capture of low-energy electrons.

2.4 Electron Attachment to Excited Molecules

In Section 2.2 we have already mentioned that electron capture by SF_6 shows a strong temperature dependence in that the dissociative channel $SF_5^- + F$ is increased by a factor of 100 when the gas temperature is increased from room temperature to 800 K. Electron capture by "hot" molecules (rotationally, vibrationally, electronically excited) is of considerable significance from both the basic view and from applica-tion. With regard to the latter, in many applied devices such as discharge switches or gaseous dielectrics the temperature may vary to some extent, thereby influencing the number density of electrons and ions. Furthermore, photon enhanced dissocia-tive attachment is considered for application in optical switches [81].

We will illustrate here the effect of temperature for the system O_2. Figure 2.23 shows that the onset and the maximum of the O^- ion yield curve is remarkably shif-ted to lower energies as the temperature is increased; the points are the experimen-tal results [82] and the solid curves are calculated values [83]. O'Malley [83] showed that the large effect can consistently be explained by the competition between auto-detachment and dissociation in the transitory negative ion. Figure 2.24 shows the survival probability used in the calculation; it decreases rapidly with increasing elec-tron energy (i. e., with decreasing internuclear separation). Population of higher vibrational levels allows transitions at larger internuclear distances, thus decreasing the appearance energy of O^- by more than 2 eV on going from room temperature to 2100 K. Since the vibrational quantum in O_2 is $\hbar\omega = 0.194$ eV, vibrational levels up to $v \gtrsim 10$ must significantly contribute to the process. On the other hand, population of

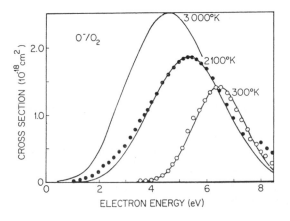

Fig. 2.23. Temperature dependence of the dissociative electron attachment cross-section for O^- from O_2. The solid curves are the calculated values [83], the points are the experimental results [82].

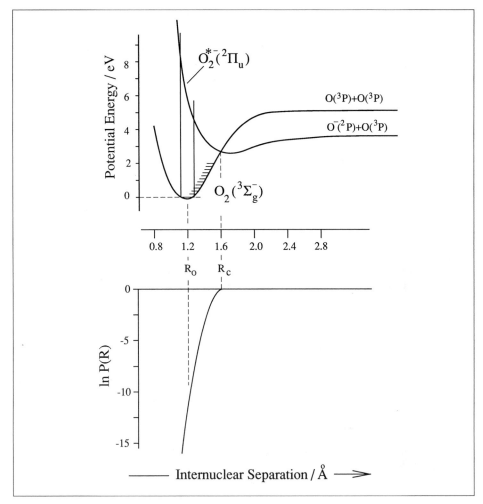

Fig. 2.24. Potential energy diagrams and survival probability illustrating the temperature dependence in dissociative electron attachment.

the $v=10$ level with respect to $v=0$ is only $2 \cdot 10^{-5}$ at 2100 K. However, due to the rapidly increasing survival probability with internuclear separation, these levels still contribute to the process, as demonstrated by the explicit calculation of O'Malley [83].

Similar arguments apply for rotational excitation which has been studied experimentally [84] and theoretically [85–87] in dissociative attachment to H_2. Here, centrifugal stretching in the neutral molecule reduces the autodetachment rate leading to an increase of cross-section with angular momentum.

The temperature dependence of electron capture has also been measured for a variety of larger molecules as summarized in the review of Christophorou [81, 91].

It has been found that for molecules were only dissociative attachment processes occur the cross-section increases with temperature as pointed out above. For molecules with pure associative electron capture, the cross-section decreases with temperature due to an increasing autodetachment rate. For molecules exhibiting both associative and dissociative processes within a given energy range, the total attachment cross section may increase or decrease, dependent on the relative distribution of associative and dissociative processes.

Results on electron attachment to electronically excited molecules are scarce, in spite of their basic and applied significance.

An experimental study on excited oxygen molecules ($O_2 \, ^1\Delta_g$) showed an increase by a factor of 4-5 [88, 92], and in $H_2^* (c^3 \Pi_u)$ a theoretical study [89] predicts an increase by 3 orders of magnitude for the cross section of dissociative electron attachment with respect to the ground state.

Electron swarm experiments on thiophenol [90, 91] showed that the *capture cross section is enhanced by 5 orders of magnitude* when the molecule is *excited to* its first triplet state (indirectly via laser excitation of the singlet state and subsequent "system intercrossing"). It is likely that the much higher polarizability of electronically excited states is responsible for this effect [93].

References

1. Oster T, Kühn A, Illenberger E (1989) Gas Phase Negative Ion Chemistry. Int J Mass Spectrom Ion Proc 89:1-72
2. Roy D (1972) Characteristics of the Trochoidal Monochromator by Calculation of Electron Energy Distribution. Rev Sci Instrum 43:535-541
3. McMillan MR, Moore, JH (1980) Optimization of the Trochoidal Electron Monochromator. Rev Sci Instrum 51:944-950
4. Stamatovic A, Schulz GJ (1979) Characteristics of the Trochoidal Electron Monochromator. Rev Sci Instrum 41:423-427
5. Stamatovic A, Schulz G.J (1968) Trochoidal Electron Monochromator. Rev Sci Instrum 39:1752-1753
6. Allan M (1989) Study of Triplet States and Short-Lived Negative Ions by Means of Electron Impact. J Electron Spectrosc Relat Phen 48:219-351
7. Jordan KD, Burrow PD (1987) Temporary Anion States of Polyatomic Hydrocarbons. Chem Rev 87:557-588
8. Giordan JC, Moore, JH, Tossell JA, Weber J (1983) Negative Ion States of 3d Metallocenes. J Am Chem Soc 105:3431-3433
9. Guerra M, Jones D, Distefano G, Scagnolari F, Modelli A (1991) Temporary Anion States in the Chloromethanes and in Monochloroalkanes. J Chem Phys 94:484-490
10. Ajello JM, Chutjian A (1979) Line Shapes for Attachment of Threshold Electrons to SF_6 and $CFCl_3$: Threshold Photoelectron (TPSA) Studies. J Chem. Phys 71:1079-1087
11. Chutjian A, Alajajian SH (1985) s-Wave Threshold in Electron Attachment: Observations and Cross Sections in CCl_4 and SF_6 at Ultralow Electron Energies. Phys Rev A 31:2885-2892
12. Orient OJ, Chutjian A (1986) Comparison of Calculated and Experimental Attachment Rate Constants for SF_6 in the Temperature Range 200-600 K. Phys Rev A 34:1841-1846

13. Christophorou LG, Pace MO (eds) (1984) Gaseous Dielectrics. Pergamon, New York

14. Kunhardt EE, Luessen LH (eds) (1983) Electrical Breakdown and Discharges in Gases. Plenum, New York

15. Bansal KM, Fessenden RW (1976) On the Oxidizing Radical Formed by Reaction of e_{aq}^- and SF_6. J Phys Chem 80:1743–1745

16. Christophorou LG (1987) Electron Attachment and Detachment Processes in Electronegative Gases. Contrib Plasma Phys 27:237–281

17. Hunter SR, Carter JG, Christophorou LG (1989) Low Energy Electron Attachment to SF_6 in N_2, Ar and Xe Buffer Gases. J Chem Phys 90:4879–4891

18. Mothes KG, Schindler RN (1971) Die Bestimmung absoluter Geschwindigkeitskonstanten für den Einfang thermischer Elektronen durch CCl_4, SF_6, C_4F_8, C_7F_{14}, N_2F_4 und NF_3. Ber Bunsenges Phys Chem 75:938–945

19. Berkowitz J (1979) Photoabsorption, Photoionization, and Photoelectron Spectroscopy. Academic Press, New York

20. Grimsrud EP, Chowdhurry S, Kebarle P (1985) Electron Affinity of SF_6 and Perfluoromethylcyclohexane. The Unusual Kinetics of Electron Transfer Reactions $A^- + B = A + B^-$ where $A = SF_6$ or Perfluorinated Cyclo-Alkanes. J Chem Phys 83:1059–1068

21. Klobukowski M, Barandiarán Z, Seijo L, Huzinaga S (1987) Towards HF SCF Value of Electron Affinity of SF6. J Chem Phys 86:1637–1638

22. Compton RN, Christophorou LG, Hurst GS, Reinhardt PW (1966) Nondissociative Electron Capture in Complex Molecules and Negative Ion Lifetimes. J Chem Phys 45:4634–4639

23. Harland PW, Thynne JCJ (1971) Autodetachment Lifetimes, Attachment Cross Sections, and Negative Ions Formed by Sulfur Hexafluoride and Sulfur Tetrafluoride. J Phys Chem 75:3517–3523

24. Odom RW, Smith, DL, Futrell HJ (1975) A Study of Electron Attachment to SF_6 and Autodetachment and Stabilization of SF_6^-. J Phys B8:1349–1366

25. Foster MS, Beauchamp JL (1975) Electron Attachment to Sulphur Hexafluoride: Formation of Stable SF_6^- at Low Pressure. Chem Phys Lett 31:482–486

26. Fenzlaff M, Gerhard R, Illenberger E (1988) Associative and Dissociative Electron Attachment to SF_6 and SF_5Cl. J Chem Phys 88:149–155

27. Hay PJ (1977) Excited States and Positive Ions in SF_6. J Am Chem Soc 99:1013–1019

28. Hay PJ (1982) The Relative Energies of SF_6^- and SF_6 as a Function of Geometry. J Chem Phys 76:502–504

29. Tang R, Callaway J (1986) Electronic Structure of SF_6. J Chem Phys 84:6854–6860

30. Stockdale JAD, Compton RN, Schweinler HC (1969) Negative Ion Formation in Selected Hexafluoride Molecules. J Chem Phys 53:1502–1507

31. Drzaic PS, Brauman JI (1982) Electron Photodetachment Study of Sulfur Hexafluoride Anion: Comments on the Structure of SF_6^-. J Am Chem Soc 104:13–19

32. Kline LE, Davies DK, Chen CL, Chantry PJ (1979) Dielectric Properties for SF_6 and SF_6 Mixtures Predicted from Basic Data. J Appl Phys 50:6789–6796

33. Chen CL, Chantry PJ (1979) Photon Enhanced Dissociative Electron Attachment in SF_6 and its Isotopic Selectivity. J Chem Phys 3897–3907

34. Babcock LM, Streit GE (1981) Ion-Molecule Reactions of SF_6: Determination of $I.P.(SF_5)$, $A.P.(SF_5^+/SF_6)$ and $D(SF_5^-F)$. J Chem Phys 74:5700–5706

35. Sieck LW, Ausloos PJ (1990) The Ionization Energy of SF_5 and the SF_5^-F Bond Dissociation Energy. J Chem Phys 93:8374–8378

36. Mead RD, Stevens AE, Lineberger WC (1984) Photodetachment in Negative Ion Beams, In: Bowers MT (ed) Gas Phase Ion Chemistry, vol 3, vol 3. Academic Press, New York

37. Oster T, Illenberger E (1988) Negative Ion Formation in SF_5NCO Following Low-Energy Electron Attachment. Int J Mass Spectrom Ion Proc 85:125–136

38. Hubrich C, Zetzsch C, Stuhl F (1977) Absorptionsspektren von halogenierten Methanen im Bereich von 275 bis 160 nm bei Temperaturen von 298 und 208 K. Ber Bunsenges Phys Chem 81:437–442

39. Huebner RH, Bushnell DL, Celotta RJ, Mielczarek SR, Kuyatt, CE (1975) Ultraviolet Photo-absorption by Halocarbons 11 and 12 from Electron Impact Measurements. Nature 257:376–378
40. Molina MJ, Rowland FS (1974) Stratospheric Sink for Chlorofluoromethanes: Chlorine Atom-Catalyzed Destruction of Ozone. Nature 249:810–812
41. Elliot S, Rowland FS (1987) Chlorofluorocarbons and Stratospheric Ozone. J Chem Educ 64:387–391
42. Solomon S (1990) Progress Towards a Quantitative Understanding of Antarctic Ozone Depletion. Nature 347:347–354
43. Schumacher R, Sprünken H-R, Christodoulides AA, Schindler RN (1978) Studies by the Electron Cyclotron Resonance Technique. 13. Electron Scavenging Properties of the Molecules CCl_3F, CCl_2F_2, $CClF_3$ and CF_4. J Phys Chem 82:2248–2252
44. Crompton RW, Haddad GN, Heberberg R, Robertson AG (1982) The Attachment Rate for Thermal Electrons to SF_6 and $CFCl_3$. J Phys B 15:L483–484
45. Illenberger E, Scheunemann H-U, Baumgärtel H (1979) Negative Ion Formation in CF_2Cl_2, CF_3Cl and $CFCl_3$ Following Low-Energy (0–10 eV) Impact with Near Monoenergetic Electrons. Chem Phys 37:21–31
46. Illenberger E, Scheunemann H-U, Baumgärtel H (1978) Fragmentierung von Halogen-methanen beim Stoß mit monochromatischen Elektronen niedriger Energie. Ber Bunsenges Phys chem 82:1154–1158
47. Ausloos P, Rebbert RE, Glasgow L (1977) Photodecomposition of Chloromethanes Adsorbed on Silica Surfaces. J Res Natl Bur Stan 82:1–8
48. Parlar H, Korte F (1981) Wie effizient ist der photoinduzierte Abbau organischer Umweltchemikalien in heterogener Phase. Chem Ztg 105:127–134
49. Dowben PA, Grunze M, Jones RG, Illenberger E (1981) Interaction of $CFCl_3$ with an Fe(100) Surface Part II: Adsorption and Decomposition at Ambient Temperatures. Ber Bunsenges Phys Chem 85:734-739
50. Buchmann L-M, Illenberger E (1987) The Interaction of $CFCl_3$ and CF_2Cl_2 with Solid Surfaces Studies by Thermal Desorption Mass Spectrometry. Ber Bunsenges Phys Chem 91:653–659
51. Dewar MJS, Thiel W (1977) Ground States with Molecules. 38. The MNDO Method. Approximations and Parameters. J Am Chem Soc 99:4899–4907
52. Olthoff JK, Tossell JA, Moore JH (1985) Electron Attachment to Haloalkanes and Halobenzenes. J Chem Phys 83:5627–5634
53. Dispert H, Lacmann K (1978) Negative Ion Formation in Collisions between Potassium and Fluoro- and Chloromethanes: Electron Affinities and Bond Dissociation Energies. Int J Mass Spectrom Ion Phys 28:49-67
54. McMillan DF, Golden DM (1982) Hydrocarbon Bond Dissociation Energies, In: Rabinovitch BS, Schurr JM, Strauss HL (eds). Annu Rev Phys Chem 33:493–532
55. Lide DR (ed) (1985) JANAF Thermochemical Tables, 3rd edn. American Chemical Society, New York
56. Harland PW, Franklin JL, Carter DE (1972) Use of Translational Energy Measurements in the Evaluation of the Energetics for Dissociative Attachment Processes. J Chem Phys 58:1430–1437
57. Franklin JL (1979) Energy Distribution in the Unimolecular Decomposition of Ions, In: Bowers MT (ed), Gas Phase Ion Chemistry, vol 1. Academic Press, New York
58. Lupo DW, Quack M (1987) IR-Laser Photochemistry, Chem Rev 87:181–216
59. Quack M, Troe J (1975), Complex Formation in Reactive and Inelastic Scattering: Statistical Adiabatic Channel Model for Unimolecular Processes III. Ber Bunsenges Phys Chem 79:170–183
60. Illenberger E (1982) Habilitationsschrift. Freie Universität Berlin, p 183–186
61. Illenberger E (1982) A Method to Determine Excess Energies in Dissociative Electron Attachment Processes. Ber Bunsenges Phys Chem 86:247-252
62. Massey HSW (1976) Negative Ions. Cambridge University Press, Cambridge

63. Rapp D, Briglia DD (1965) Total Cross Sections for Ionization and Attachment in Gases by Electron Impact. II. Negative Ion Formation. J Chem Phys 43:1480–1489
64. Chantry PJ, Schulz GJ (1967) Kinetic Energy Distribution of Negative Ions Formed by Dissociative Attachment and the Measurement of the Electron Affinity of Oxygen. Phys Rev 156:134–141
65. Robin MB (1985) Higher Excited States of Polyatomic Molecules, vol III. Academic, Orlando, Florida
66. Suto M, Washida N, Akimoto H, Nakamura M (1983) Emission Spectra of CF_3 Radicals. III. Spectra and Quenching of CF_3 Emission Band Produced in the VUV Photolyses of CF_3Cl and CF_3Br. J Chem Phys 78:1019–1024
67. Fenzlaff M, Illenberger E (1989) Energy Partitioning in the Unimolecular Decomposition of Cyclic Perfluororadical Anions. Chem Phys 136:443–452
68. Scheunemann H-U, Heni E, Illenberger E, Baumgärtel H (1982) Dissociative Attachment and Ion Pair Formation in CF_4, CHF_3, CH_2F_2 and CH_3F under Low Energy (0-20 eV) Electron Impact. Ber Bunsenges Phys Chem 86:321–326
69. Illenberger E (1982) Energetics of Negative Ion Formation in Dissociative Attachment. Ber Bunsenges Phys Chem 86:252–261
70. Heni M, Illenberger E (1986) Dissociative Electron Attachment to CF_3I: An Example of a Completely Unbalanced Excess Energy Distribution. Chem Phys Lett 131:314–318
71. Klein R; McGinnis RP, Leone SR (1983) Photodetachment Threshold of CN^- by Laser Optogalvanic Spectroscopy. Chem Phys Lett 100:475–478
72. Heni M, Illenberger E, Lentz D (1986) The Isomers CF_3NC and CF_3CN. Formation and Dissociation of the Anions Formed on Electron Attachment. Int J Mass Spectrom Ion Proc 71:199–210
73. Heni M, Illenberger E (1986) Electron Attachment by Saturated Nitriles, Acrylonitrile (C_2H_3CN), and Benzonitrile (C_6H_5CN). Int J Mass Spectrom Ion Proc 73:127–144
74. Riley SJ, Wilson KR (1972) Excited Fragments from Excited Molecules: Energy Partitioning in the Photodissociation of Alkyl Iodides. Faraday Discuss Chem Soc 53:132–153
75. Krajnovich D, Butler LJ, Lee YT (1984) Photodissociation of C_2F_5Br, C_2F_5I and 1,2-C_2F_4BrI. J Chem Phys 81:3031–3047
76. Fenzlaff HP, Kühn A, Illenberger E (1988) Formation and Dissociation of Negative Ion Resonances in Methanol and Allylalcohol. J Chem Phys 88:7453–7458
77. Weast RC (ed) (1985) Handbook of Chemistry and Physics, 66th edn. CRC Press, Boca Raton, Florida
78. von Trepka L, Neuert H (1963) Über die Entstehung von negativen Ionen aus einigen Kohlenwasserstoffen und Alkoholen durch Elektronenbeschuß. Z Naturforsch 18:1295–1303
79. Illenberger E, Baumgärtel H, Süzer S (1984) Electron Attachment Spectroscopy: Formation and Dissociation of Negative Ions in the Fluoroethylenes. J Electron Spectrosc Relat Phen 33:123–139
80. Illenberger E (1981) Measurement of the Translational Excess Energy in Dissociative Electron Attachment Processes. Chem Phys Lett 80:153–158
81. Christophorou LG (1987) Electron Attachment and Detachment Processes in Electronegative Gases. Contrib Plasma Phys 27:238–281
82. Henderson WR, Fite WL, Brackmann RT (1969) Dissociative Attachment to Hot Oxygen. Phys Rev 183:157–166
83. O'Malley TF (1966) Calculation of Dissociative Attachment in Hot O_2. Phys Rev 155:59–63
84. Allan M, Wong SF (1978) Effect of Vibrational and Rotational Excitation on Dissociative Attachment in Hydrogen. Phys Rev Lett 41:1791–1794
85. Chen JCY, Peacher JL (1967) Survival Probability in Dissociative Attachment. Phys Rev 163:103–111
86. Wadehra JM, Bradsley JN (1978) Vibrational- and Rotational-State Dependence of Dissociative Attachment in e–H_2 Collisions. Phys Rev Lett 26:1795–1798

87. Domcke W, Mündel C (1985) Calculation of Cross Sections for Vibrational Excitation and Dissociative Attachment in HCl and DCl Beyond the Local Complex-Potential Approximation. J Phys B 18:4491–4509
88. Burrow PD (1973) Dissociative Attachment from the $O_2(a^1\Delta_g)$ State. J Chem Phys 59:4922–4931
89. Bottcher C, Buckley BD (1979) Dissociative Electron Attachment to the Metastable $c^3\Pi_u$ State of Molecular Hydrogen. J Phys B 12:L 497–500
90. Pinnaduwage LA, Christophorou LG, Hunter SR (1989) Optically Enhanced Electron Attachment to Thiophenol. J Chem Phys 90:6275–6289
91. Christophorou LG (1991) Electron-Excited Molecule Interactions, Invited General Lecture, XX International Conference on Phenomene in Ionized Gases. Il Ciocco, Italy (in press)
92. Jaffke T, Meinke M, Hashemi R, Christophoron LG, Illenberger E (1992) Electron Attachment to Singlet Oxygen. Chem Phys Lett (in press)
93. Christophoron LG, Illenberger E (1992) Scattering of Slow Electrons from Excited Atoms: The Dominant Role of the Polarization Potential. Chem Phys Lett (submitted)

3 Negative Ions Observed in Electron Transmission and Electron Attachment Spectroscopy

So far, we have studied those temporary negative parent ions formed on resonant electron attachment, which can be observed by mass spectrometry either directly or indirectly through thermodynamically stabel ionic decomposition products. As was pointed out in Chapter 1, this is not possible for a large number of molecules such as nitrogen, ethylene or benzene. Temporary negative anions of these compounds are observable in ETS. We will now show how ETS can be used for ethylene and several fluoroethylenes in order to study the effect of fluorination on the attachment energy. Experimental techniques such as *photoelectron* and *photoion mass spectroscopy* provide information about the energy of *occupied MOs*. *Electron transmission*, on the other hand, gives information about the energy of the normally *unoccupied MOs*. Since ethylene possesses a negative electron affinity, the LUMO is directly accessible on electron impact. Various scattering experiments strongly support the picture of electron attachment into the $b_{2g} (\pi^*)$ MO [1-3]. Whereas $C_2H_4^-$ ions can only be observed in electron transmission, negative ion formation in numerous ethylene derivatives can be observed in EAS also. In the case of difluoroethylene, for example, the dissociated channel

$$C_2H_2F_2^- \, (^2\Pi) \rightarrow F^- + C_2H_2F \qquad\qquad 3.1$$

is now accessible from the energy of TNI, thus allowing the observation of electron attachment in both transmission and attachment spectroscopy.

3.1 Experimental Considerations

Figure 3.1 shows a schematic diagram of the electron transmission spectrometer [4]. The instrument is based on the concept introduced by Sanche and Schulz [5-8], who extensively studied resonances in atoms and diatomic molecules.

The apparatus consists of a TEM similar to that used in EAS: a collision chamber $(D_1 - D_3)$ and an electron retarding $(P_1 - P_3)$ and collection system (A). All parts of the transmission spectrometer are (as in the electron attachment spectrometer) made of molybdenum; the electron monochromator and the collision chamber are pumped separately in order to keep the pressure low in the monochromator region.

The energy-selected electrons (fwhm < 0.1 eV) are aligned by the homogeneous magnetic field and enter the scattering chamber (D_1, D_2, D_3) where they undergo collisions with the gas molecules. Electrons which reach the exit of the collision region

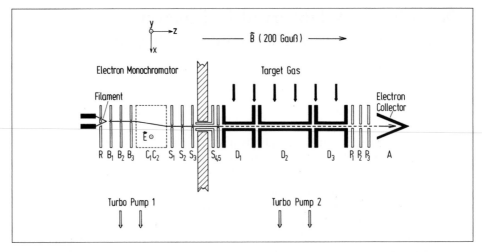

Fig. 3.1. Schematic representation of the electron transmission spectrometer (from [4]).

(P_1) are decelerated in such a way that only unscattered electrons are able to overcome the retarding potential (between P_2 and P_3) in order to be accelerated to the electron collector (A). The retarding procedure eliminates all electrons which have reoriented their initial velocity vectors. The electron collector thus measures the unscattered part (I) of the incident electron beam intensity (I_0). I and I_0 are related by the Lambert-Beer law

$$I(\varepsilon) = I_0 \exp\left(- n\sigma(\varepsilon)\, l\right), \qquad\qquad 3.2$$

where n is the gas density in the scattering chamber, $\sigma(\varepsilon)$ is the total scattering cross-section, and l the scattering path length.

Since the formation of a TNI is characterized by rapid changes of the cross-section with energy, the derivative of the transmitted current with respect to energy,

$$\frac{dI}{d\varepsilon} = -\frac{d\sigma}{d\varepsilon}\, I_0\, n\sigma l \exp\left(- n\sigma l\right) \qquad\qquad 3.3$$

is measured. Resonant scattering is superimposed on a slowly varying background of non-resonant scattering. If the magnitude of the latter is large compared with resonant scattering, the exponential term in Eq. 3.3 remains essentially constant and to a first approximation, $dI/d\varepsilon \approx d\sigma/d\varepsilon$.

The derivative is recorded by applying a small modulation voltage (700 Hz, 0.03 V) to the collision chamber, thereby modulating the electron energy. The resulting modulation in the transmitted current is measured and amplified with a phase sensitive (lock-in) detector.

The optimum value of the gas density yielding the largest modulation signal can be obtained by setting

$$\frac{d\,(dI/d\varepsilon)}{dn} = 0\,,$$

which gives $n\sigma l = 1$. With the present configuration ($l = 10$ cm) and typical scattering cross-sections of the order of 10^{-16}-10^{-15} cm^2, gas pressures in the range p $\approx 10^{-3}$-10^{-2} mbar are required. The experiments indeed show that the largest signal is observed near that gas density.

The many advantages of this technique have been described by Sanche and Schulz [5, 6] and, more recently, by Jordan and Burrow [9] and Allan [10]. The review article of Jordan and Burrow [9] includes a bibliography of all studies in electron transmission spectroscopy up to 1987.

In the present chapter, we will apply ETS to study the formation of TNIs in ethylene and the fluorinated ethylenes. In addition, it will be shown that application of EAS can provide additional information on the fate of the TNI observed in ETS.

3.2 Ethylene and Fluoroethylenes. Energy of π^* MOs and Decomposition of Transient Negative Ions

Figure 3.2 shows the transmission spectrum for N_2 and C_2H_2. The peak near zero eV reflects the onset behavior of the electron current in the presence of the gas. In the nitrogen spectrum we see pronounced structures at higher energies due to the

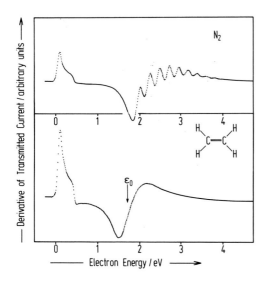

Fig. 3.2. Electron transmission spectra of nitrogen and ethylene.

formation of N_2^- ($^2\Pi_g$) ions in several vibrational states. In the case of $C_2H_4^-$ ions, fine structure is no longer observable. The steepest slope in the derivative corresponds to the peak in the scattering cross-section. It represents the vertical attachment energy ($\varepsilon_0 = 1.75$ eV) associated with $C_2H_4^-$ in its electronic ground state [1–3].

The transmission spectra of the six fluoroethylenes are plotted in Figs. 3.3 and 3.4. Like ethylene, each of the fluorinated compounds exhibits one single pronounced enhancement in the scattering cross-section, thus signalizing the formation of temporary ions. Only $C_2F_4^-$ exhibits an additional, sharp structure below 0.5 eV which overlaps with the peak representing the onset of the electron current. This low-energy structure may be an artifact caused by electrons backscattered by the retarding system ("retarding cusps"). Such effects have been analyzed in detail by Johnston and Burrow [11]. However, since such low-energy structure is visible only in C_2F_4 and since its position is not changed by varying the retarding voltage, it may indeed be a result of resonant scattering (see below). In Figs. 3.3 and 3.4 the steepest slope of the derivative (i.e., the resonance maximum) is indicated by an arrow. It can clearly be seen from the transmission spectra that the vertical attachment energy increases on fluorination, from 1.75 eV in ethylene to 3.0 eV in perfluoroethylene, indicating that the anions (and hence the π^* MOs) are continuously destabilized as a function of fluorination. The filled π MOs, in contrast, are essentially insensitive to fluorination as monitored by the ionization energies from photoelectron spectroscopy [12].

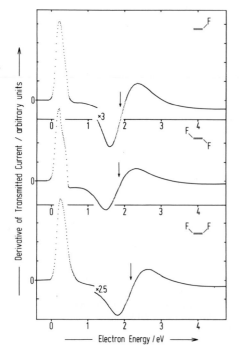

Fig. 3.3. Electron transmission spectra of fluoroethylene, trans-1,2-difluoroethylene and cis-1,2-difluoroethylene (from [4]).

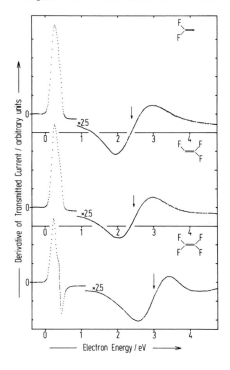

Fig. 3.4. Electron transmission spectra of 1,1-difluoroethylene, trifluoroethylene, and tetrafluoroethylene (from [4]).

Burrow et al. [13] have discussed this differing behavior in terms of the stabilizing inductive effect and the destabilizing resonance interaction introduced by the fluorine atoms.

Superficially, the continuous destabilization of the anions upon fluorination seems surprising since perfluoroethylene and difluoroethylene are known to exist as stable anionic compounds in low temperature matrices [14] while the $C_2H_4^-$ ion has never been observed. In the ETS experiment, however, we measure the *vertical attachment energy* as opposed to the (adiabatic) energy difference between the neutral and anionic ground states. Furthermore, the spectra clearly show that, not only the resonance maximum is shifted to higher energy, but its width increases on fluorination. The latter effect reflects Franck-Condon transitions to repulsive potential energy surfaces which become more repulsive as the degree of fluorination increases. Therefore, the equilibrium geometries of the fluoroethylene anions are expected to differ substantially from those of the neutrals. Theoretical calculations, in fact, predict that the anions are unstable with respect to out-of-plane distortion [15] so that the vertical attachment energy may be considerably different from the adiabatic electron affinity. Although the vertical attachment energy increases with the extent of fluorination, the relaxation energy can overcompensate for this increase so that the (adiabatic) electron affinity of these compounds decreases on fluorination. It should be noted that in the chloroethylenes the vertical attachment energy decreases with chlorination, from 1.3 eV in monochloroethylene to near 0 eV in tetrachloroethylene [16, 17].

Figure 3.5 illustrates the situation in terms of the total energy for ethylene and cis-1,2-difluoroethylene. While the $C_2H_4^-$ $(^2\Pi)$ resonance lies distinctly below dissociation channels involving thermodynamically stable fragments, this is no longer the case for the fluorinated species. The high electron affinity of fluorine and the high stability of HF allows the temporary anion to decompose into the thermodynamically stable negatively charged and neutral fragments, thus permitting the observation of the temporary $^2\Pi$ anions via dissociative electron attachment.

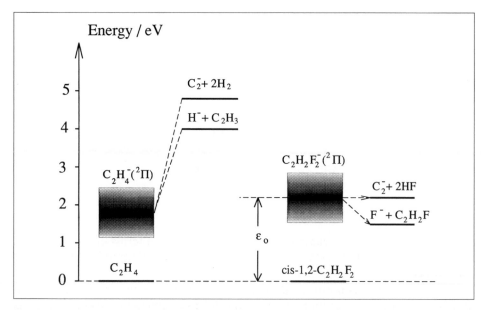

Fig. 3.5. Energy diagram illustrating the lowest dissociation channels for ground state anions of ethylene and cis-1,2-difluoroethylene.

In contrast to the relatively broad resonances reflecting transitions to repulsive potential energy surfaces, the sharp structure below 0.5 eV in C_2F_4 indicates a transition to an anionic surface near the equilibrium configuration. For energetic reasons this electronic state cannot decompose into stable negative and neutral fragments and is thus not visible in dissociative electron attachment. A similar observation in CF_4 was made by Brongersma et al. [18]. These authors interpreted their result as the temporary accommodation of the incident electron into a low-lying Rydberg-type MO. This may also be the case in C_2F_4 as the low energy structure suggests that the additional charge does not induce any significant geometry change in the molecule.

The F^- ion yield curves for the six fluoroethylenes are plotted in Figs. 3.6 and 3.7. The absolute positions of the peak maxima of the first resonance essentially coincide with those observed in electron transmission. In addition, the width of the reso-

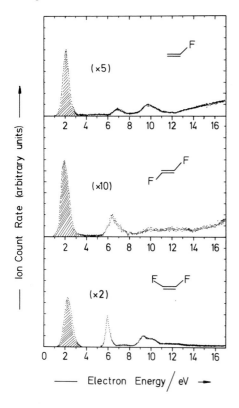

Fig. 3.6. F^- formation from fluoroethylene, trans-1,2-difluoroethylene, and cis-1,2-difluoroethylene (adapted from [20]).

nance increases with fluorination. The $(^2\Pi)$ fluoroethylene radical anions decompose into F^- ions and the corresponding radical (M−F). Since the fragments in their electronic ground states are F^- (^1S) and (M−F) $(^2\Sigma)$, the associated potential energy curves should have Σ symmetry. However, as mentioned above, the temporary anions are expected to suffer an out-of-plane distortion which can be interpreted as arising from a mixing of π^* MOs with high lying σ^* MOs, permitting the temporary anion to decompose by the cleavage of a σ bond.

The ion yield curves all show an additional, pronounced resonance in the region of 6–7 eV. A correlation of these energies with electronically excited states in the neutral fluoroethylenes makes it seem plausible that this is a core-excited resonance associated with the 3s Rydberg state converging to the first (π) ionization energy [19].

Figures 3.8 and 3.9 show the various fragmentation products observed in cis-1,2-difluoroethylene and 1,1-difluoroethylene. It can be seen from the scale factors that in both cases F^- formation is not the most abundant channel.

On inspecting the decomposition products and the intensity with which they are formed some intriguing features of the unimolecular decomposition reactions become manifest: the C_2^- ion is formed from the $^2\Pi$ resonance in cis-1,2-difluoroethylene (Fig. 3.8) and trans-1,2-difluoroethylene (not shown here), but *not* in

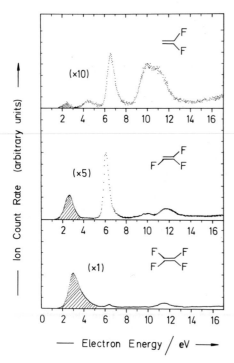

Fig. 3.7. F⁻ formation from 1,1-difluoroethylene, trifluoroethylene, and perfluoroethylene (adapted from [20]).

1,1-difluoroethylene (Fig. 3.9). The energetically lowest lying dissociation channels for the formation of the C_2^- ion are

$$C_2H_2F_2^- \,(^2\Pi) \rightarrow C_2^- + 2HF \qquad\qquad 3.4$$

$$\rightarrow C_2^- + H_2 + F_2. \qquad\qquad 3.5$$

The energetic threshold for channel 3.4 is 2.5 eV in the case of cis- and trans-difluoroethylene, and 3.0 eV for 1,1-difluoroethylene (as calculated from established thermodynamic data [19, 20]. Channel 3.5 lies by 5.6 eV above channel 3.4, so the $^2\Pi$ resonance in the difluoroethylenes can decay into C_2^- ions only by formation of two HF molecules.

The capture of electrons into the π^* MO and the subsequent excitation of vibrational modes containing the C–C stretch and HCF scissoring component explains the observed preferential formation of HF when the hydrogen and fluorine atoms originate from the same carbon atom.

Similar arguments apply to the formation of the C_2HF^- ion which is, by far, the predominant fragment from cis-1,2-difluoroethylene (Fig. 3.8) and also trans-1,2-difluoroethylene [19] (not shown here). This strongly suggests that C_2HF^- is a fluorovinylidene-like species. In fact, for the $C_2H_2^-$ ion it is well established experimentally

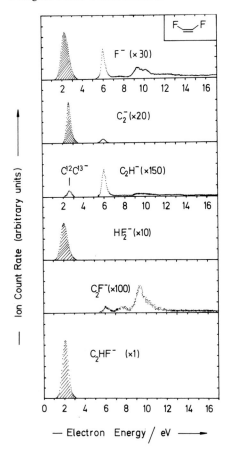

Fig. 3.8. Dissociation products in cis-1,2-difluoroethylene (from [20]).

[21, 22] and theoretically [23–25] that the stable configuration is the vinylidene structure. Acetylene, on the other hand, possesses a strongly negative electron affinity, as shown by electron transmission spectroscopy [26].

The bifluoride ion (FHF$^-$) observed from 1,2-difluoroethylene (Fig. 3.8) is known to possess an unusually strong hydrogen bond, namely, D (FH–F) $\geq$ 1.5 eV [27, 28]. Its observation from the $^2\Pi$ resonance in some fluoroethylenes suggests a remarkably high electron affinity for the neutral radical, EA (FHF) $\geq$ 4.8 eV [29].

We remember that molecules of the type XHY (X,Y: halogen atom) have recently been studied as prototype systems to probe the transition state in hydrogen transfer reactions of the form

$$X + HY \rightarrow XHY \rightarrow XH + Y.$$

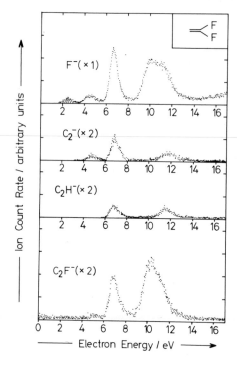

Fig. 3.9. Dissociation products in 1,1-difluoroethylene (from [19]).

Upon photodetachment of the stable anion $(XHY)^-$ one generates the neutral system in a configuration close to the transition state, which can be probed by photoelectron spectroscopy (see also Chapter 2, Part I).

For the symmetric systems XHX experiments have so far been reported for ClHCl [30], IHI [31], and BrHBr [32]. A further photodetachment experiment was reported on the system FH_2^- as a probe for the transition state of the reaction $F + H_2 \rightarrow HF + H$ [33].

References

1. Walker IC, Stamatovic A, Wong SF (1978) Vibrational Excitation of Ethylene by Electron Impact: 1-11 eV. J Chem Phys 69:5532–5537
2. Bowman CR, Miller WD (1965) Excitation of Methane, Ethane, Ethylene, Propylene, Acetylene, Propyne, and 1-Butyne by Low-Energy Electron Beams. J Chem Phys 42:681–686
3. Burrow PD, Jordan KD (1975) On the Electron Affinities of Ethylene and 1,3-Butadiene. Chem Phys Lett 36:594–598
4. Kwiatkowski G, Illenberger E (1984) Negative Ions in the Gas Phase Observed in Electron Transmission and Electron Attachment Spectroscopy. Ber Bunsenges Phys Chem 88:670–675
5. Sanche L, Schulz GJ (1972) Electron Transmission Spectroscopy: Rare Gases. Phys Rev A5:1672–1683
6. Sanche L, Schulz GJ (1972) Electron Transmission Spectroscopy: Core Excited Resonances in Diatomic Molecules. Phys Rev A6:69–86

7. Schulz GJ (1973) Resonances in Electron Impact on Atoms. Rev Mod Phys 45:378–422
8. Schulz GJ (1973) Resonances in Electron Impact on Diatomic Molecules. Rev Mod Phys 45:423–486
9. Jordan KD, Burrow PD (1987) Temporary Anion States of Polyatomic Hydrocarbons. Chem Rev 87:557–588
10. Allan M (1989) Study of Triplet States and Short-Lived Negative Ions by Means of Electron Impact Spectroscopy. J Electron Spectrosc Relat Phen 48:219–351
11. Johnston AR, Burrow PD (1982) Scattered Electron Rejection in Electron Transmission Spectroscopy. J Electron Spectrosc Relat Phen 25:119–133
12. Brundle CR, Robin MB, Kuebler NA, Basch H (1972) Perfluoro Effect in Photoelectron Spectroscopy. I. Nonaromatic Molecules. J Am Chem Soc 94:1451–1465
13. Chiu NS, Burrow PD, Jordan KD (1979) Temporary Anions of the Fluoroethylenes. Chem Phys Lett 68:121–126
14. Wang JT, Williams F (1989) Radical Anion of 1,1-Difluoroethylene. EPR Evidence for a Perpendicular Geometry. J Am Chem Soc 103:2902–2904
15. Paddon-Row MN, Rondan NG, Houk KN, Jordan KD (1982) Geometries of the Radical Anions of Ethylene, Fluoroethylene, 1,1-Difluoroethylene, and Tetrafluoroethylene. J Am Chem Soc 104:1143–1145
16. Burrow PD, Modelli A, Chiu NS, Jordan KD (1981) Temporary Σ and Π Anions of the Chloroethylenes and Chlorofluoroethylenes. Chem Phys Lett 82:270–276
17. Kaufel R, Illenberger E, Baumgärtel H (1984) Formation and Dissociation of the Chloroethylene Anions. Chem Phys Lett 106:342–346
18. Verhaart GJ, van der Hart WJ, Brongersma HH (1978) Low Energy Electron Impact on Chlorofluoromethanes and CF_4: Resonances, Dissociative Attachment and Excitation. Chem Phys 34:161–167
19. Heni M, Illenberger E (1986) The Unimolecular Decomposition of the Fluoroethylene Radical Anions Formed by Electron Attachment. J Electron Spectrosc Relat Phen 41:453–466
20. Heni M, Illenberger E, Baumgärtel H, Süzer S (1982) The Dissociation of the $^2\Pi$ Fluoroethylene Anions. Chem Phys Lett 87:244–248
21. Burnett SM, Stevens AM, Feigerle CS, Lineberger WC (1983) Observation of X^1A Vinylidene by Photoelectron Spectroscopy of the $C_2H_2^-$ Ion. Chem Phys Lett 100:124–128
22. Ervin KM, Ho J, Lineberger WC (1990) A Study of the Singlet and Triplet States of Vinylidene by Photoelectron Spectroscopy of $H_2C=C^-$, $D_2C=C^-$ and $HDC=C^-$. Vinylidene-Acetylene Isomerization. J Chem Phys 91:5974–5992
23. Rosmus P, Botschwina P, Maier JP (1981) On the Ionic States of Vinylidene and Acetylene. Chem Phys Lett 84:71–75
24. Frenking G (1983) The Neutral and Ionic Vinylidene-Acetylene Rearrangement. Chem Phys Lett 100:484–487
25. Carrington T, Hubbard LM, Schaefer III HF, Miller WH (1984) Vinylidene: Potential Energy Surface and Unimolecular Reaction Dynamics. J Chem Phys 80:4347–4354
26. Jordan KD, Burrow PD (1978) Studies of the Temporary Anion States of Unsaturated Hydrocarbons by Electron Transmission Spectroscopy. Acc Chem Res 11:341–348
27- Pimentel GC; McClellan AL (1971) Hydrogen Bonding. Annu Rev Phys Chem 22:347–385
28. Jiang GJ, Anderson GR (1973) A Semiempirical Study of Hydrogen Bonding in the Bifluoride Ion. J Phys Chem 77:1764–1768
29. Heni M, Illenberger E (1985) The Stability of the Bifluoride Ion (HF_2^-) in the Gas Phase. J Chem Phys 83:6056–6057
30. Metz RB, Kitsopoulos TN, Weaver A, Neumark DM (1988) Study of the Transition State Region in the Cl+HCl Reaction by Photoelectron Spectroscopy of ClHCl$^-$. J Chem Phys 88:1463–1465
31. Weaver A, Metz RB, Bradforth SE, Neumark, DM (1988) Spectroscopy of the I+HI Transition-State Region by Photodetachment of IHI$^-$. J Phys Chem 92:5558–5560

32. Metz RB, Weaver A, Bradforth SE, Kitsopoulos TN, Neumark DM (1990) Probing the Transition State with Negative Ion Photodetachment: The Cl+HCl and Br—HBr Reactions. J Phys Chem 94:1377–1388
33. Weaver A, Metz RB, Bradforth SE, Neumark DM (1990) Investigation of the F+H_2 Transition State Region via Photoelectron Spectroscopy of the FH_2^- Anion. J Chem Phys 93:5352-5353

4 Electron Attachment to Molecular Clusters and Condensed Molecules

The term "cluster" is used in many different disciplines, from astronomy to chemistry. In the latter, structures in solid or liquid materials or in gases are generally labeled with this title. Webster' Dictionary [1] defines a cluster as "any group of persons, animals or things close together". In this sense, we will assign a group of molecules in the gas phase which are close together as a free molecular cluster or a free molecular aggregate. It is well known since the work of van der Waals [2] that the existence of condensed phases of matter stems from the attractive forces between molecules. These forces are commonly assigned as van der Waals forces: a group of molecules which are held together by these forces is labeled as a van der Waals cluster. Although at the time of van der Waals, the origins of such intermolecular forces were not understood, the fundamental connection between the *macroscopic properties of matter* and the *forces between the constituent molecules* was already evident.

Traditionally, science has concentrated on understanding the properties of the two separate forms of matter, single molecules in the gas phase and bulk liquids or solids. However, within the last 15 years or so there has been an enormous increase of experimental and theoretical work concerning the properties of clusters [3–7] which are considered to represent a link between the gas phase and the condensed phase. New experimental techniques are now available that allow the preparation of clusters in the gas phase under fairly definite conditions. Fundamental questions emerged, like "how many metal atoms are necessary to evolve an electric conduction band?", or "how small is the number of atoms or molecules to support a bulk crystal structure?". In the case of metal atoms like Hg, the transition from van der Waals binding over covalent to metallic binding as the cluster size increases has extensiveley been studied and (sometimes controversially) discussed [8, 9].

In this contribution, we will focus on some aspects concerning electron attachment to van der Waals clusters of small organic molecules, and in the last section we will briefly discuss some experiments dealing with electron capture by molecules in the condensed phase. The general area of non-covalent or van der Waals interactions between molecules plays a key role in many fields of present day natural science, in chemistry and physics, as well as in biology. Van der Waals complexes occur in virtually all bimolecular reactions and the role of intermolecular forces in the field of biological structures and reactivity may be compared with that of covalent binding in chemistry. For a detailed study of intermolecular forces, their origins and determination, the interested reader is referred to the monograph of Maitland, Rigby, Smith and Wakeham [10].

Negatively charged clusters are of particular interest since they are considered to provide models for excess electrons in liquids [3, 11]. Although solvated electrons have been known for more than 100 years [12], it was only recently that negatively charged clusters have been observed in supersonic beam experiments [13-19]. Anionic free clusters in the gas phase can be formed either by injection of electrons into the expansion zone of a supersonic jet [17, 18], or by attachment of electrons to a beam of pre-existing clusters [13–16, 19, 20].

While electron attachment to isolated molecules is considered to occur via accommodation of the extra electron into normally unfilled MOs, this is no longer necessarily the case in clusters. The free water molecule, for example, captures electrons within a series of overlapping resonances between 6.5 and 12 eV, and the TNI decomposes instantly into various dissociative attachment channels [21, 22]. Water clusters, in contrast, additionally absorb electrons near 0 eV to form mass spectrometrically observable $(H_2O)_n^-$ aggregates with $n \geq 11$ [19]. Landman et al. [23] have made theoretical studies of electron attachment to water clusters using quantum path-integral molecular dynamics (QUPID) calculations, and photodetachment experiments have been performed on $(H_2O)_n^-$, $n = 2$–69 [24]. For larger clusters it is likely that, in the relaxed configuration of $(H_2O)_n^-$, the excess electron is trapped in the field of the oriented water dipoles, thus representing a "solvated electron".

We will present results on electron capture by some simple halogenated hydrocarbons and elucidate the behavior of the basic quantities (electron attachment energy, products ultimately formed, and distribution of excess energy) when proceeding from the isolated molecule to the respective molecular aggregate.

4.1 Experimental Considerations

Electron attachment to clusters is studied by crossing a supersonic molecular beam (containing a distribution of clusters of different size) with an electron beam and detecting the negative ions by a quadrupole mass spectrometer (Fig. 4.1).

The supersonic beam is generated by adiabatic expansion of the gas under consideration, seeded in Ar $(1:10)$ through a 80 μm nozzle, followed by a skimmer which separates the expansion chamber from the main chamber.

Approximately 7 cm downstream from the skimmer the molecular beam is crossed at right angles with an electron beam. The electron beam is either produced by a simple device consisting of a hairpin tungsten filament followed by a series of electrodes or by a trochoidal electron monochromator similar to that described in Section 2.1. If a simple tungsten filament is used, the beam possesses a comparatively low energy resolution (≈ 0.7 eV). In any case, the beam is aligned by a homogeneous magnetic field (50-100 Gauss) and transmitted into the reaction volume defined by the crossing of the electron beam with the molecular beam.

The resulting ions are extracted, analyzed with a commercial quadrupole mass filter, and then detected by standard pulse-counting electronics.

As mentioned, the supersonic beam contains a *distribution* of clusters. The average size of the clusters can be varied to some extent by the experimental conditions.

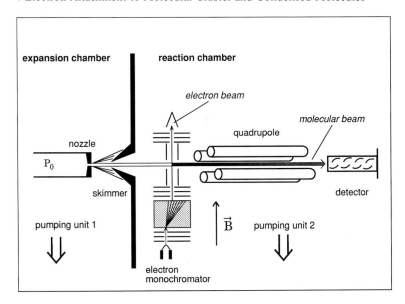

Fig. 4.1. Schematic representation of the apparatus for the study of electron attachment to clusters.

We shall, at this point, briefly characterize such a supersonic molecular beam in relation to an effusive molecular beam.

In a volume of gas such as inside a source or reservoir, the molecules have a Maxwell-Boltzmann velocity distribution:

$$f(v)\, dv \sim v^2 \exp\left(\frac{-Mv^2}{2kT}\right) dv, \qquad\qquad 4.1$$

with the mean velocity

$$\bar{v} = \int_0^\infty vf(v)\, dv = \left(\frac{8kT}{\pi M}\right)^{1/2} = 1.60\left(\frac{kT}{M}\right)^{1/2}. \qquad\qquad 4.2$$

If the gas effuses from the reservoir through an orifice into the vacuum, we have an *effusive molecular beam*. Since the probability of a molecule emerging from the orifice is proportional to its velocity, the intensity distribution in the molecular beam is the velocity weighted Boltzmann distribution:

$$I(v)\, dv \sim v^3 \exp\left(\frac{-Mv^2}{2kT}\right) dv, \qquad\qquad 4.3$$

with the mean velocity

$$\bar{v} = \left(\frac{9\pi\, kT}{8M}\right)^{1/2} = 1.88\left(\frac{kT}{M}\right)^{1/2}, \qquad\qquad 4.4$$

and the most probable velocity (maximum of the distribution function):

$$v_0 = \left(\frac{3kT}{M}\right)^{1/2} = 1.73 \left(\frac{kT}{M}\right)^{1/2}.$$

4.5

Equation 4.3 holds if the spatial and velocity distributions of the molecules inside the reservoir are not affected by the effusion of the molecules. This is the case when the diameter of the orifice (d) is much smaller than the mean free collision path λ within the source:

$$\lambda \gg d.$$

4.6

For nitrogen at room temperature the mean free path is ≈ 6 m at 10^{-5} mb and 0.06 mm at 1 mb. In actual practice it is found that effusive sources are effective when $\lambda \geq d$ [26]; the flow from the orifice is molecular since there are virtually no collisions between the gas molecules within the beam.

In 1951, Kantrowitz and Grey [27] proposed the "supersonic jet" as a molecular beam source. In this case, the reservoir pressure and the orifice diameter is changed to a point where

$$d \gg \lambda.$$

4.7

Now, there are many collisions between the molecules as the gas flows through the orifice and downstreams from the orifice. Such a flow is called hydrodynamic. *A hydrodynamic expansion converts the enthalpy associated with random molecular motion into directed mass flow* [27–31].

The conversion of random motion causes a decrease of (translational) temperature in the beam, associated with a decrease of the speed of sound, c, given by

$$c = (\varkappa k T/M)^{1/2},$$

4.8

with $\varkappa = C_p/C_v$ the heat capacity ratio. The properties of a supersonic jet are often expressed in terms of the Mach number M_a, which is defined as

$$M_a = \frac{v}{c},$$

4.9

where v is the (average) velocity of the particles in the direction of the expansion.

When the gas has expanded through the nozzle it will have a narrow velocity distribution in the direction of the mass flow, as illustrated in Fig. 4.2. For a monoatomic gas, the average velocity is given by

$$v_t = \left(\frac{5kT_0}{M}\right)^{1/2},$$

4.10

where T_o is the temperature of the source. The (translational) temperature of the beam is determined by the relative velocity of the particles in the beam, i.e., by the width of the velocity distribution Δv which decreases as the Mach number increases (Fig. 4.2). A high Mach number suggests that the gas is moving at a very high velocity, but this is *not* the case. The velocity of the particles in a supersonic beam is only moderately higher than the average velocity in an effusive beam, i.e., $(5kT/M)^{1/2} \approx 2.24\,(kT/M)^{1/2}$, as opposed to $(9\pi kT/8M)^{1/2} \approx 1.88\,(kT/M)^{1/2}$.

How cold can a jet become? Under adiabatic reversible flow conditions (absence of viscous forces, shock waves, heat sources, and sinks like in chemical reactions, and heat conductivity) the expansion is isentropic. Under these assumptions, the following expression holds [32–34]:

$$\frac{T_o}{T} = \left(\frac{p_o}{p}\right)^{\frac{1-\varkappa}{\varkappa}} = 1 + \frac{1-\varkappa}{2}\,M_a^2, \qquad\qquad 4.11$$

where T_o and p_o are the temperature and pressure in the source, and T and p the same quantities in the beam.

For distances greater than some nozzle diameters, the Mach number is given by

$$M_a = A\,(x/d)^{\varkappa - 1}, \qquad\qquad 4.12$$

with A being a constant depending on $\varkappa$, and x the distance from the nozzle. For a monoatomic gas, Eq. 4.12 becomes

$$M_a = 3.26\,(x/d)^{2/3}.$$

Equation 4.12 says that, downstream the beam, the Mach number increases, while at the same time, the temperature and pressure decreases. At some distance, however,

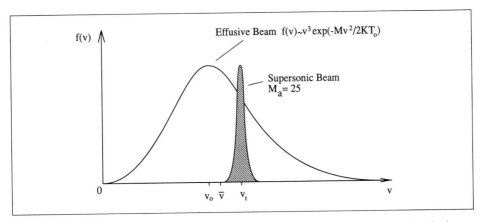

Fig. 4.2. Velocity distribution in an effusive molecular beam and in a supersonic molecular beam.

the density drops to a value at which the collision rate can no longer maintain the redistribution of velocity required by the hydrodynamic equations of the flow and, accordingly, the Mach number and temperature approach some terminal values.

Andersen and Fenn [35] found that, for a monoatomic gas, the terminal Mach number can be expressed as

$$M_a = 1.33\,(p_o d)^{0.4},$$

4.13

when p_o is expressed in bars and d in cm. For the present configuration ($d=80$ μm), Eq. 4.13 predicts a terminal Mach number of 25 at a stagnation pressure of 2 bar, and Eq. 4.11 a beam temperature of 1.4 K (at $T_o = 300$ K) for a monoatomic gas.

Mach numbers up to 250 corresponding to terminal (translational) temperatures of 0.015 K have been achieved [36].

From Eq. 4.11, we see that cooling is less effective for polyatomic molecules ($\varkappa \approx 1$). Cooling of large molecules is achieved with the "seeded beam" technique. Here, the molecules of interest are diluted in an inert carrier gas (He, Ar). The expansion cools the translational degrees of freedom of the gas mixture, particularly the monoatomic carrier gas which acts as a *refrigerant* for the *other degrees of freedom*. The actual state of the beam then depends on the rate at which the other degrees of freedom come into equilibrium with the cold translational bath. As mentioned, at some distance the collision rate drops to such a value that the energy flow between translational and internal degrees of freedom becomes negligible, thus leaving the system in an eventually highly *non-equilibrium state*. Vibrational cooling is generally much less effective than rotational cooling, and we have the relation $T_{trans} < T_{rot} < T_{vib}$.

At some point, atoms and molecules in the beam will start to polymerize. If one is interested in the spectroscopy of cold molecules, cluster formation is an unwanted effect [33, 34]. Since cooling of the beam requires two-body collisions, and formation of clusters requires many-body collisions, cluster formation can be maximized or minimized by an appropriate choice of p_o and d.

4.2 Electron Capture by Clusters of CF₃Cl and CF₄

It was shown in Section 2.3.3 that trifluoromethane captures electrons within two separated resonances centered around 1.4 eV and 4–5 eV. While the low energy state exclusively yields Cl⁻, the TNI generated near 4–5 eV decomposes into a variety of negatively charged fragments (Cl⁻, F⁻, CF₃⁻ etc.). TOF experiments revealed that the 1.4 eV TNI decomposes into Cl⁻ + CF₃ via electronic dissociation along a repulsive potential energy surface.

How does the situation change on proceeding to CF₃Cl clusters? As a first approach, Fig. 4.3 shows the lower part of a negative ion mass spectrum obtained in electron impact to a cluster beam of CF₃Cl at a stagnation pressure of 2 bar. The diagram represents a superposition of two spectra recorded at electron energies of 1.4

and 4.5 eV, respectively. For a clearer representation, Fig. 4.3 only shows peaks corresponding to ^{35}Cl. Apart from the fragments known from the isolated molecule, one detects a variety of larger charged complexes with the series M_n^-, $M_n \cdot Cl^-$, $M_n \cdot F^-$, $M_n \cdot CF^-$ and $M_n \cdot CF_3^-$. We note that the product ion intensity distribution does not directly reflect the size distribution of the neutral clusters in the beam, since the ions generally result from decompositions of larger clusters. In addition, the transmission probability of a quadrupole mass filter decreases significantly with increasing mass number. At what incident electron energies are these various products formed?

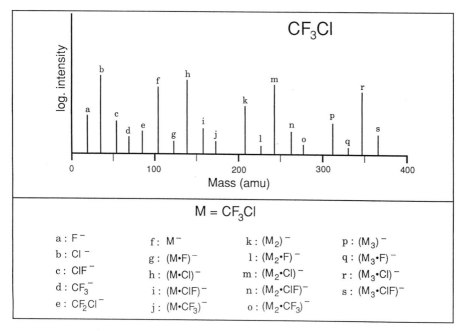

Fig. 4.3. Negative ion mass spectrum obtained in electron impact (0–10 eV) to a beam of CF_3Cl clusters (from [16]).

Figure 4.4 shows a few representative ion yield curves taken at 2 bar stagnation pressure. These spectra have been recorded with the unmonochromatized electron beam. In spite of the low energy resolution inherent in this experiment, we see that all the ions are formed within a low energy resonance and/or a resonance between 4–5 eV already known from the isolated compound. The shapes and energetic positions of these ion yield curves do not depend on the stagnation pressure, at least not to an extent observable by the poor resolution. The only significant effect is the observation that the relative intensity of the low-energy Cl^- signal with respect to the one at 4.8 eV increases with stagnation pressure. This behavior is explained by a dramatic change in the Cl^- kinetic energy release (see below).

The ions M_n^- and $M_n \cdot Cl^-$ are predominantly formed from the resonance of low energy and $M_n \cdot F^-$ is solely associated with the second resonance, and closely resem-

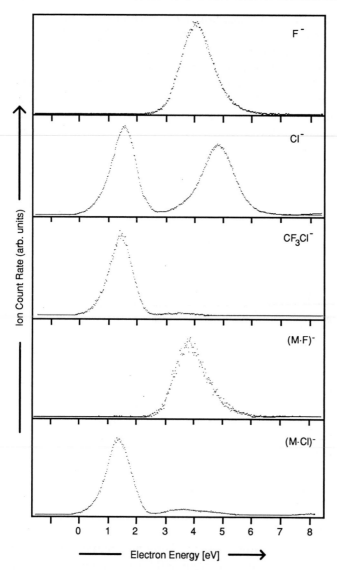

Fig. 4.4. Ion yield curves for some products from CF_3Cl clusters.

bles the F^- ion yield. The shapes of the ion yield curves are virtually independent of n. All products appear within a resonance near 1.4 eV and/or a resonance of higher energy (4–5 eV) already known in the isolated molecule. This indicates that electron capture by CF_3Cl clusters proceeds via initial formation of an individual anion (CF_3Cl^-) in the aggregate whose initial state is not much affected by the surrounding constituents of the target cluster. While the isolated molecule decomposes into the different fragments, the ionized cluster gives rise to the many different product ions apparent in the mass spectrum (Fig. 4.3).

We will now restrict the discussion to low-energy attachment associated with the electronic ground state of CF_3Cl^-. In the isolated molecule this resonance decomposes by direct electronic dissociation into $Cl^- + CF_3$. This situation becomes more complex when the molecular ion is formed in an aggregate

$$e^- (1.4 \text{ eV}) + (CF_3Cl)_n \rightarrow (CF_3Cl^-) \cdot (CF_3Cl)_{n-1}. \qquad 4.14$$

The ionized aggregate decomposes according to

$$(CF_3Cl^-)(CF_3Cl)_{n-1} \rightarrow (CF_3Cl^-)_m + (CF_3Cl)_{n-m} \qquad 4.15$$

$$\rightarrow Cl^- + (CF_3) \cdot (CF_3Cl)_{n-1} \qquad 4.16$$

$$\rightarrow (CF_3Cl)_m \cdot Cl^- + (CF_3) \cdot (CF_3Cl)_{n-m-1}. \qquad 4.17$$

In Eq. 4.15–4.17, each neutral channel is assigned to consist of only one compound. Of course, the reactions may generally release enough excess energy so that many further fragmentation processes can occur. Channel 4.15, m = 1, leaves the parent radical anion in its relaxed configuration. This implies that the potential energy surface of M^- in its electronic ground state must *possess a minimum*, as illustrated in Fig. 2.7. In this case, a relaxation energy of more that 1.3 eV (at resonance maximum) must be distributed among the neutral part of the reacting system.

By applying the TOF technique introduced in Section 2.3, we will now study how the kinetic energy release of Cl^- changes when proceeding to clusters.

As mentioned in the experimental section, the supersonic beam consits of a *distribution* of clusters. This distribution cannot be observed directly by the mass spectrometer since the ions are generally results of fragmentation reactions following ionization of the target aggregate. We can, however, witness the kinetic energy release of Cl^- by recording TOF spectra at increasing stagnation pressures and, thus, increasing Mach numbers. Since the TOF spectra show only a very weak dependence on electron energy, we restrict the analysis to spectra taken at the peak maximum (1.4 eV).

At low stagnation pressure (0.5 bar, Fig. 4.5), one observes a separated TOF doublet similar to that in Fig. 2.19 for the isolated system. At that pressure, there is no indication of cluster formation and Cl^- arises solely from electron capture by monomers. By increasing the stagnation pressure (1 bar) an additional feature near time zero which dominates the spectrum at 2 bar becomes apparent. Any further increase in pressure leaves the TOF distribution virtually unchanged.

Figure 4.5d shows the result of a graphical subtraction. It indicates that the TOF spectrum consists of two components, one due to Cl^- ions with considerable kinetic energy and an additional one yielding low energy ions.

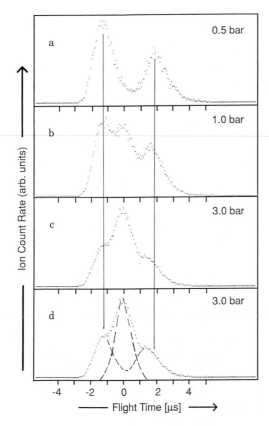

Fig. 4.5. TOF spectra of Cl⁻ from CF₃Cl clusters recorded at different stagnation pressures.

A closer look at the TOF distribution reveals that, at pressures above 1 bar, the distribution is slightly shifted to lower time values, with the turn-around peak more affected than the direct one. This is, in fact, what one calculates for a Cl⁻ ion emitted from a target moving with the seeded beam velocity

$$v_t = (5kT/M_c)^{1/2};$$

(M_c is the mass of the carrier gas and $T = 300$ K) towards the detector. For argon as carrier gas, we have $v_t \approx 560$ m s⁻¹. The mean velocity of a modified Maxwell-Boltzmann distribution describing the effusion from the nozzle at low pressures is given by Eq. 4.4. For the mass of CF₃Cl, we calculate $\bar{v} \approx 290$ m s⁻¹. In comparison, the velocity a Cl⁻ ion gains from the dissociation process is $v_d \approx 1600$ m s⁻¹.

The appearance of low energy Cl⁻ ions at stagnation pressures above 1 bar is responsible for the overproportional increase in the Cl⁻ intensity between 1 and 3 bar (Fig. 4.4). For thermal ions, we no longer have a discrimination against perpendicular velocity components resulting in a higher detection probability.

So far, we have demonstrated that cluster formation coincides with the appearance of low energy Cl^- ions. The question concerning the origin of the high-energy component (which is still present in the TOF spectra at high stagnation pressures), however, is still open. This feature may be due to Cl^- emission from monomers (in the beam or from back-ground gas) or from particular clusters.

If target clusters exist with structures where the F_3C-Cl axis points away from the other molecules, one would expect the free emission of Cl^- ions and so a behavior similar to the isolated case.

On the other hand, for configurations in which the F_3C-Cl axis of the temporary ion points toward other members of the target clusters, one expects either the emission of a low-energy Cl^- ions, slowed down by scattering events or the ultimate formation of the other products observed at that energy (CF_3Cl^-, M_n^- or $M_n \cdot Cl^-$). We have carried out TOF experiments with helium as carrier gas which showed that the high energy component is, in fact, dominantly due to Cl^- emission from scattered background monomers, and only a small amount due to targets in the supersonic beam (monomers and/or clusters) .

It should be noted that the Cl^- TOF spectra do not show evidence for metastable decompositions, i.e., reactions occurring on the microsecond scale. Such decays would cause complicated and asymmetric TOF profiles in contrast to the "clean" distributions obtained in the present study.

In conclusion, for the present system, one can see that:

a) electron capture by clusters of CF_3Cl proceeds through the initial formation of an individual molecular ion CF_3Cl in the aggregate;

b) although the temporary negative ion (at 1.4 eV) is formed by a Franck-Condon transition in a repulsive $\sigma^*(C-Cl)$ state, one observes a strong contribution of thermal Cl^- emission from clusters;

c) among the many product ions formed at low energy, one observes the parent radical anion CF_3Cl^- not accessible in electron capture by the isolated molecule; in that case the target aggregate acts as a *heating bath* absorbing more than 1 eV of relaxation energy.

This "built-in" many-body stabilization mechanism allows to prepare anions in their relaxed configuration. In the case of oxygen, for example, it has been shown that ground state O_2^- can also be observed following electron attachment to oxygen clusters (see the following section).

The effect of excess energy randomization becomes much more dramatic in the case of CF_4 clusters. As shown in Section 2.3.2, tetrafluoromethane captures electrons within a broad resonance centered around 7–8 eV, which is associated with the formation of F^- and CF_3^- (Fig. 2.8).

On proceeding to clusters, we find a variety of additional products such as F_2^-, M_n^-, $M_n \cdot F^-$, and $M_n \cdot CF_3^-$. Among these, the appearance of CF_4^- is most surprising. To our knowledge, this ion has not been observed before.

The ion yield curves show that all products are formed within the broad resonance region known from electron capture by the monomer [37]. Their individual shape, however, varies significantly from one product to another. All members of the series

$M_n \cdot CF_3^-$ virtually coincide with the CF_3^- profile, while the maximum of the formation probability of CF_4^- (and the other members of the series M_n^-) is shifted to significantly lower energies. As an example, Fig. 4.6 shows the ion yield cuves for CF_4^- compared with F^-. These spectra are recorded with a *monochromatized* electron beam at a stagnation pressure of 4 bar.

Since the energy at which a product is formed reflects the attachment energy in the initial target compound, we can conclude (in analogy to CF_3Cl) that electron cap-

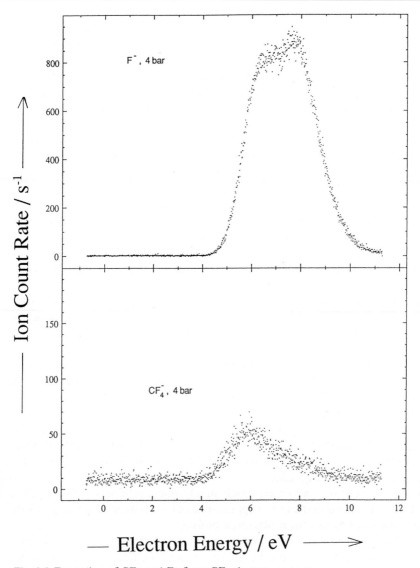

Fig. 4.6. Formation of CF_4^- and F^- from CF_4 clusters.

ture by CF_4 clusters also proceeds via formation of a localized CF_4^- in the cluster. As mentioned above, this temporary negative ion is formed in a strongly repulsive state. If it is formed within an aggregate, we have to consider the reactions

$$e^- + (CF_4)_n \rightarrow (CF_4^-) \cdot (CF_4)_{n-1} \rightarrow F^- + CF_3 \cdot (CF_4)_{n-1} \qquad 4.18$$

$$\rightarrow CF_3^- + F \cdot (CF_4)_{n-1} \qquad 4.19$$

$$\rightarrow (CF_4)_m^- + (CF_4)_{n-m} \qquad 4.20$$

$$\rightarrow (CF_4)_m \cdot F^- + CF_3 \cdot (CF_4)_{n-m-1} \qquad 4.21$$

$$\rightarrow (CF_4)_m \cdot CF_3^- + F \cdot (CF_4)_{n-m-1}. \qquad 4.22$$

Observation of CF_4^- implies that its potential energy surface must possess a minimum at an energy below the lowest dissociation channel, $F^- + CF_3$, which lies 2.26 ± 0.2 eV above the neutral ground state of CF_4.

A possible potential energy diagram for CF_4 and CF_4^- in their respective electronic ground states is shown in Fig. 4.7. If CF_4^- is ultimately formed in its equilibrium configuration, a relaxation energy of up to 7–8 eV has to be distributed among the target aggregate! This can only occur by a substantial evaporation of the initial cluster. It is not clear whether or not CF_4 possesses a *positive adiabatic electron affinity*. Figure 4.7 illustrates the ion in a metastable state. Although the energy of the relaxed ion is

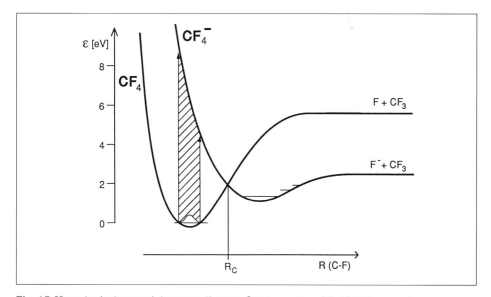

Fig. 4.7. Hypothetical potential energy diagram for the system CF_4/CF_4^- (see text).

above that of ground state CF_4 (negative adiabatic electron affinity) autodetachment is a Franck-Condon forbidden transition. It is further likely that CF_4^- represents a weakly bound adduct $CF_3 \cdot F^-$ (with one bond significantly weakened) rather than a tetrahedral CF_4^-.

Returning to the ion yield curve (Fig. 4.6), the most significant feature here is the asymmetric profile of the M^- yield curve toward lower energies within the resonance.

As introduced in Section 1.3, the cross-section for the formation of a specific ion X^- can be written as

$$\sigma(X^-) = \sigma_0 \cdot P(X^-),$$

with σ_0 the attachment cross-section of the target compound and $P(X^-)$ the probability for the formation of X^- with respect to other competitive channels, including autodetachment. If $P(X^-)$ does not depend on energy, we can write $\sigma(X^-) \sim \sigma_0$ and the line shape of the ion yield curve will have a Gaussian profile based on the reflection principle. For more complex reactions $P(X^-)$ generally depends on the energy of the precursor ion which will influence the peak shape of the ion yield curve. It is plausible that CF_4^- formed at the lower energy side of the resonance has a higher chance of being stabilized, since less relaxation energy has to be distributed among the target aggregate. This results in an ion yield curve with an *asymmetric* profile.

A TOF analysis of the fragments F^- and CF_3^- reveals that they are effectively slowed down when proceeding from the monomers to clusters [38] in analogy to Cl^- ejection discussed above [37].

4.3 Electron Capture Induced Processes in $(C_2F_4)_n$ and $(CF_3Cl)_m \cdot (H_2O)_n$

Tetrafluoroethylene shows a qualitative new feature concerning the state of an excess electron within an aggregate.

Temporary negative ions in the fluoroethylenes have been discussed in Section 3.2. From these studies it is well known that electron capture in C_2F_4 gives rise to an isolated resonance peaking at 3 eV. This electronic state is described by accommodation of the excess electron into the lowest virtual orbital having π^* character. The $C_2F_4^-$ ($^2\Pi$) resonance then decomposes into channels yielding F^-, CF_3^-, CF_2^- and CF_3^- [39–41].

In clusters, we observe the expected resonance near 3 eV yielding all the fragments known from the monomer and, additionally, various larger complexes such as $M_n^-, M_n \cdot F^-$, etc. The ion yield curves for the larger products are shifted by 0.2–0.5 eV toward lower energies.

The most striking feature is the appearance of a new absorption band at low energy. This electronic state is coupled exclusively with the formation of M_n^- ($n \geq 2$)!

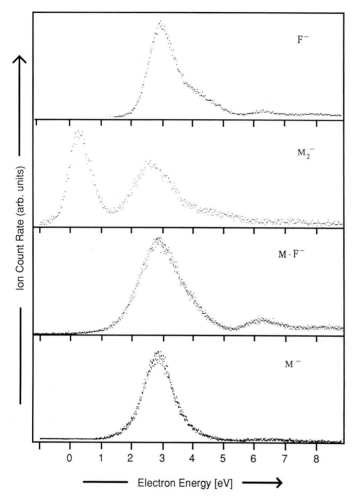

Fig. 4.8. Ion yield curves obtained in electron attachment to C_2F_4 clusters.

Figure 4.8 gives a few ion yield curves recorded at a stagnation pressure of 2 bar. We note that F^- formation is not expected from the low energy resonance for simple energetic reasons, since the C−F bond energy exceeds the electron affinity of F by approximately 1.5 eV. It is, however, surprising that the radical anion M^- is not formed at low energy although the adiabatic electron affinity of C_2F_4 is expected to be near 0 eV [41, 42].

It is also striking that, among all larger compounds, M_2^- is formed with an extraordinary high intensity and one might think that the sample contains some impurity of perfluorocyclobutane which, in fact, captures low energy electrons to form a long-lived radical anion like other cyclic perfluoro compounds [43, 44]. However, experiments at low stagnation pressure did not show any indication of an M_2^- signal, thus, we conclude that this ion must be a product of electron capture by clusters of perfluoroethylene.

What is the nature of this new electronic state? It is likely that M_2^- in its *relaxed* form represents a stable cyclobutane anion, but this says nothing about the initial state of the excess electron in the target aggregate at the instant of the electronic transition. In the present system, the response of the charge distribution toward the incoming electron obviously changes when proceeding from monomer to aggregate.

As discussed in Section 1.2, the formation of a temporary negative ion in an atom or molecule is usually described in terms of the long range interaction between the electron and the neutral particle:

$$V(r) = -\frac{\alpha e^2}{2r^4} + \frac{\hbar^2 l(l+1)}{2mr^2}.$$

4.23

In Eq. 4.23 α denotes the polarizability and l the angular momentum quantum number of the incoming electron associated with the orbital into which it is captured. The combined interaction of the attractive polarization and the repulsive centrifugal potential creates a centrifugal barrier. If the energy of the incident electron is appropriate, i.e., if its wavefunction fits into the potential well inside the barrier, it can be trapped in a quasi bound state, thus forming a temporary negative ion.

The energy of the anion or the vertical electron attachment energy is largely determined by the polarizability of the neutral compound. Accordingly, if the aggregate or a specific structure in it possesses a larger polarizability than the isolated compound, there will always be the possibility of trapping an excess electron at a lower energy. In a somewhat simplified view, one would expect for clusters, in which the intermolecular distances are large compared with the bond length in a molecule, that, the incident electron to interact primarily with one individual molecule. The localized anion M^- is formed in an electronic state not significantly perturbed by the other molecules.

Conversely, if particular clusters or structures within them exist that allow a more collective response of the charge distribution with respect to the incoming electron, one would expect electron capture to occur at lower energy.

The low-energy absorption band shows an intriguing pressure dependence when monitoring the M_2^- ion yield curve [16]. By increasing the stagnation pressure one first observes the low-energy absorption band, and only at higher stagnation pressures does the $^2\Pi$ resonance (shifted to lower energies by 0.5 eV) become apparent. The same behavior can be noted for all members of the series M_n^- ($n \geq 2$), with the general trend that the signal onset arises at higher stagnation pressures with increasing n. This suggests that small clusters (probably dimers due to their significantly high intensity) represent the specific structure that allows low-energy electron absorption.

The occurrence of additional absorption bands is not a general feature in aggregates consisting of larger polyatomics. Clusters of perfluoroethane, for example, only absorb electrons near the energy known from the isolated molecule [45]. Studies in trifluoroethylene [46], on the other hand, also reveal an additional low energy peak associated with M_n^- ($n \geq 2$) and this might indicate that the π system is responsible for the effect.

Finally, we shall illustrate negative ion formation following resonant electron capture in heterogeneous clusters composed of H_2O and CF_3Cl. The individual molecules possess resonances at considerably different energies. As shown above, CF_3Cl exhibits two clearly separated resonances near 1.4 eV and 4–5 eV, while H_2O captures electrons within three overlapping resonances between 6.5 and 12 eV which are associated with the formation of H^-, O^- and OH^- [21, 22]. For the channel $H^- + OH$ it has been established that most of the excess energy is released as kinetic energy to the fragments [47]. The obvious question arises as to whether product ions composed of both constituents are formed.

The heterogeneous clusters are prepared by flowing CF_3Cl seeded in Ar $(1:10)$ at a pressure of 1.8 bar through a stainless steel vessel containing liquid water near room temperature and expanding the mixture through the nozzle as described in the Experimental Section.

Although the clusters under consideration are composed of simple molecules, the assignment of the mass spectrum is not in any case straightforward. In the 0–10 eV region the negative ion mass spectrum shows evidence for the solvated ions $((H_2O)_n \cdot Cl^-$ ($n \geq 1$), $(H_2O)_n \cdot F^-$ ($n \geq 1$), $(H_2O)_n \cdot OH^-$ ($n \geq 2$)) and all the products obtained in electron capture by homogeneous CF_3Cl aggregates described above, as well as O^- and OH^-. We have restricted our experiments to mass numbers below 150 so that the possible formation of $(H_2O)_n^-$, ($n \geq 11$) [19] has not been a subject of the present investigation. The assignment of the mass spectrum is sometimes ambiguous since different possible compounds have identical mass numbers. For example, $CF_3Cl \cdot Cl^-$ (mass numbers 139, 141, 143) overlaps with $(H_2O)_6 \cdot Cl^-$ (143, 145), $(H_2O)_7 \cdot OH^-$ (143) and $(H_2O)_7 \cdot F^-$ (145). They can, however, sometimes be distinguished by their different energy dependences.

Figure 4.9 shows the lower part of a mass spectrum recorded at 1.4 eV electron energy. Only peaks due to ^{35}Cl are shown. We ascribe the mass numbers 53, 71, 89, 107, 125, and 143 to $(H_2O)_n \cdot {}^{35}Cl^-$ ($n = 1$–6, see below).

In Fig. 4.10 we have recorded the ion yields corresponding to mass numbers 35, 53, 55 and 19 in the energy range 0–12 eV. Above 11 eV one observes a continuously rising signal which cannot be discriminated by mass filter or any electrostatic potential. This signal is easily identified as arising from electronically excited carrier gas atoms in their well-known metastable states, Ar* ($^3P_2$, 11.55 eV), Ar* ($^3P_0$, 11.72 eV) [48] causing electron emission at the first dynode of the electron multiplier.

Figure 4.10 gives the $^{35}Cl^-$ yield which is similar to that obtained from homogeneous CF_3Cl clusters (Fig. 4.4, p 318). In the ion yield curve recorded for $M = 53$ (Fig. 4.10b) we assign the signals near 1.4 and 4.5 eV to $H_2O \cdot {}^{35}Cl^-$ and the weak contribution above 7 eV to $(H_2O)_2 \cdot OH^-$. Solvated hydroxyl ions $(H_2O)_n \cdot OH^-$ ($n \geq 2$) have

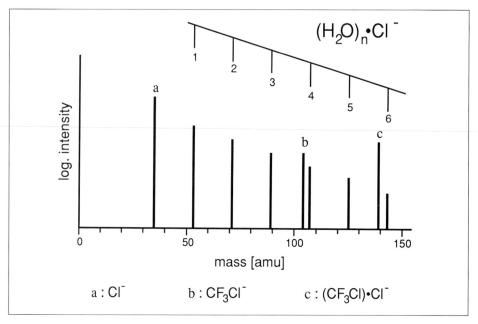

Fig. 4.9. Negative ion mass spectrum observed in electron impact (1.4 eV) to heterogeneous clusters composed of H_2O and CF_3Cl.

also been detected between 7 and 10 eV in electron impact to H_2O clusters [19]. As expected, the ion yield recorded at $M = 55$ (Fig. 4.10 c) gives not contribution near 7 eV; the intensity ratio between the two low energy resonances, however, differs substantially from that in Fig. 4.10 b. We interpret the signal near 4–5 eV as being composed of $(H_2O) \cdot {}^{37}Cl^-$ and $(H_2O)_2 \cdot F^-$. Solvated ions of the type $(H_2O)_n \cdot F^-$ ($n \geq 1$) can, in fact, be observed in clusters composed of H_2O and perfluorinated compounds arising at energies close to that of F^- formation. For comparison, Fig. 4.10 d shows the F^- yield.

The present example illustrates that the formation of the solvated ions $(H_2O)_n \cdot Cl^-$ proceeds predominantly through the low energy resonance of CF_3Cl at 1.4 eV. From the results presented above, we know that this molecular anion is generated in a strongly repulsive $C–Cl^-$ electronic state.

Apart from free electron attachment to clusters considered here, considerable progress has recently been made in studying electron transfer from Rydberg noble gas atoms to molecular clusters, hereby generating negatively charged (cluster) ions [49, 50]. When state selected Rydberg atoms are used, a surprisingly strong dependence of cluster ion spectra on principal quantum number has been observed for the system N_2O, [51]. The explanation of this effect is by no means straightforward; it has been discussed in terms of the different decay pathways of the Coulombic collision complex $[Ar^+ \cdot (N_2O)^{*-}]$.

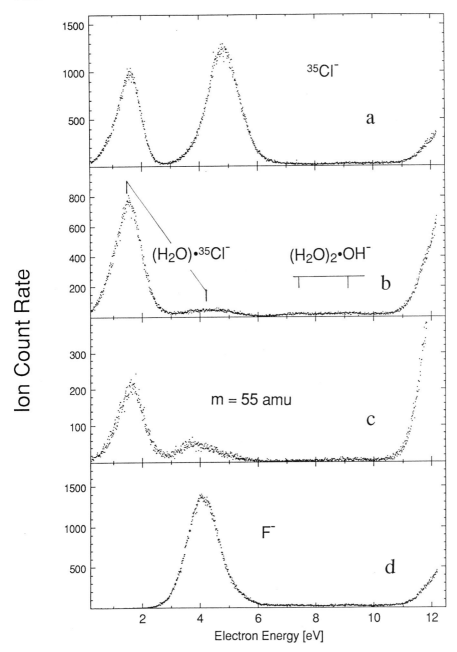

Fig. 4.10. Ion yield curves for some products from heterogeneous clusters composed of H_2O and CF_3Cl (see text).

4.4 Dissociative Electron Attachment from Condensed Molecules

We have, so far, considered resonant electron scattering from isolated molecules and free molecular aggregates and focused on reactions leading to stable anions. Here we will consider some exciting results recently obtained in low energy electron scattering from oxygen molecules condensed on a solid surface. We will then finally compare these studies with recent experiments on free oxygen clusters.

One may generally distinguish between two major classes of experiments which provide information on electron scattering from condensed molecules: those in which the energy loss (and eventually the angular dependence) of the *scattered electrons* is measured [52–54], and others by which the products induced by electron impact are recorded.

Desorption of neutrals and ions induced by electronic excitation on condensed molecules has received considerable attention within the past decade and the interested reader is referred to a series of monographs [56–59] and review articles [60, 61]. The vast majority of results in this area has been obtained at higher energies (≥ 20 eV). Desorption studies induced by low energy impact (≤ 20 eV) are essentially restricted to the work of Sanche and coworkers, which is summarized in a review article [62]. It has been shown that elastic scattering can only be explained in terms of solid state concepts while *inelastic interactions* can be interpreted by invoking *resonant* and *direct scattering at a molecular site*. Out of the various systems studied so far [63–67], we shall illustrate here results for condensed oxygen.

Figure 4.11 shows the ion yield curves of O^- desorbed from a three monolayer O_2 film ((a) and (b)) and a 20 monolayer Ar film containing 10 % (c) and 2 % volume O_2 (d). For comparison, O^- formation from gaseous O_2 is plotted in Fig. 4.11 (e) (see also Fig. 2.15). The molecules are condensed on a clean, electrically isolated polycrystalline platinum (Pt) ribbon press fitted on the cold tip of a closed-cycle refrigerated cryostat. All components (sample, electron gun and mass spectrometer) are housed in an ultrahigh vacuum (UHV) chamber reaching pressures below $5 \cdot 10^{-11}$ mb [68–70].

Curve (b) in Fig. 4.11 was recorded by retarding the ions with a potential of -1.5 eV prior to entering the mass spectrometer.

These results can be explained by considering the potential energy curves of the system O_2^- [71, 72]. Although a variety of molecular O_2^- states can be formed (24 for example, from ground state O (^3P) and O^- (^2P)), only few possess the proper characteristics to allow substantial decay via dissociative attachment. Conservation of multiplicity requires a doublet or quartet state, the TNI should be repulsive in the Franck-Condon region, and its lifetime should be sufficiently long to avoid autodetachment.

Molecular orbital analysis of the O_2^- states correlated with the lowest channels, O (^3P) $+ O^-$ (^2P) and O (^1D) $+ O^-$ (^2P) then gives the three repulsive states shown in Fig. 4.12 having the proper characteristics to produce O^- with a considerable cross

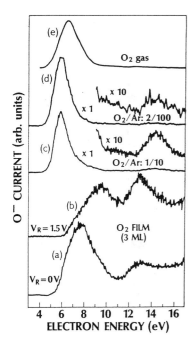

Fig. 4.11. O$^-$ stimulated desorption yields from O$_2$ and Ar/O$_2$ films, see text (from [62]).

section [69,73]. All three repulsive states are core excited $(2p - 1h)$ resonances, namely O$_2^{*-}$ $(1\pi_u^{-1} 1\pi_g^2)$ $^2\Pi_u$, O$_2^{*-}$ $(3\sigma_g^{-1} 1\pi_g^2)$ $^2\Sigma_g^+$ and O$_2^{*-}$ $(2\sigma_u^{-1} 1\pi_g^2)$ $^2\Sigma_u^+$.

Out of these states, the Σ^+ configuration cannot be populated in electron attachment to isolated O$_2$ $(^3\Sigma_g^-)$ ground state since the isolated target-electron frame of reference excludes $\Sigma^- \leftrightarrow \Sigma^+$ transitions [73]. This selection rule holds for electron attachment and autoionization processes; it is based on the fact that the one electron wavefunction must principally be σ^+. Thus, dissociative electron attachment to single O$_2$ results only in one peak in the O$^-$ ion yield curve (Fig. 4.11 (e) and Fig. 2.15) which is due to reaction

$$e^- + O_2\,(^3\Sigma_g^-) \rightarrow O_2^-\,(^2\Pi_u) \rightarrow O^-\,(^2P) + O\,(^3P)\,. \qquad 4.24$$

The same process dominates in O$_2$ doped Ar matrices (Fig. 4.11).

However, as the concentration of O$_2$ molecules increases (the molecules are then less isolated from each other) additional features at higher energy appear and the 6 eV peak broadens and shifts to higher energies. These new structures are relatively amplified when only higher energy ions are transmitted to the mass spectrometer (Fig. 4.11 b). The new peaks at 9 and 13–14 eV are ascribed to the $^2\Sigma_g^+$ and $^2\Sigma_u^+$ states now involved in dissociative attachment. This is a clearcut indication of the breakdown of the σ^- selection rule in the condensed phase.

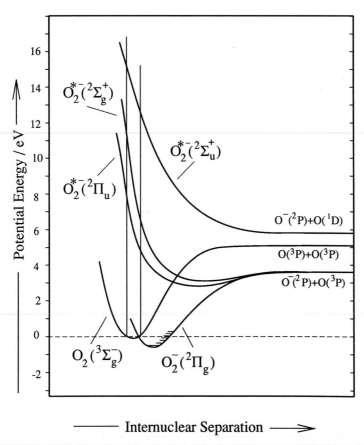

Fig. 4.12. Potential energy curves of O_2 and O_2^-. Only those negative ion states are shown which are expected to generate significant O^- yields via dissociative attachment.

The present example clearly demonstrates that electron scattering from condensed molecules is reasonably described in terms of potential energy diagrams at a molecular site. However, selection rules operative under isolated conditions are violated in the condensed phase.

How is the situation in free oxygen clusters which represent a link between isolated and condensed molecules? Figure 4.13 shows a very recent result obtained with the supersonic beam configuration described in Section 4.4. Pure oxygen is expanded through an 80-μm nozzle which is kept at $-120\,°C$. At low stagnation pressure (no cluster formation) we observe one single O^- peak around 6.2 eV. The continuously rising signal above 17 eV is due to the ion pair formation process:

$$e^- + O_2(^3\Sigma_g^-) \rightarrow O_2^{**} + e^-$$
$$ \rightarrow O^+(^4S) + O^-(^2P). \qquad\qquad 4.25$$

At higher stagnation pressure the O^- signal exhibits two additional features near 8 eV and 14.5. eV which we ascribe to reactions proceeding via the $O_2^- \, (^2\Sigma_g^+)$ and $O_2^- \, (^2\Sigma_u^+)$ states introduced above.

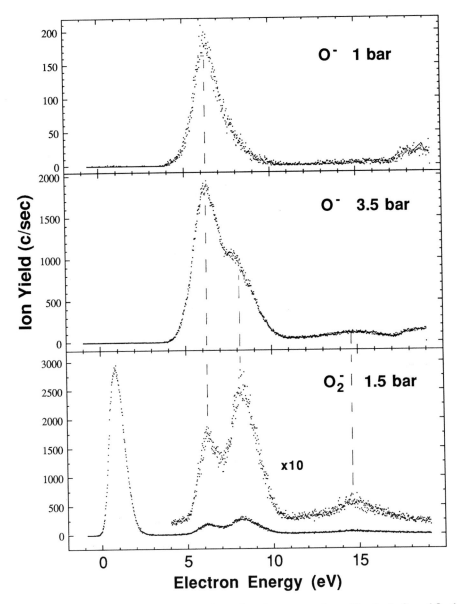

Fig. 4.13. Ion yield curves for O^- and O_2^- following electron attachment by single O_2 and O_2 clusters (see text).

Interestingly, these features are also visible in the O_2^- product, pointing toward complex scattering events ultimately leaving O_2^- in a thermodynamically stable state, i.e. O_2^- ($^2\Pi_g$, $v < 4$) below ground state O_2 ($^3\Sigma_g^-$). At that point, however, it must be emphasized that any reaction product can principally be the result of (1) the evolution of a temporary molecular ion within the cluster or (2) capture by an electron which has been *scattered* within the cluster. In the case of O_2^- formation (and its larger homologues) it is likely that an inelastically scattered slow electron is captured, generating O_2^- in its electronic ground state which is then stabilized to a vibrational level ($v < 4$) thereby evaporating the target cluster. In fact, the threshold electron impact excitation spectrum of oxygen (only secondary electrons are recorded which have lost their total primary energy) shows two pronounced peaks at 6.2 eV and 8.2 eV [74]. These inelastically scattered electrons are due to excitation of oxygen molecules through direct or resonant scattering. On the other hand, Fig. 4.13 shows that the dominant channel for O_2^- formation is through *direct* attachment of slow electrons and subsequent collisional stabilization.

Apart from O^- and O_2^- shown in Fig. 4.13, electron attachment to oxygen clusters yields the homologous series $(O_2)_n^-$, $n \geq 1$ and $(O_2)_n \cdot O^-$, $n \geq 0$ with energy profiles similar to O_2^- an O^-, respectively [75] (not shown here).

The present study is a textbook example of the "built in" three body stabilization mechanism in clusters; it additionally shows the breakdown of the σ^- selection rule in clusters and the occurrence of products arising from intracluster electron scattering processes.

References

1. Webster's New World Dictionary (1979). Simon and Schuster, New York
2. Van der Waals JD (1873) Doctoral Dissertation, Leiden
3. Jortner J, Pullman A, Pullman B (eds) (1987) Large Finite Systems. Reidel, Dordrecht
4. Benedek G, Martin TP, Pacchioni G (eds) (1988) Elemental and Molecular Clusters. Springer, Berlin
5. Maier JP (ed) (1989) Ion and Cluster Ion Spectroscopy and Structure. Elsevier, Amsterdam
6. Z Physik D (1991) vols 19 and 20, Special Issue, Proceedings ISSPIC 5, 5th International Symposium on Small Particles and Inorganic Clusters, Konstanz
7. Bowers MT, Jarrold MF, Stace AC (eds) (1990) Gas Phase Cluster Ions, Special Issue. Int J Mass Spectrom Ion Proc 102
8. Haberland H, Kornmeier H. Langosch H, Oschwald M, Tanner G (1990) Experimental Study of the Transition from van der Waals over Covalent to Metallic Bonding in Mercury Clusters. J Chem Soc Faraday Trans 86:2437–2481
9. Rademann K (1989) Photoionization Mass Spectrometry and Valence Photoelectron-Photoion Coincidence Spectroscopy of Isolated Clusters in a Molecular Beam. Ber Bunsenge Phys Chem 93:653-670
10. Maitland GC, Rigby M, Smith EB, Wakeham WA (1987) Intermolecular Forces. Oxford Scientific Publication, Oxford

11. Pikaev AK (1971) The Solvated Electron in Radiation Chemistry. Academy of Science of the USSR, Translated from Russian. Israel Program for Scientific Translations, Jerusalem
12. Lepoutre G (1984) Colloque Weyl: A Short History. J Phys Chem 88:3699–3700
13. Klots CE, Compton RN (1977) Electron Attachment to Carbon Dioxide Clusters in a Supersonic Beam. J Chem Phys 67:1779–1780
14. Märk TD, Leiter K, Ritter W, Stamatovic A (1985) Low-Energy-Electron Attachment to Oxygen Clusters Produced by Nozzle Expansion. Phys Rev Lett 55:2559–2562
15. Stamatovic A, Scheier P, Märk TD (1988) Electron Attachment and Electron Impact Ionization of SF_6 and SF_6/Ar Clusters. J Chem Phys 88:6884–6888
16. Hashemi R, Kühn A, Illenberger E (1990) Electron Capture Induced Processes in Molecules and Molecular Aggregates. Int J Mass Spectrom Ion Proc 100:753–784
17. Armbruster M, Haberland H, Schindler H-G (1981) Negatively Charged Water Clusters, or the First Observation of Free Hydrated Electrons. Phys Rev Lett 47:323–326
18. Haberland H, Langosch H, Schindler H-G, Worsnop DR (1984) Negatively Charged Water Clusters: Mass Spectra of $(H_2O)_n^-$ and $(D_2O)_n^-$. J Phys Chem 88:3903–3904
19. Knapp M, Echt O, Kreisle D, Recknagel E (1987) Electron Attachment to Water Clusters under Collision-Free Conditions. J Phys Chem 91:2601–2607
20. Mitsuke K, Tada H, Misaizu F, Kondow T, Kuchitsu K (1988) Negative Ion Formation from CCl_4 Clusters in Collision with Highly Excited Rydberg Atoms and Slow Electrons. Chem Phys Lett 143:6–12
21. Melton CE (1972) Cross Sections and Interpretation of Dissociative Attachment Reactions Producing OH^-, O^- and H^- in H_2O. J Chem Phys 57:4218–4225
22. Jungen M, Vogt J, Staemmler V (1979) Feshbach-Resonances and Dissociative Electron Attachment to H_2O. Chem Phys 37:49–55
23. Landman U, Barnett RN, Cleveland CL, Scharf D, Jortner J (1987) Electron Excitation Dynamics, Localization, and Solvation in Small Clusters. J Phys Chem 91:4890–4899
24. Coe JV, Lee GH, Eaton JG, Arnold ST, Sarkas HW, Bowen KH, Ludewigt C, Haberland H (1990) Photoelectron Spectroscopy of Hydrated Cluster Anions, $(H_2O)_{n=2-69}^-$. J Chem Phys 92:3980–3982
25. Haberland H (1990) Solvated-Electron Clusters. In: Scoles G (ed) The Chemical Physics of Atomic and Molecular Clusters. North Holland, Amsterdam
26. Ramsey NF (1985) Molecular Beams. Oxford University Press, Oxford
27. Kantrowitz A, Grey J (1951) A High Intensity Source for the Molecular Beam. Part I. Theoretical. Rev Sci Instrum 22:328–332
28. Anderson JB (1974) Molecular Beams from Nozzle Sources. In: Wegner PP (ed) Molecular Beams and Low Density Gas Dynamics. Marcel Dekker, New York
29. Miller DR (1988) Free Jet Sources. In: Scoles G (ed) Atomic and Molecular Beam Methods, vol 1. Oxford University Press
30. Becker E (1965) Gasdynamik. Teubner, Stuttgart
31. Fricke J (1973) Kondensation in Düsenstrahlen. Physik in unserer Zeit 4:21–27
32. Liepmann HW, Roshko A (1957) Elements of Gas Dynamics. Wiley, New York
33. Smalley RE, Wharton L, Levy DH (1977) Molecular Optical Spectroscopy with Supersonic Beams and Jets. Accounts Chem Res 10:139–145
34. Levy DH (1980) Laser Spectroscopy of Cold Gas-Phase Molecules. Ann Rev Phys Chem 31:197–225
35. Anderson JB, Fenn JB (1965) Velocity Distributions in Molecular Beams from Nozzle Sources. Phys Fluids 8:780–787
36. Toennies JP, Winkelmann K (1977) Theoretical Studies of Highly Expanded Free Jets: Influence of Quantum Effects and a Realistic Intermolecular Potential. J Chem Phys 66:3965–3979
37. Lotter J, Illenberger E (1990) Electron Capture Induced Reactions in CF_4 Clusters. J Phys Chem 94:8951–8956
38. Kühn A, Illenberger E (1990) Low Energy (0-10 eV) Electron Attachment to CF_3Cl Clusters: Formation of Product Ions and Analysis of Excess Translational Energy. J Chem Phys 93:357–364

39. Heni M, Illenberger E, Baumgärtel H, Süzer S (1982) The Dissociation of the $^2\Pi$ Fluoroethylene Anions. Chem Phys Lett 87:244–248
40. Heni M, Illenberger E (1986) The Unimolecular Decomposition of the Fluoroethylene Radical Anions Formed by Electron Attachment. J Electron Spectrosc Relat Phen 41:453–466
41. Oster T, Kühn A, Illenberger E (1989) Gas Phase Negative Ion Chemistry. Int J Mass Spectrom Ion Proc 89:1–72
42. Paddon-Row MN, Rondan NG, Houk KN, Jordan KD (1982) Geometries of the Radical Anions of Ethylene, Fluoroethylene, 1,1-Difluoroethylene, and Tetrafluoroethylene. J Am Chem Soc 104:1143–1145
43. Fenzlaff M, Illenberger E (1989) Energy Partitioning in the Unimolecular Decomposition of Cyclic Perfluororadical Anions. Chem Phys 136:443–452
44. Süzer S, Illenberger E, Baumgärtel H (1984) Negative Ion Mass Spectra of Hexafluoro-1,3-butadiene, Hexafluoro-2-butyne and Hexafluorocyclobutene. Identification of the Structural Isomers. Org Mass Spectrom 19:292–293
45. Hashemi R (1989) Elektroneneinfanginduzierte Reaktionen in van der Waals-Aggregaten halogenierter Kohlenwasserstoffe. Diplomarbeit. Freie Universität Berlin
46. Echt O, Knapp M, Schwarz C, Recknagel E (1987) Electron Attachment to Clusters. In: Jortner J, Pullman A, Pullman B (eds) Large Finite Systems. Reidel, Dordrecht
47. Belic DS, Landau M, Hall RI (1981) Energy and Angular Dependence of $H^-(D^-)$ Ions Produced by Dissociative Electron Attachment to H_2O (D_2O). J Phys B 14:175–190
48. Moore CE (1971) Atomic Energy Levels. US Dept Commerce NSRDS-NBS 35, Washington, DC
49. Mitsuke K, Kondow T, Kuchitsu K (1986) Collisional Electron Transfer to CH_3CN Clusters from High-Rydberg Krypton Atoms. J Phys Chem 90:1505–1506
50. Kraft T, Ruf M-W, Hotop H (1991) Electron Transfer from State-Selected Rydberg Atoms to $(N_2O)_m$ and $(CF_3Cl)_m$ Clusters. Z Phys D
51. Kraft T, Ruf M-W, Hotop H (1990) Strong Dependence of Negative Cluster Ion Spectra on Principal Quantum Number n in Collisions of State-Selected Ar^{**} (nd) Rydberg Atoms with N_2O Clusters. Z Phys D 17:37–43
52. Ertl G, Küppers J (1985) Low Energy Electrons and Surface Chemistry. VCH Verlagsgesellschaft mbH, Weinheim
53. Willis RF (ed) (1990) Vibrational Spectroscopy of Adsorbates. Springer, Berlin
54. Christmann K (1991) Introduction to Surface Physical Chemistry. Topics in Physical Chemistry, vol I. Steinkopff, Darmstadt
55. Ibach H, Mill DL (1982) Electron Energy Loss Spectroscopy and Surface Vibrations. Academic Press, New York
56. Tolk NH, Traum MM, Tully JC, Madey TE (eds) (1983) Desorption Induced by Electronic Transitions DIET I. Springer, Berlin
57. Brenig W, Menzel D (eds) (1985) Desorption Induced by Electronic Transitions DIET II. Springer, Berlin
58. Stulen RH, Knotek ML (eds) (1988) Desorption Induced by Electronic Transitions DIET III. Springer, Berlin
59. Betz G, Varga P (eds) (1990) Desorption Induced by Electronic Transitions DIET IV. Springer, Berlin
60. Menzel D (1986). Desorption Induced by Electronic Transitions. Nucl Instrum Methods B 13:507–517
61. Avouris P, Walkup RE (1989) Fundamental Mechanisms of Desorption and Fragmentation Induced by Electronic Transitions at Surfaces. Ann Rev Phys Chem 40:173–206
62. Sanche L (1990) Low-Energy Electron Scattering from Molecules on Surfaces. J Phys B: At Mol Opt Phys 23:1597–1624
63. Sanche L, Parenteau L (1986) Dissociative Attachment in Electron-Stimulated Desorption from Condensed NO and N_2O. J Vac Sci Technol A 4:1240–1242
64. Azria R, Parenteau L, Sanche L (1987) Dynamics of Dissociative Attachment Reactions in Electron-Stimulated Desorption: Cl^- from Condensed Cl_2. J Chem Phys 87:2292–2296

65. Sanche L, Parenteau L (1987) Ion-Molecule Surface Reaction Induced by Slow (5–20 eV) Electrons. Phys Rev Lett 59:136–139
66. Sambe H, Ramaker DE, Parenteau L, Sanche L (1987) Electron-Stimulated Desorption Enhanced by Coherent Scattering. Phys Rev Lett 59:505–508
67. Sambe H, Ramaker DE, Deschênes M, Bass AD, Sanche L (1990) Absolute Cross Section for Dissociative Electron Attachment in O_2 Condensed on Kr Film. Phys Rev Lett 64:523–526
68. Sanche L (1984) Dissociative Attachment in Electron Scattering from Condensed O_2 and CO. Phys Rev Lett 53:1638–1641
69. Azria R, Parenteau L, Sanche L (1987) Dissociative Attachment from Condensed O_2: Violation of the Selection Rule $\Sigma^- \leftrightarrow \Sigma^+$. Phys Rev Lett 59:638–640
70. Sanche L, Parenteau L, Cloutier P (1989) Dissociative Attachment Reactions in Electron-Stimulated Desorption from Condensed O_2 and O_2^- Doped Rare-Gas Matrices. J Chem Phys 91:2664–2674
71. Krupenie PH (1972) The Spectrum of Molecular Oxygen. J Phys Chem Rev Data 1:423–434
72. Krauss M, Neumann D, Wahl AC, Das G, Zemke W (1973). Excited Electronic States of O_2^-. Phys Rev A 7:69–77
73. Sambe H, Ramaker DE (1987) The σ^- Selection Rule in Electron Attachment and Autoionization of Diatomic Molecules. Chem Phys Lett 139:386–388
74. Schulz GJ, Dowell JT (1962) Excitation of Vibrational and Electronic Levels in O_2 by Electron Impact. Phys Rev 128:174–177
75. Hashemi R, Illenberger E (1991) Violation of the σ^- Selection Rule in Electron Attachment to Oxygen Clusters. Chem Phys Lett 187:623–627

Subject Index

K. **Christmann**, Berlin

Introduction to
Surface Physical Chemistry

(Topics in Physical Chemistry. Eds. H. Baumgärtel, E.U. Franck, W. Grünbein on behalf of Deutsche Bunsen-Gesellschaft für Physikalische Chemie. **Vol. 1**)

1991. X, 274 pages. 135 figures.
Hardcover DM 64,–.
ISBN 3-7985-0858-5 (FRG). ISBN 0-387-91405-6 (USA).
(Reduced price for members of the Deutsche Bunsen-Gesellschaft: DM 44,80)

Contents: Introduction – Macroscopic Treatment of Surface Phenomena: Thermodynamics and Kinetics of Surfaces – Microscopic Treatment of Surface Phenomena – Some of the Surface Scientist's Tools – Surface Reactions and Model Catalysis – General Conclusions.

This first volume in the series **Topics in Physical Chemistry** is devoted to the physical chemistry of surfaces. It underscores the importance of this subject in both fundamental and applied physical chemistry. The work is particularly benficial for a more basic understanding of heterogeneously catalyzed (surface) processes leading to, among others, improved industrial fabrication processes.

The general scope of this book is to foster an understanding of the physical basis of surface processes, and to elucidate the close relationship between classical surface physics and applied interface chemistry.

Available at your local bookstore

Steinkopff **Dr. Dietrich Steinkopff Verlag**
Saalbaustraße 12, W-6100 Darmstadt/FRG